PLASTICITY AND GEOMECHANICS

Plasticity theory is widely used to describe the behaviour of soil and rock in many engineering situations. *Plasticity and Geomechanics* presents a concise introduction to the general subject of plasticity with a particular emphasis on applications in geomechanics. Metal plasticity is described and elementary theories are discussed before attention is focused specifically on geomaterials. The greater part of the book is devoted to the classical aspects of plasticity, particularly the use of upper and lower bound theorems and slip line theory. Critical state theory is introduced and Cam Clay is described in detail.

Derived from the authors' own lecture notes, this book is written with students firmly in mind; the main body of the work is concerned with applications, while the more theoretical aspects such as proofs of theorems are placed in appendices. Excessive use of mathematical methods is avoided in the main body of the text and, where possible, physical interpretations are given for important concepts. In this way the authors present a clear introduction to the complex ideas and concepts of plasticity as well as demonstrating how this developing subject is of critical importance to geomechanics and geotechnical engineering.

Although entirely self-contained, this book constitutes a companion volume to the acclaimed *Elasticity and Geomechanics* by the same authors, and will appeal to students and researchers in the fields of civil, mechanical, material and geological engineering. It may be used as a text for senior-level undergraduate and graduate courses in soil mechanics, foundation engineering and geomechanics.

R. O. DAVIS is Professor of Civil Engineering at the University of Canterbury. A. P. S. SELVADURAI is Professor of Civil Engineering and Applied Mechanics at McGill University. Both authors are dedicated educators and researchers in the fields of geotechnical engineering and geomechanics with a combined experience exceeding 50 years. They are joint authors of the well-received *Elasticity and Geomechanics* published in 1996 by Cambridge University Press, and Professor Selvadurai is also the author of the two-volume monograph *Partial Differential Equations in Mechanics* (2000).

PLASTICITY AND GEOMECHANICS

R. O. DAVIS
University of Canterbury

A. P. S. SELVADURAI
McGill University

CAMBRIDGE UNIVERSITY PRESS
Cambridge, New York, Melbourne, Madrid, Cape Town, Singapore, São Paulo

Cambridge University Press
The Edinburgh Building, Cambridge CB2 2RU, UK

Published in the United States of America by Cambridge University Press, New York

www.cambridge.org
Information on this title: www.cambridge.org/9780521818308

First published 2002
This digitally printed first paperback version 2005

A catalogue record for this publication is available from the British Library

ISBN-13 978-0-521-81830-8 hardback
ISBN-10 0-521-81830-3 hardback

ISBN-13 978-0-521-01809-8 paperback
ISBN-10 0-521-01809-9 paperback

Contents

Preface

Plasticity and Geomechanics follows on from our earlier book *Elasticity and Geomechanics*. Like the earlier book, this one is very much a textbook rather than a treatise or reference book. It has grown from lecture notes and is written with students firmly in mind. Hopefully it will provide an easy, accessible introduction to a subject which, while being widely used in engineering practice, is often difficult for students to assimilate. The plasticity of metals is itself a subject of some complexity. When, instead of metals, the material we are concerned with is either soil or rock, the level of complexity is increased significantly. We have attempted here to untangle the ideas and concepts, and to lay out as clear a picture as possible of a subject area that is still in a state of development and discovery.

The book is organised as follows. Chapters 1 and 2 review some of the basic elements of stress and strain as well as the fundamentals of elasticity. Chapter 2 also presents a general discussion of inelastic response in soil, emphasising the defining characteristics of yield under isotropic compression and dilatancy as a result of shearing. Chapters 3 and 4 set out the fundamental ideas of yield surface and flow rules. The geometry of principal stress space is developed in detail. Yield loci for metals, for Coulomb materials and for some modifications of Coulomb materials are all presented. The Cam Clay and Modified Cam Clay surfaces are summarised. Chapter 4 develops the basic ideas of normality and the associated flow rule as well as non-associated flow. The concepts of perfect plasticity and work hardening are introduced and a complete stress–strain relationship for a general material with non-associated flow is derived. Whenever possible, important concepts such as normality are demonstrated by simple examples. A more complex but practically important example involving cavity expansion is also considered. Chapter 5 introduces the collapse load theorems and limit analysis. This is the longest chapter. In it we attempt to provide a clear introduction to what might be termed the art of finding useful stress (lower bound) and deformation (upper bound) fields for practical problems. Chapter 6 presents an introduction to slip line fields. In the interest of simplicity, the topic is developed initially for purely cohesive materials. Frictional materials are introduced as an embellishment of the purely cohesive case and complicated mathematics is avoided wherever possible. Finally, in Chapter 7, work hardening and critical state soil mechanics are described. As in the preceding chapters we try to avoid excessive detail, but endeavour to demonstrate important concepts by appealing to examples. The fundamentals of critical state theory are developed using Cam Clay together with a simple example problem. A micromechanical theory for normal or virgin compression of an idealised soil is also presented in this chapter. Throughout the book our choice of material is guided by a belief in the importance of simplicity and a desire to make fundamental ideas accessible to students.

Each chapter is followed by a short reading list detailing original sources for the material presented, complemented by references to additional reading of a more general nature. Also, following each chapter is a selection of problems that may be used to help develop the reader's understanding and skill.

The book concludes with a collection of appendices. These expand or elaborate on topics that do not fit easily with the flow of writing in the main text. Most aspects of a more mathematical nature are placed here. In particular, proofs associated with the important theorems of limit analysis as well as a complete development of Mohr's circle, virtual work and uniqueness of solutions are given. The appendices provide rigour for those readers who wish it without interrupting the more physical development in the chapters.

The bulk of the book is devoted to perfectly plastic materials. This may seem odd in light of the current interest in critical state theories for soils, but in our view it is essential knowledge. A firm understanding of basic principles is the foundation for expertise in any subject, and plasticity is no exception. We share in a growing concern that the demands on engineering curricula in current times are such that many students have had little opportunity to gain an adequate background in what might be termed the more 'classical' aspects of plasticity theory. This occurs because of two recent developments. The first is critical state soil mechanics. Critical state theory has become the new paradigm for the analysis of geotechnical problems. This is quite proper but, as with any rapidly developing paradigm, there is a tendency for a gold-rush attitude to infiltrate and subvert the normal course of study. The second development is the advent of computer methods in engineering. The widespread availability of powerful, inexpensive computers together with commercial software has revolutionised all aspects of engineering design over the last 20 years. This all too often creates a culture of uninspired thought, sometimes lacking in judgement. Numerical solutions now proliferate where once thoughtful, critical analysis was the 'only game in town'. Of course, there is no doubt that both critical state soil mechanics and numerical solutions are positive developments, but, to use them safely and efficiently, these advances must be underpinned by a well-developed understanding of both basic plasticity and elements of continuum mechanics. Our purpose here is to provide an introduction to the basic concepts in as painless a way as possible.

There are of course many excellent books on the theory of plasticity. For beginning students, Calladine's monograph *Engineering Plasticity* (full citation given at the end of Chapter 4) is a superb introduction to aspects of metal plasticity. Nadai's treatise, *Theory of Flow and Fracture in Solids* (cited in Chapter 6), is not only a reference work of great depth and scope but is also notable for taking pains to develop a variety of ideas in the context of modern soil mechanics, together with strong links to continuum mechanics. In the realm of geotechnical literature, nearly all modern textbooks contain varying amounts of material related to both the theory and the application of plasticity. A number of books more or less devoted to critical state theory have appeared since the seminal work *Critical State Soil Mechanics* by Schofield and Wroth (cited in Chapter 3). Our book in no way competes with any of these. Indeed, the exact opposite is true. We delve into critical state theory but only in the most elementary way and only after we have dealt with the classical topics of limit analysis and slip line theory. We merely wish to expose the reader to the potential of critical state analysis in the hope of encouraging further study. Among the more specialised geotechnical literature, two books deserve special mention. Chen's *Limit Analysis and Soil Plasticity* (cited in Chapter 5) contains a wealth of solutions in limit analysis covering many topics of practical interest to geotechnical engineering, and Sokolovski's *Statics of Soil Media* (cited in Chapter 6) presents the most thorough development of slip line analysis. Both books are dedicated to specific aspects of plasticity and could be regarded as required reading for research students. Neither, however, would be especially suitable as an introductory text. Our aim in this book is to fill the gap between elementary soil mechanics and more specialised books such as those by Chen or Sokolovski, as well as the books devoted to critical state theory.

Finally, there are several individuals and organisations to whom we express thanks for their assistance in preparing this book. We are indebted to the Institut A für Mechanik, University of Stuttgart, Germany and Ecole National des Ponts et Chaussées in Paris, where parts of the work were researched and written. One of us (APSS) thanks the University of Canterbury for the award of a Visiting Erskine Fellowship and the Canadian Council for the Arts for the award of a Killam Research Fellowship. The libraries of Cambridge University were a great help in obtaining original references, both early and modern. Professor Malcolm Bolton of Cambridge University first showed us how simply the Cam Clay model can be developed in the context of simple shearing. His development is reiterated in Chapter 7. Dr Glenn McDowell of the University of Nottingham and Professor Jim Hill of the University of Wollongong reviewed parts of the manuscript and made many constructive comments. Much of the writing was done in a Hertfordshire cottage belonging to Mr and Mrs K.A. Maclean. Their hospitality is acknowledged with gratitude. The friendship and encouragement of Mr Norman Travis is also acknowledged with special thanks. Finally, we thank Anne and Sally for their patience and understanding throughout the trials and tribulations of lost files, crashing hard disks, jammed printers, headaches, backaches and all the other joys of writing, and Sally is specially thanked for compiling the index for this volume.

R. O. Davis
Christchurch
June 2002

A. P. S. Selvadurai
Montreal
June 2002

1
Stress and strain

1.1 Introduction

How a material responds to load is an everyday concern for civil engineers. As an example we can consider a beam that forms some part of a structure. When loads are applied to the structure the beam experiences deflections. If the loads are continuously increased the beam will experience progressively increasing deflections and ultimately the beam will fail. If the applied loads are small in comparison with the load at failure then the response of the beam may be proportional, i.e. a small change in load will result in a correspondingly small change in deflection. This proportional behaviour will not continue if the load approaches the failure value. At that point a small increase in load will result in a very large increase in deflection. We say the beam has failed. The mode of failure will depend on the material from which the beam is made. A steel beam will bend continuously and the steel itself will appear to flow much like a highly viscous material. A concrete beam will experience cracking at critical locations as the brittle cement paste fractures. Flow and fracture are the two failure modes we find in all materials of interest in civil engineering. Generally speaking, the job of the civil engineer is threefold: first to calculate the expected deflection of the beam when the loads are small; second to estimate the critical load at which failure is incipient; and third to predict how the beam may respond under failure conditions.

Geotechnical engineers and engineering geologists are mainly interested in the behaviour of soils and rocks. They are often confronted by each of the three tasks mentioned above. Most problems will involve either foundations, retaining walls or slopes. The loads will usually involve the weight of structures that must be supported as well as the weight of the soil or rock itself. Failure may occur by flow or fracture depending on the soil or rock properties. The geo-engineer will generally be interested in the deformations that may occur when the loads are small, the critical load that will bring about failure and what happens if failure does occur.

When the loads are smaller than a critical value, the geotechnical engineer will often represent the soil or rock as an elastic material. This is an approximation but it can be used effectively to provide answers to the first question: what deformations will occur when loads are small? The approximation of soil as a linear elastic material has been explored in a number of textbooks including our own – *Elasticity and Geomechanics*.* For convenience we will refer to this book as *EG*. In *EG* we outlined the fundamentals of the classical or linear theory of elasticity and we investigated some simple applications useful in geotechnical engineering. The book you now hold is meant to be a logical progression from *EG*. *Plasticity and Geomechanics* carries the reader forward into the area of failure and flow. We will outline the mathematical theory of plasticity and consider some simple questions concerning collapse loads, post-failure deformations and why soils behave as they do when stresses become too severe. Like *EG* this book is not meant to be a treatise. It will hopefully provide a concise introduction to the fundamentals of the theory of plasticity and will provide some relatively simple applications that are relevant in geo-engineering.

As a matter of necessity some of the material from *EG* must be repeated here in order that this book may be self-contained. In the present chapter we will cover some fundamental ideas concerning deformation, strain and stress, together with the concept of equilibrium. Chapter 2 then outlines basic elastic behaviour and discusses aspects of inelastic behaviour in respect to soil and rock. The nomenclature used here is similar to that adopted in *EG*. Readers who feel they have a firm grasp of stress, strain and elasticity, especially those who may have spent some time with *EG*, may wish to omit this chapter, and parts of the next, and move more quickly to Chapter 3. In Chapter 3 the concept of *yielding* is introduced. This is the state at which the failure process is about to commence. In Chapter 4 we investigate the process of *plastic flow*. That is, we try to determine the rules that govern deformations occurring once yield has taken place. Chapter 5 considers two important theorems that provide bounds on the behaviour of a plastically deforming material. These theorems may be extremely useful in approximating the response of geotechnical materials in realistic loading situations without necessitating any elaborate mathematics. Chapter 6 briefly touches on the mathematics of finding exact solutions for a limited class of problems and, finally, Chapter 7 introduces certain modern developments in the use of plasticity specifically for soils. The main body of

* Complete references to cited works are given at the end of the chapter where they first appear.

the book is followed by appendices that offer a more rigourous development of several important aspects.

1.2 Soil mechanics and continuum mechanics

Even the most casual inspection of any real soil shows clearly the random, particulate, disordered character we associate with natural materials of geologic origin. The soil will be a mixture of particles of varying mineral (and possibly organic) content, with the pore space between particles being occupied by either water, or air, or both. There are many important virtues associated with this aspect of a soil, not least its use as an agricultural medium; but, when we approach soil in an engineering context, it will often be desirable to overlook its particulate character. Modern theories that model particulate behaviour directly do exist and we will discuss one in Chapter 7, but in nearly all engineering applications we idealise soil as a continuum: a body that may be subdivided indefinitely without altering its character.

The treatment of soil as a continuum has its roots in the eighteenth century when interest in geotechnical engineering began in earnest. Charles Augustus Coulomb, one of the founding fathers of soil mechanics, clearly implied the continuum description of soil for engineering purposes in 1773. Since then nearly all engineering theories of soil behaviour of practical interest have depended on the continuum assumption. This is true of nearly all the soil plasticity theories we discuss in this book.

Relying on the continuum assumption, we can attribute familiar properties to all points in a soil body. For example, we can associate with any point $\boldsymbol{x}$ in the body a mass density ρ. In continuum mechanics we define ρ as the limiting ratio of an elemental mass ΔM and volume ΔV

$$\rho = \lim_{\Delta V \to 0} \frac{\Delta M}{\Delta V} \tag{1.1}$$

Of course we realise that were we to shrink the elemental volume ΔV to zero in a real soil we would find a highly variable result depending on whether the point coincides with the position occupied by a particle, or by water, or by air. Thus we interpret the density in (1.1) as a representative average value, as if the volume remains finite and of sufficient size to capture the salient qualities of the soil as a whole in the region of our point. Similar notions apply to other quantities of engineering interest. For example, there will be forces acting in the interior of the soil mass. In reality they will be unwieldy combinations of interparticle contact forces and hydrostatic forces. We will consider appropriate average forces and permit them to be supported by continuous surfaces. We can then consider the ratio of an elemental force on an elemental area and define stresses within the soil. It is elementary concepts such as these that we wish to elaborate in this chapter.

Although the concept of a continuum is elementary, it represents a powerful artifice, which enables the mathematical treatment of physical and mechanical phenomena in materials with complex internal structure such as soils. It allows us to take advantage of many mathematical tools in formulating theories of material behaviour for practical engineering applications.

1.3 Sign conventions

Before launching into our discussion of stress and strain, we will first consider the question of how signs for both quantities will be determined. In nearly all aspects of solid mechanics, tension is assumed to be positive. This includes both tensile stress and tensile strain. In geomechanics, on the other hand, most practitioners prefer to make compression positive, or at least to have compressive stress positive. This reflects the fact that particulate materials derive strength from confinement and confinement results from compressive stress. We will adopt the convention of compression being positive throughout this textbook.

Naturally, if compressive stress is considered to be positive then so must be compressive strain, and that requirement introduces an awkward aspect to the mathematical development of our subject. We can see the reason for this by considering a simple tension test as shown in Figure 1.1. In the figure a bar of some material is stretched by tensile forces T applied at each end. The axis of the bar is aligned with the coordinate axis x, and the end of the bar at the origin is fixed so that it cannot move. If the bar initially has length L, then application of the force T will be expected to cause an elongation of, say, Δ. Let the *displacement* of the bar be a function of x defined by $u = u(x) = \Delta(x/L)$. Physically the displacement tells us how far the particle initially located at x has moved, due to the force T. The extensional strain in the bar may be written as $\varepsilon = du/dx = \Delta/L$. If we were to adopt the solid mechanics convention of tension being positive, then the force T would be positive and so would be

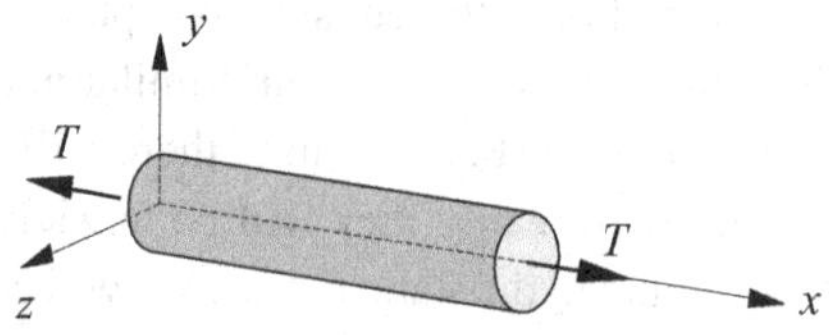

Figure 1.1. Prismatic bar in simple tension.

the extensional strain. Obviously all is well. On the other hand, if we wish to use the geomechanics convention that compression is positive, then the tensile force T is negative; but the strain, defined by $\varepsilon = du/dx$ remains positive. We could simply prescribe ε as a negative quantity, but that would not provide a general description for all situations. Instead we need some general method to correctly produce the appropriate sign for the strain.

There are two possible solutions to our problem. One approach is to redefine the extensional strain as $\varepsilon = -du/dx$. This will have the desired effect of making compressive strain always positive, but will have the undesirable effect of introducing negative signs in a number of equations where they may not be expected by the unwary and hence may cause confusion. The second solution is to agree from the outset that *positive displacements will always act in the negative coordinate direction*. If we adopt this convention, then the displacement of the bar is given by $u = u(x) = -\Delta(x/L)$. This second solution is the one we will adopt throughout the book. As a result nearly all the familiar equations of solid mechanics can be imported directly into our geomechanics context without any surprising negative signs. Moreover, there will be few opportunities where we must refer directly to the sign of the displacements, and so the convention of a positive displacement in the negative coordinate direction will mostly remain in the background. Specific comments will be made wherever we feel confusion might arise.

1.4 Deformation and strain

We begin by considering a continuum body with some generic shape similar to that shown in Figure 1.2. The body is placed in a reference system that we take to be a simple three-dimensional, rectangular Cartesian coordinate

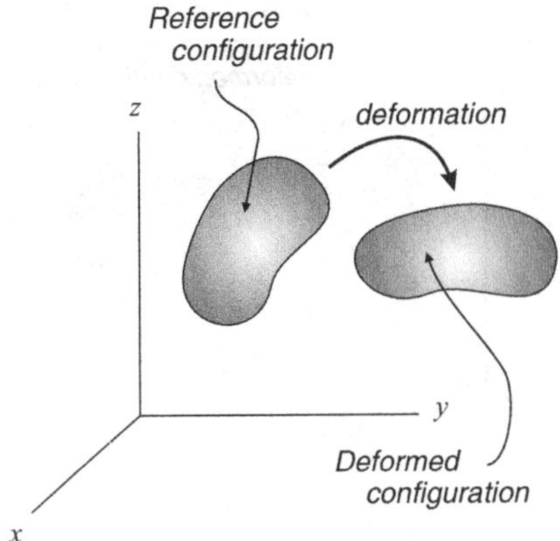

Figure 1.2. Reference and deformed configurations of body.

frame as shown in the figure. A deformation of the body results in it being moved from its original *reference configuration* to a new *deformed configuration*.

All deformations of a continuum are composed of two distinct parts. First there are *rigid motions*. These are deformations for which the shape of the body is not changed in any way. Two categories of rigid motion are possible, *rigid translation* and *rigid rotation*. A rigid translation simply moves the body from one location in space to another without changing its attitude in relation to the coordinate directions. A rigid rotation changes the attitude of the body but not its position.

The second part of our deformation involves all the changes of shape of the body. It may be stretched, or twisted, or inflated or compressed. These sorts of deformations result in *straining*. Strains are usually the most interesting aspect of a deformation.

One way to characterise any deformation is to assign a *displacement vector* to every point in the body. The displacement vector joins the position of a point in the reference configuration to its position in the deformed configuration. We represent the vector by

$$\boldsymbol{u} = \boldsymbol{u}(\boldsymbol{x}, t) \tag{1.2}$$

where $\boldsymbol{x}$ denotes the position of any point within the body and t denotes time. A typical displacement vector is shown in Figure 1.3. Since there is a displacement vector associated with every point in the body, we say there is a *displacement vector field* covering the body. In our x, y, z coordinate frame, $\boldsymbol{u}$ has components denoted by u_x, u_y, u_z. Each component is, in general, a function of position

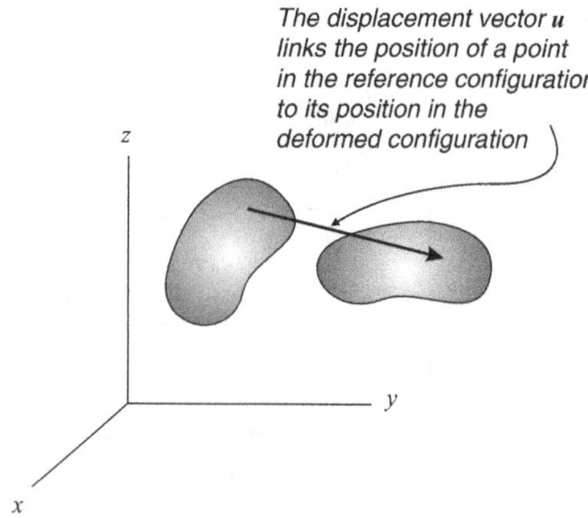

Figure 1.3. The displacement vector.

and time, and, according to our sign convention, components acting in negative coordinate directions will be considered to be positive.

If we know the displacement vector field, then we have complete knowledge of the deformation. Of course, part of the displacement field may be involved with rigid motions while the remainder results from straining. Our first task is to separate the two.

We begin by taking spatial derivatives of the components of the displacement vector. We arrange the derivatives into a 3×3 matrix called the displacement gradient matrix, $\nabla \boldsymbol{u}$.* If we are working in a three-dimensional rectangular Cartesian coordinate system we can represent $\nabla \boldsymbol{u}$ in an array as follows:

$$\nabla \boldsymbol{u} = \begin{bmatrix} \dfrac{\partial u_x}{\partial x} & \dfrac{\partial u_x}{\partial y} & \dfrac{\partial u_x}{\partial z} \\ \dfrac{\partial u_y}{\partial x} & \dfrac{\partial u_y}{\partial y} & \dfrac{\partial u_y}{\partial z} \\ \dfrac{\partial u_z}{\partial x} & \dfrac{\partial u_z}{\partial y} & \dfrac{\partial u_z}{\partial z} \end{bmatrix} \tag{1.3}$$

Note the use of partial derivatives. Note also that the derivatives of $\boldsymbol{u}$ will *not* be affected by rigid translations. This might suggest we could use (1.3) as a measure of strain. But rigid rotations will give rise to non-zero derivatives of $\boldsymbol{u}$, so we need to introduce one more refinement. We use the *symmetric part* of $\nabla \boldsymbol{u}$. Let

$$\boldsymbol{\varepsilon} = \frac{1}{2}[\nabla \boldsymbol{u} + (\nabla \boldsymbol{u})^T] \tag{1.4}$$

We call $\boldsymbol{\varepsilon}$ the *strain matrix*. Note that the superscript T indicates the transpose of the displacement gradient matrix. Also note that $\boldsymbol{\varepsilon}$ is a symmetric matrix. As its name implies, $\boldsymbol{\varepsilon}$ represents the straining that occurs during our deformation. Just as is the case with the displacement vector, $\boldsymbol{\varepsilon}$ is also a function of both position $\boldsymbol{x}$ and time t.

We write the components of $\boldsymbol{\varepsilon}$ as follows:

$$\boldsymbol{\varepsilon} = \begin{bmatrix} \varepsilon_{xx} & \varepsilon_{xy} & \varepsilon_{xz} \\ \varepsilon_{yx} & \varepsilon_{yy} & \varepsilon_{yz} \\ \varepsilon_{zx} & \varepsilon_{zy} & \varepsilon_{zz} \end{bmatrix} \tag{1.5}$$

The diagonal components of $\boldsymbol{\varepsilon}$ are referred to as *extensional strains*,

$$\varepsilon_{xx} = \frac{\partial u_x}{\partial x}, \qquad \varepsilon_{yy} = \frac{\partial u_y}{\partial y}, \qquad \varepsilon_{zz} = \frac{\partial u_z}{\partial z} \tag{1.6}$$

* We use the symbol ∇ to denote the del operator $\frac{\partial}{\partial x}\hat{\boldsymbol{\imath}} + \frac{\partial}{\partial y}\hat{\boldsymbol{\jmath}} + \frac{\partial}{\partial z}\hat{\boldsymbol{k}}$, where $\hat{\boldsymbol{\imath}}, \hat{\boldsymbol{\jmath}}, \hat{\boldsymbol{k}}$ denote the triad of unit base vectors.

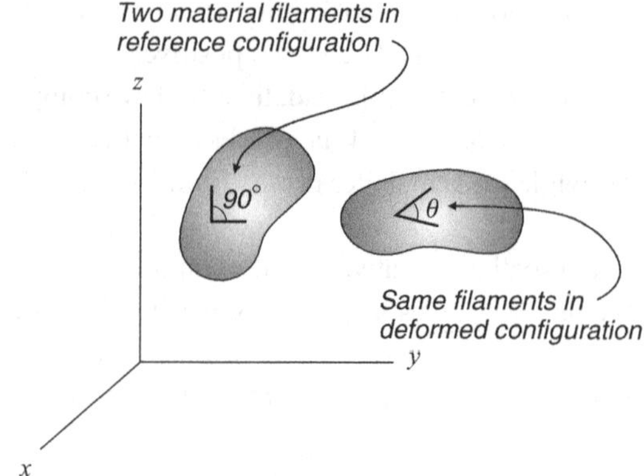

Figure 1.4. Physical meaning of shearing strain.

Each of these represents the change in length per unit length of a material filament aligned in the appropriate coordinate direction.

The off-diagonal components of $\boldsymbol{\varepsilon}$ are called *shear strains*

$$\begin{aligned} \varepsilon_{xy} = \varepsilon_{yx} &= \frac{1}{2}\left(\frac{\partial u_x}{\partial y} + \frac{\partial u_y}{\partial x}\right) \\ \varepsilon_{yz} = \varepsilon_{zy} &= \frac{1}{2}\left(\frac{\partial u_y}{\partial z} + \frac{\partial u_z}{\partial y}\right) \\ \varepsilon_{zx} = \varepsilon_{xz} &= \frac{1}{2}\left(\frac{\partial u_z}{\partial x} + \frac{\partial u_x}{\partial z}\right) \end{aligned} \tag{1.7}$$

These strains represent one-half the increase* in the initially right angle between two material filaments aligned with the appropriate coordinate directions in the reference configuration. For example, consider two filaments aligned with the x- and y-directions in the reference configuration as shown in Figure 1.4. After the deformation the attitude of the filaments may have changed and the angle between them is now θ. Then $2\varepsilon_{xy} = 2\varepsilon_{yx} = \theta - \pi/2$. The presence of the factor of $\frac{1}{2}$ in (1.7) is important to ensure that the strain matrix will give the correct measure of straining in different coordinate systems. Often the change in an initially right angle (rather than one-half the change) is referred to as the *engineering shear strain*. It is usually denoted by the Greek letter gamma, γ. Obviously if we know one of the shear strains defined in (1.7), then we can determine the corresponding engineering shear strain.

* In solid mechanics the shear strain represents the *decrease* in the right angle. We have the *increase* because of the assumption that compression is positive and our sign convention for displacements.

An important aspect of the definition of the strain matrix in (1.4) is the requirement that the displacement derivatives remain small during the deformation. Sometimes the matrix $\boldsymbol{\varepsilon}$ is referred to as the *small* strain matrix. The name is meant to imply that the components of $\boldsymbol{\varepsilon}$ are only a correct measure of the actual straining so long as the components of $\nabla \boldsymbol{u}$ are much smaller in magnitude than 1. More complex definitions of strain are required in the case where deformation gradient components have large magnitudes. If the components of $\nabla \boldsymbol{u}$ are $\ll 1$, then products of the components can be ignored and the small-strain definition (1.4) results. There are substantial advantages associated with the small-strain matrix $\boldsymbol{\varepsilon}$ because it is a linear function of the displacement derivatives, while the large-strain measures are not. Because of this fact we may find that $\boldsymbol{\varepsilon}$ is used in some situations where it is not strictly applicable. Simple solutions are often good solutions, even if they are technically only approximations, and in geotechnical engineering the virtue of simplicity may justify a considerable loss of rigour.

Arising from the small-strain approximation is another measure of strain, the *volumetric strain, e*. It represents the change in volume per unit volume of the material in the reference configuration. It is defined as the sum of the three extensional strains:

$$e = \varepsilon_{xx} + \varepsilon_{yy} + \varepsilon_{zz} = \nabla \cdot \boldsymbol{u} \tag{1.8}$$

Here $\nabla \cdot \boldsymbol{u}$ represents the divergence of the vector $\boldsymbol{u}$.* There are a number of instances where the sum of the diagonal terms of a matrix gives a useful result. Because of this we define an operator called the *trace*, abbreviated as *tr*, which gives the sum. Thus (1.8) could also be written as $e = tr(\boldsymbol{\varepsilon})$.

In classical plasticity theory where metals are the primary material of interest, it is usual to assume that the material is incompressible and hence e is always zero. This is often not the case for soils, at least when they are permitted to drain. In undrained situations a fully saturated soil may be nearly incompressible, but if drainage can occur volume change is likely. In keeping with our definition of extensional strain, compressive volumetric strain will be considered to be positive.

Finally, note that all of the development above is based on the assumption that we are using a rectangular or Cartesian coordinate frame. At times it may be more convenient to use cylindrical or spherical coordinates. In that case there will be some subtle differences in many of the results given thus far. Appendix A

* It is the scalar quantity defined by $\nabla \cdot \boldsymbol{u} = (\frac{\partial}{\partial x}\hat{\boldsymbol{i}} + \frac{\partial}{\partial y}\hat{\boldsymbol{i}} + \frac{\partial}{\partial z}\hat{\boldsymbol{j}}) \cdot (u_x\hat{\boldsymbol{i}} + u_y\hat{\boldsymbol{j}} + u_z\hat{\boldsymbol{k}}) = \frac{\partial u_x}{\partial x} + \frac{\partial u_y}{\partial y} + \frac{\partial u_z}{\partial z}$.

outlines how one moves from rectangular to cylindrical or spherical coordinates and summarises the main results in non-Cartesian coordinate frames.

1.5 Strain compatibility

An important concept with regard to deformation and strain is the idea of *strain compatibility*. In simplest terms this is the physically reasonable requirement that when an intact body deforms, it does so without the development of *gaps or overlaps*. To be a little more precise, consider a point in the reference configuration, and construct some small neighbourhood of surrounding points. If we examine that same point in the deformed configuration, then we would hope to find the same neighbouring points surrounding it and, moreover, we would expect them to have similar relationships to the central point. That is, if neighbouring points α and β are arranged in the reference configuration so that α is closer and β more distant from the central point, then that arrangement should prevail in the deformed configuration as well.

Another way to look at this concept is to consider the definition of the strain matrix itself (1.4). We see that six independent components of strain are obtained from three independent components of displacement. If the displacement vector field is fully specified, then there is clearly no difficulty in determining the strains, but what if the problem is turned around? Suppose the six components of strain are specified. Is it then possible to integrate (1.4) to determine the three displacements uniquely? In general it is not. Moving from strains to displacements we find that the problem is over-determined, i.e. we have more equations than unknowns.

The great French mathematician Barré de Saint-Venant solved the general problem of strain compatibility in 1860. He showed that the strain components must satisfy a set of six *compatibility equations* shown in (1.9). A derivation of these equations may be found in Appendix A of *EG*. The derivation shows how equations (1.9) given below ensure that (1.4) can be integrated to yield single-valued and continuous displacements:

$$\frac{\partial^2 \varepsilon_{xx}}{\partial y^2} + \frac{\partial^2 \varepsilon_{yy}}{\partial x^2} = 2\frac{\partial^2 \varepsilon_{xy}}{\partial x \partial y}$$

$$\frac{\partial^2 \varepsilon_{yy}}{\partial z^2} + \frac{\partial^2 \varepsilon_{zz}}{\partial y^2} = 2\frac{\partial^2 \varepsilon_{yz}}{\partial y \partial z}$$

$$\frac{\partial^2 \varepsilon_{zz}}{\partial x^2} + \frac{\partial^2 \varepsilon_{xx}}{\partial z^2} = 2\frac{\partial^2 \varepsilon_{xz}}{\partial x \partial z}$$

$$\begin{aligned}
\frac{\partial^2 \varepsilon_{xx}}{\partial y \partial z} &= -\frac{\partial^2 \varepsilon_{yz}}{\partial x^2} + \frac{\partial^2 \varepsilon_{zx}}{\partial x \partial y} + \frac{\partial^2 \varepsilon_{xy}}{\partial x \partial z} \\
\frac{\partial^2 \varepsilon_{yy}}{\partial z \partial x} &= -\frac{\partial^2 \varepsilon_{zx}}{\partial y^2} + \frac{\partial^2 \varepsilon_{xy}}{\partial y \partial z} + \frac{\partial^2 \varepsilon_{yz}}{\partial y \partial x} \\
\frac{\partial^2 \varepsilon_{zz}}{\partial x \partial y} &= -\frac{\partial^2 \varepsilon_{xy}}{\partial z^2} + \frac{\partial^2 \varepsilon_{yz}}{\partial z \partial x} + \frac{\partial^2 \varepsilon_{zx}}{\partial z \partial y}
\end{aligned} \tag{1.9}$$

Finally, it is perhaps worth noting that the compatibility conditions impose a kinematic constraint on the strains in a continuum where the mechanical behaviour is as yet unspecified.

1.6 Forces and tractions

We approach the concept of stress through considering the forces that act on an exterior boundary or inside the body. We are aware that there are two distinct types of forces: *contact forces* and *body forces*. Body forces are forces caused by outside influences such as gravity or magnetism. They are associated with the volume or mass of the body and they are fully specified at the outset of any problem. Contact forces are associated with surfaces, either surfaces inside the body or segments of the exterior bounding surface of the body. Contact forces result from the action of the body on itself, such as the tension that exists inside a stretched rubber band or from specified boundary conditions such as an applied load on the upper surface of a beam.

For the time being we will concentrate our attention on contact forces. Every contact force is associated with a surface, so we consider a small element of surface dA embedded somewhere inside our continuum body. If we magnify the element as shown in Figure 1.5 then we can see its associated contact force as a vector $d\boldsymbol{F}$. Presumably $d\boldsymbol{F}$ results from the action of the body on itself since dA lies in the interior of the body. We then define the *surface traction vector*, $\boldsymbol{T}$, as the limiting value of the ratio of force and area.

$$\boldsymbol{T} = \lim_{dA \to 0} \frac{d\boldsymbol{F}}{dA} \tag{1.10}$$

We are aware of course that in the context of a real soil the limiting process must be treated with considerable care. We are concerned with a continuum, or at least a continuum approximation of the real material. In a real soil we would not wish to shrink dA to zero area, rather to terminate the limiting process at some point giving a reasonable representation of the soil structure.

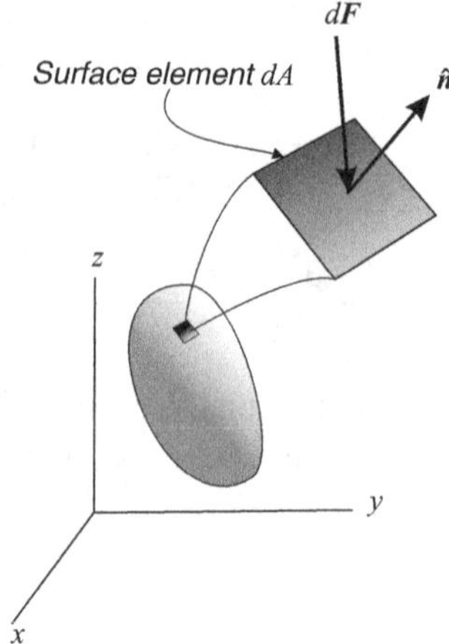

Figure 1.5. Traction vector acting on a surface element.

Note that the traction vector $\boldsymbol{T}$ is directly associated with the particular surface element we have chosen. If we choose a different surface element at the same point in the body, we will generally find a different traction vector. Therefore we see that the *orientation* of the surface element plays an important role. Since there are infinitely many possible orientations for our surface, there are infinitely many traction vectors operating at any given point in the body. This fact raises significant problems with regard to the description of stress. A number of eminent researchers in the eighteenth century were unsure of how stress might be easily characterised in all but simple problems. As it turns out, the problem is not difficult. We will only need to know tractions on three surfaces in order to fully prescribe the traction on any other surface.

1.7 The stress matrix

In 1823 the French mathematician Augustin Cauchy showed how we may solve the problem of determining the traction vector for a given surface. First we need to identify the orientation of the surface we are interested in. This is accomplished by the construction of a unit vector $\hat{\boldsymbol{n}}$ normal to the surface as shown in Figure 1.5. Then Cauchy showed that the product of a 3×3 square matrix $\boldsymbol{\sigma}^T$ with the vector $\hat{\boldsymbol{n}}$ gives the traction $\boldsymbol{T}$ acting on the surface,

$$\boldsymbol{T} = \boldsymbol{\sigma}^T \hat{\boldsymbol{n}} \tag{1.11}$$

This equation is derived in detail in Appendix C of *EG*. In equation (1.11) the superscript T indicates the transpose of the matrix. The matrix $\boldsymbol{\sigma}$ is called the

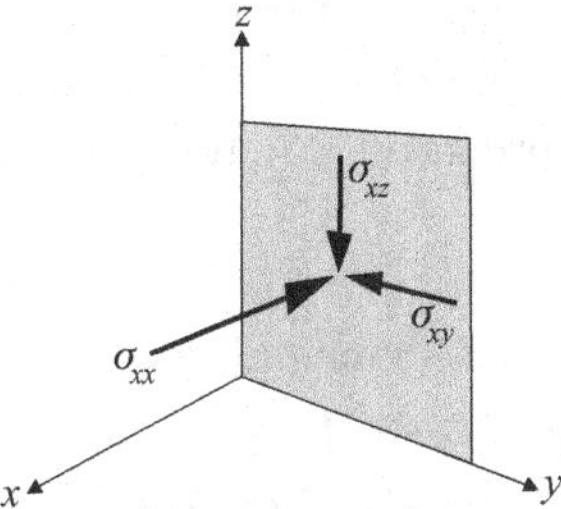

Figure 1.6. Components of the stress matrix acting on a surface perpendicular to the x-direction.

stress matrix. Its component form looks like this

$$\boldsymbol{\sigma} = \begin{bmatrix} \sigma_{xx} & \sigma_{xy} & \sigma_{xz} \\ \sigma_{yx} & \sigma_{yy} & \sigma_{yz} \\ \sigma_{zx} & \sigma_{zy} & \sigma_{zz} \end{bmatrix} \tag{1.12}$$

Each of the components, σ_{xx}, σ_{xy}, etc. is a component of a particular surface traction vector. For example, the components of the first row of $\boldsymbol{\sigma}$ are precisely the components of the traction vector that acts on a surface which is perpendicular to the x-axis as shown in Figure 1.6. This follows immediately if we note that the unit normal vector to the surface is $\hat{\boldsymbol{n}} = [1, 0, 0]^T$. Similarly, the second and third rows of the $\boldsymbol{\sigma}$ matrix are composed of, respectively, the components of traction vectors acting on surfaces perpendicular to the y- and z-axes. The subscripts of the stress matrix components identify which component of which surface traction is being represented. The xx-component, σ_{xx}, is the x-component of the traction acting on the surface perpendicular to the x-direction. Similarly, σ_{xy} is the y-component of that same traction. The yz-component, σ_{yz}, is the z-component of the traction acting on the surface perpendicular to the y-direction.

Note that in Figure 1.6 the stress matrix components are drawn pointing in the opposite direction to the coordinate axes. Because of this σ_{xx} appears to be a compressive stress. This is the usual sign convention in geomechanics where compression is positive.

The diagonal components of $\boldsymbol{\sigma}(\sigma_{xx}, \sigma_{yy}, \sigma_{zz})$ are called the normal stress components, or simply the normal stresses. They act normal to the three surfaces perpendicular to the three coordinate directions. The off-diagonal components, $\sigma_{xy}, \sigma_{yz}, \ldots$ are called the shear stress components, or simply shear stresses. They act tangential to the three surfaces. Cauchy also showed that, in the absence of internal couples, the shear stresses must be complementary and hence the stress matrix is symmetric, i.e. $\sigma_{xy} = \sigma_{yx}, \sigma_{xz} = \sigma_{zx}, \sigma_{zy} = \sigma_{yz}$.

Because of this fact, the transpose of $\boldsymbol{\sigma}$ in (1.11) is not really important. We choose not to omit it, however, since the understanding of the physical meaning of the stress components springs directly from the equation.

1.8 Principal stresses

At any point in the body there will always be at least three surfaces on which the shear stresses $\sigma_{xy}, \sigma_{yz}, \ldots$ will vanish. These are the *principal surfaces* or *principal planes*. To see how this comes about note that if there is no shear stress on a surface the traction vector $\boldsymbol{T}$ must be parallel to the unit normal vector $\hat{\boldsymbol{n}}$. Then using (1.11) we see that

$$\boldsymbol{T} = \boldsymbol{\sigma}^T \hat{\boldsymbol{n}} = \alpha \hat{\boldsymbol{n}} \tag{1.13}$$

where α is a scalar multiplier. We can rearrange this result to obtain

$$(\boldsymbol{\sigma} - \alpha \mathbf{I}) \hat{\boldsymbol{n}} = 0 \tag{1.14}$$

where $\mathbf{I}$ denotes the identity matrix and we have used the fact that $\boldsymbol{\sigma}$ is a symmetric matrix. Equation (1.14) gives three homogeneous linear equations. We know from linear algebra that there will either be no solutions, infinitely many solutions or a unique solution for any system of homogeneous linear equations. The condition for the existence of a unique solution is

$$\det(\boldsymbol{\sigma} - \alpha \mathbf{I}) = 0 \tag{1.15}$$

So we have an eigenvalue problem. If we expand the determinant in (1.15) we find the following *characteristic equation*:

$$-\alpha^3 + I_1 \alpha^2 - I_2 \alpha + I_3 = 0 \tag{1.16}$$

where the coefficients I_1, I_2 and I_3 are functions of the stress matrix components $\sigma_{xx}, \sigma_{xy}, \ldots$. This cubic equation will have three roots (or three eigenvalues) for the multiplier α. Referring back to (1.13) we see that the roots will be the physical magnitudes of the traction $\boldsymbol{T}$ on each of the surfaces where there is no shear stress. We call these the *principal stresses* and denote them by σ_1, σ_2 and σ_3. Compression is taken to be positive here as everywhere in our development.

The greatest and least principal stress are called the *major principal stress* and *minor principal stress*, respectively. The remaining stress is called the *intermediate principal stress*. In some applications it is convenient to agree to number the principal stresses so that σ_1 is the major principal stress while σ_3 is the minor principal stress. This is a common convention but it may not always be the preferred option and we will not apply any particular rule to how σ_1, σ_2

and σ_3 may be related. In some circumstances we may have the conventional definition of $\sigma_1 \geq \sigma_2 \geq \sigma_3$, but at other times it may be more convenient to have $\sigma_3 \geq \sigma_2 \geq \sigma_1$, or one of the other four possible permutations of the three indices.

If we now substitute each of σ_1, σ_2 or σ_3 back in (1.13) to replace α, we can solve for the corresponding eigenvectors $\hat{\boldsymbol{n}}_1, \hat{\boldsymbol{n}}_2, \hat{\boldsymbol{n}}_3$. These three vectors are called the principal directions. They define the three *principal surfaces*, i.e. the surfaces on which $\boldsymbol{T}$ and $\hat{\boldsymbol{n}}$ are parallel and therefore the surfaces that support no shear. A theorem from linear algebra assures us that the eigenvectors will be mutually orthogonal, hence the principal surfaces will also be mutually orthogonal. This can be a particularly useful result. It means that we can always find *some* coordinate system, say x', y', z', such that the coordinate directions are parallel to the principal directions (or perpendicular to the principal surfaces). In that coordinate system the stress matrix will have this simple form

$$\boldsymbol{\sigma} = \begin{bmatrix} \sigma_1 & 0 & 0 \\ 0 & \sigma_2 & 0 \\ 0 & 0 & \sigma_3 \end{bmatrix} \tag{1.17}$$

That is, in this particular coordinate system, surfaces that are perpendicular to the coordinate axes support no shear. They are the principal surfaces.

Another interesting point arises here. Note that regardless of what coordinate system we happen to use, the principal stresses are independent entities. The components of $\boldsymbol{\sigma}$ at a point will, in general, be different in different coordinate systems, but the three principal stresses that we determine by finding the roots of (1.16) will always be the same. They are unique quantities associated with the particular point of interest in the continuum. We say that the principal stresses are *invariant* under a coordinate transformation. Invariants are often useful quantities owing to their independence from our choice of coordinate directions. This can be especially useful when it comes to creating descriptions of how materials behave. Obviously a material cannot know what coordinate directions we have chosen to use for its description. Therefore it would be unwise to create a model for the material stress–strain response that depended on the coordinate axis directions. But if we model the material using invariant quantities such as principal stresses, then there is no connection between the material model and the chosen frame of reference.

Of course σ_1, σ_2 and σ_3 are not the only invariant quantities associated with the stress matrix. It also follows from (1.16) that the three coefficients I_1, I_2 and I_3 must also be invariants. This must be true since, if we were to substitute one of the invariant principal stresses for the quantity α, the equation would be satisfied. If the principal stresses do not depend on the choice of coordinates, then neither can the coefficients I_1, I_2 and I_3. We call I_1, I_2 and I_3 the *principal stress invariants*. They are related to the components of the stress matrix by the following equations:

$$\begin{aligned} I_1 &= tr(\boldsymbol{\sigma}) \\ I_2 &= \frac{1}{2}[(tr\boldsymbol{\sigma})^2 - tr(\boldsymbol{\sigma}^2)] \\ I_3 &= \det(\boldsymbol{\sigma}) \end{aligned} \tag{1.18}$$

where we recall that the trace operator *tr* gives the sum of the diagonal components of the matrix. In the event that our coordinate system happened to align with the principal directions, and the stress matrix had the simple form shown in (1.17), the above equations would become

$$\begin{aligned} I_1 &= \sigma_1 + \sigma_2 + \sigma_3 \\ I_2 &= \sigma_1\sigma_2 + \sigma_2\sigma_3 + \sigma_3\sigma_1 \\ I_3 &= \sigma_1\sigma_2\sigma_3 \end{aligned} \tag{1.19}$$

Of course these equations are always true regardless of the choice of coordinate system. The sum of two invariant quantities will itself be invariant, as will the product of two invariants. For that matter any combination of invariants will also be an invariant. Equation (1.19) is simply the universal relationship between the principal stresses and the principal stress invariants. Note that the dimensions of the three principal invariants are [*stress*], [$stress^2$] and [$stress^3$].

One other invariant quantity that is often defined is the *mean stress* or *pressure*, denoted by p. It is equal to one-third of the first invariant, $I_1/3$. Thus $p = (\sigma_1 + \sigma_2 + \sigma_3)/3 = (\sigma_{xx} + \sigma_{yy} + \sigma_{zz})/3$. In the theory of elasticity, tensile stress is commonly taken as positive and the pressure is defined as the negative of $I_1/3$ so that positive pressure is compressive. We have no need of that definition since we have made compressive stress positive from the outset.

1.9 Mohr circles

Next, suppose we want to consider the stress state in a body at a specified point. Let us assume that the components of the stress matrix are known. In that case (1.11) applies and we can determine the traction $\boldsymbol{T}$ acting on any surface passing through the point. We could characterise the stress state by simply writing out the stress matrix, or we could list the principal stresses and the principal directions. In either case six independent numbers would be required.* If we

* Why are only six numbers needed to describe the three principal stresses and three principal directions? See Exercise 1.5.

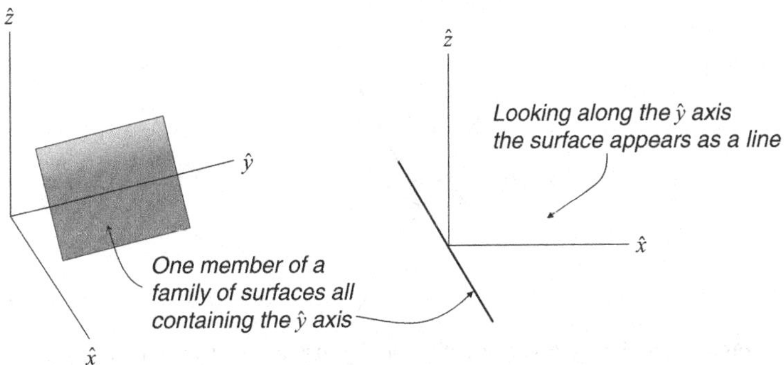

Figure 1.7. One of infinitely many surface elements generated by the $\hat{y}$-axis.

wished, we could visualise the stress state as a point in a six-dimensional space. However, there is another way to characterise the stress. We can create a simple graphical representation called the Mohr stress circle. The Mohr stress circle, or simply Mohr circle, is so important in relation to the theory of plasticity that Appendix B is completely devoted to its development. Only the major points will be described here to ensure that this introductory chapter remains brief.

Again suppose that the components of the stress matrix are known for some particular point in the body. Then we could solve the eigenvalue problem (1.14) to find the principal directions $\hat{\boldsymbol{n}}_1, \hat{\boldsymbol{n}}_2, \hat{\boldsymbol{n}}_3$. These three vectors form the basis for a coordinate system that we might represent by $\hat{x}$, $\hat{y}$, $\hat{z}$ as shown in Figure 1.7. We know that the principal surfaces must be perpendicular to these coordinate directions. Suppose we now consider a family of surfaces composed of all the surfaces that are perpendicular to the $(\hat{x}, \hat{z})$-plane. One particular surface is shown in Figure 1.7. Any other member of the family could be obtained by rotating that surface about the $\hat{y}$-axis. The $\hat{y}$-axis is called a *generator* for the family of surfaces. We can use (1.11) to ascertain the traction vector $\boldsymbol{T}$ for each surface of our family. This will give us infinitely many traction vectors, but we won't worry about that point for the moment. Each traction vector $\boldsymbol{T}$ will have components in the $\hat{x}$- and $\hat{z}$-directions, but the component in the $\hat{y}$-direction will always be zero. This is a consequence of using the principal directions as our coordinate system.

To obtain a Mohr stress circle, we now plot the components of all the traction vectors for all the surfaces of our family. However, we do not plot the traction components acting in the $\hat{x}$ and $\hat{z}$ directions. Instead we plot the components that act normal and tangential to the surface on which $\boldsymbol{T}$ acts. To be more precise, consider the surface shown in Figure 1.7. If we arrange our view point so that we look directly down the $\hat{y}$-axis, we see the situation shown in Figure 1.8. In that figure the $\hat{y}$-axis is perpendicular to the plane of the figure and we see

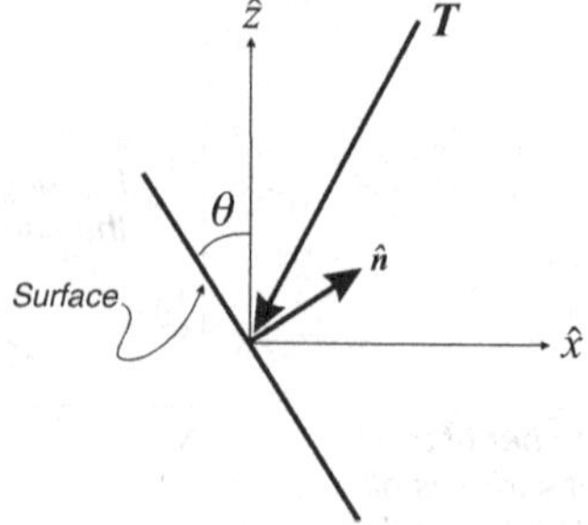

Figure 1.8. Traction vector acting on the surface element in Figure 1.7.

it as a point at the origin. Our surface appears as a line. Both the normal vector to the surface $\hat{\boldsymbol{n}}$ and the traction vector $\boldsymbol{T}$ are shown and both lie in the plane of the figure.

If we use the angle θ shown in Figure 1.8 to identify the particular surface, then the unit normal vector components can be written as

$$\hat{\boldsymbol{n}} = \begin{bmatrix} \sin\theta \\ \cos\theta \end{bmatrix} \tag{1.20}$$

Also, since the coordinate axes are parallel to the principal directions, the stress matrix will have the form (1.17). Then (1.11) gives the following result for the components of $\boldsymbol{T}$ in the $\hat{x}$- and $\hat{z}$-directions:

$$\boldsymbol{T} = \begin{bmatrix} \sigma_1 \sin\theta \\ \sigma_3 \cos\theta \end{bmatrix} \tag{1.21}$$

Now let σ and τ identify the components of $\boldsymbol{T}$ that act normal and tangential to our surface. We find σ by taking the inner product of $\boldsymbol{T}$ and $\hat{\boldsymbol{n}}$

$$\sigma = \boldsymbol{T} \cdot \hat{\boldsymbol{n}} = \sigma_1 \sin^2\theta + \sigma_3 \cos^2\theta \tag{1.22}$$

It is similarly easy to show that

$$\tau = (\sigma_1 - \sigma_3)\sin\theta\cos\theta \tag{1.23}$$

The final step is to plot τ against σ for all the surfaces as θ varies between 0 and π.* The result is a circle, the Mohr circle. A typical Mohr circle is shown in Figure 1.9. Each point on the circumference of the circle identifies the normal and tangential components of the traction vector acting on one particular member of our family of surfaces. We refer to the points on the circle circumference as *stress points*.

* Note that there is no need to let θ run to 2π since a rotation of only π radians brings us back to our starting surface.

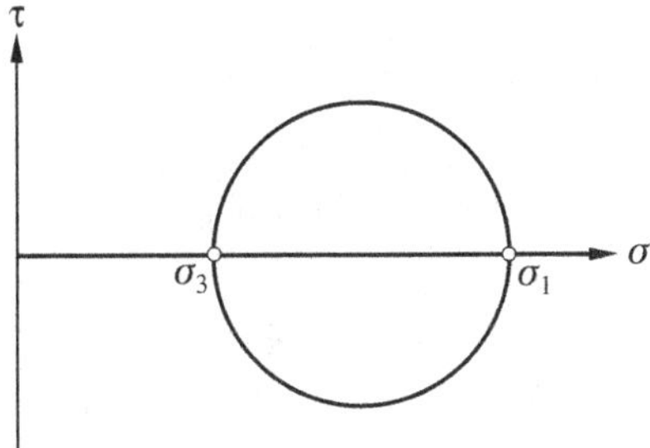

Figure 1.9. Mohr stress circle.

The centre of the circle must lie on the σ-axis. The circle crosses the σ-axis at the points that correspond to the two surfaces that support zero shear: the principal surfaces. As a result the diameter of the circle is the principal stress difference, in this case $(\sigma_1 - \sigma_3)$. The greatest and least shear stresses are equal to the positive and negative values of the circle radius $(\sigma_1 - \sigma_3)/2$. If once again we think of physically rotating the surface shown in Figure 1.8, then a rotation of π radians will result in the corresponding stress point moving completely around the circle and returning to its original starting point. In Appendix B the exact relationship between any surface and its corresponding stress point is developed in full.

The Mohr circle in Figure 1.9 contains all the stress information for all the surfaces of our family. Obviously, however, there are many other surfaces we have not yet considered. We could easily go through the same procedures for surfaces generated by the $\hat{x}$-axis and this would give another Mohr circle. Since the $\hat{x}$-axis corresponds to the $\hat{\boldsymbol{n}}_1$ principal direction, the resulting circle would cross the σ-axis at the principal stresses σ_2 and σ_3. Similarly, if we considered surfaces generated by the $\hat{z}$-axis, we would obtain a third circle spanning the principal stresses σ_2 and σ_1. The three circles might look like those sketched in Figure 1.10. Note how the circles join at the principal stresses and how each circle spans two of the principal stress values. We have drawn the figure as if $\sigma_1 > \sigma_2 > \sigma_3$, i.e. the usual convention used for numbering principal stresses, but we realise that any other numbering, such as $\sigma_2 > \sigma_3 > \sigma_1$, is equally possible.

Now we have exhausted all the obvious possibilities for surfaces. We have considered all the surfaces that are generated by each of the three principal directions and this has led to three Mohr circles. What about all of the other possible surfaces that are not generated by the principal directions but instead are oriented at non-right angles to the principal surfaces? These surfaces will generally have traction vectors that have non-zero components in all three of

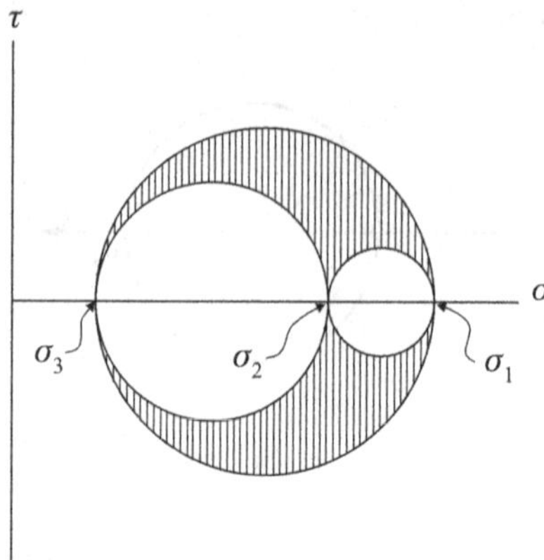

Figure 1.10. Mohr stress circle for the three-dimensional stress state.

the coordinate directions $\hat{x}$, $\hat{y}$ and $\hat{z}$. If we determine their components normal and tangential to their respective surfaces, and plot the components σ and τ we find that the resulting points exactly fill the regions between the three Mohr circles. That is, the stress points associated with these remaining surfaces all fall within the hatched regions in Figure 1.10. The three circles plus the interior points represent the entire stress state graphically.

Often, because of symmetry about the σ-axis, only the upper half of the Mohr stress circle is drawn. Also only the outermost circle is frequently shown. This reflects the fact that the most extreme stress states are represented by points on the outermost circle. Regardless of these details, the Mohr circle is an extremely useful tool. It allows one to visualise the entire stress state at any point in a body easily and it permits an intuitive grasp of stress that is not possible by considering formal equations such as (1.11). Later in the book when we consider yield criteria the Mohr circle will be a very valuable tool.

1.10 The effective stress principle

A concept familiar to all geotechnical engineers is the effective stress principle. It was formulated by one of the founding fathers of soil mechanics, Karl Terzaghi, in 1925. Terzaghi realised that in a saturated soil the solid particle skeleton must play a much more important role than the pore water. This is particularly true in regard to shearing stresses since the pore water can carry no shear stress at all. All shearing stresses are supported by the solid particle skeleton. The situation with normal stresses, however, is not quite so clear.

Let us consider a fully saturated soil that has been subjected to loads of one kind or another. For any surface within the soil we can determine σ, the component of the traction vector $\boldsymbol{T}$ that acts normal to the surface. The effective stress principle tells us that we may view the normal stress σ as if it were composed of two parts. The stress in the pore water will be an isotropic stress called the *pore pressure*. It is usually denoted by u. The remaining normal stress is the part of σ that is supported by the solid particle skeleton. It is called the *effective normal stress* or simply the *effective stress*. We denote it by σ'. The overall stress σ is often referred to as the *total stress*. The effective stress principle states that

$$\sigma = \sigma' + u \tag{1.24}$$

In other words, Terzaghi suggested that we can decompose the total stress into two parts: the effective stress and the pore pressure, each associated with a different soil constituent.

The total stress σ is the stress that has global significance in the sense that it conforms to the requirement for equilibrium. We discuss equilibrium in the next section, but, from an intuitive standpoint, we can view σ as being the stress necessary to ensure that the soil mass remains in equilibrium with whatever forces may be acting on it. In contrast, neither u or σ' is directly related to equilibrium of the soil mass. Generally u results simply from hydrostatic forces within the pore fluid. The effective stress σ' is, in some undefined way, an average normal stress acting within the solid particle skeleton and associated with the particular surface in question.

We can generalise the effective stress concept to define an *effective stress matrix* $\boldsymbol{\sigma}'$. It is given by

$$\boldsymbol{\sigma}' = \boldsymbol{\sigma} - u\mathbf{I} \tag{1.25}$$

Now consider the state of stress at some point within a body of saturated soil. Suppose we wish to plot the Mohr stress circle for this point. It turns out that we can use either σ or σ' as the normal stress component. That is, we can plot the *total stress* Mohr circle, using σ, or we can plot the *effective stress* Mohr circle, using σ'. Note that since u is an isotropic stress, its Mohr circle is simply a point. The pore water supports no shear stress. Equation (1.24) then holds the key to how we can plot two Mohr circles. For any surface the total and effective stresses differ by the pore pressure u. Therefore the two Mohr circles will be separated by an amount equal to u. The effective stress circle will lie to the left of the total stress circle, so long as $u > 0$. A typical example is shown in Figure 1.11. Every corresponding stress point on the two circles is separated horizontally by an amount u.

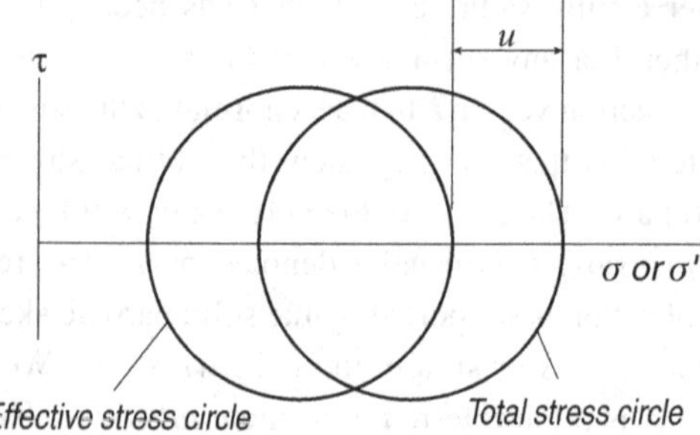

Figure 1.11. Total stress and effective stress Mohr circles.

1.11 Equilibrium

To conclude this chapter we need to introduce the concept of equilibrium of forces in relation to the body forces and contact forces that may be found inside the body. We will consider static equilibrium only, although the generalisation to dynamic conditions is not difficult. To begin, consider some region of the body with volume V and surface S such as that shown in Figure 1.12. This might be any part of the body or even all of it.

Both contact and body forces will act on V. We can think of the body forces simply as being gravity. Other body forces are possible, but gravity will generally be the only body force of interest to geotechnical engineers. Let the body force per unit volume be represented by a vector $\boldsymbol{b}$. We can think of $\boldsymbol{b}$ as being a vector with magnitude equal to the product of the acceleration of gravity times the mass density ρ, and direction pointing to the centre of the Earth. Then the total body force acting on V will be given by the integral of $\boldsymbol{b}$ over the

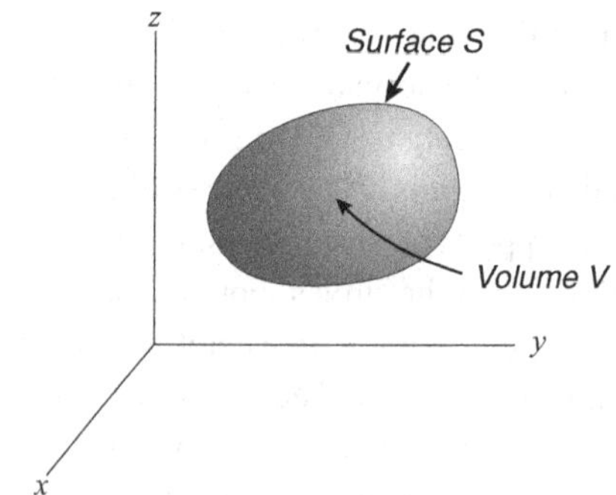

Figure 1.12. An arbitrary region with surface S and volume V.

volume V:

$$\text{total body force} = \int_V \boldsymbol{b}\, dV \tag{1.26}$$

Note that $\boldsymbol{b}$ may be a function of position inside V, assuming the mass density changes from place to place.

There will be contact forces associated with surfaces throughout the volume V, but nearly all of them will make no difference to equilibrium. The reason is, for any surface inside V, there will be equal and opposite traction vectors acting on either side of the surface and they will equilibrate each other. It is only on the surface S that we find non-equilibrating tractions or contact forces. The surface S separates V from the remainder of the body. That part of the body outside S will exert contact forces on V and we must take them into account when working out equilibrium. The contact force on some element of surface dS will be given by the product $\boldsymbol{T}\, dS$. The total contact force acting on V is therefore

$$\text{total contact force} = \int_S \boldsymbol{T}\, dS = \int_S \boldsymbol{\sigma}^T \hat{\boldsymbol{n}}\, dS \tag{1.27}$$

where (1.11) has been used. The vector $\hat{\boldsymbol{n}}$ is the *outward* unit normal vector to the surface S. We must take the outward-pointing normal vector in order to obtain the traction that acts *from* the remainder of the body *on to* the volume V. There is a powerful theorem from vector calculus that is useful here. It is called the divergence theorem and it can be used to convert the integral over S in (1.27) into an integral over the volume enclosed by S,

$$\int_S \boldsymbol{\sigma}^T \hat{\boldsymbol{n}}\, dS = \int_V \boldsymbol{\nabla} \cdot \boldsymbol{\sigma}^T\, dV \tag{1.28}$$

where $\boldsymbol{\nabla} \cdot \boldsymbol{\sigma}^T$ represents the divergence of the stress matrix transpose. Recalling that the divergence of a vector gives a scalar, it is similarly true that the divergence of a matrix (in fact a tensor) gives us a vector. In rectangular Cartesian coordinates the components of $\boldsymbol{\nabla} \cdot \boldsymbol{\sigma}^T$ are

$$\boldsymbol{\nabla} \cdot \boldsymbol{\sigma}^T = \begin{bmatrix} \dfrac{\partial \sigma_{xx}}{\partial x} + \dfrac{\partial \sigma_{yx}}{\partial y} + \dfrac{\partial \sigma_{zx}}{\partial z} \\ \dfrac{\partial \sigma_{xy}}{\partial x} + \dfrac{\partial \sigma_{yy}}{\partial y} + \dfrac{\partial \sigma_{zy}}{\partial z} \\ \dfrac{\partial \sigma_{xz}}{\partial x} + \dfrac{\partial \sigma_{yz}}{\partial y} + \dfrac{\partial \sigma_{zz}}{\partial z} \end{bmatrix} \tag{1.29}$$

Finally, if we assume that the body is in static equilibrium, we must set the contact forces and body forces equal. This gives

$$\int_V \nabla \cdot \boldsymbol{\sigma}^T \, dV = \int_V \boldsymbol{b} \, dV \tag{1.30}$$

or

$$\int_V (\nabla \cdot \boldsymbol{\sigma}^T - \boldsymbol{b}) \, dV = 0 \tag{1.31}$$

But the volume V was completely general in that it may be any region of the body. If (1.31) must be zero for *any* region the only conclusion we can draw is that the integrand itself must be zero everywhere inside the body. So equilibrium is expressed by the vector equation*

$$\nabla \cdot \boldsymbol{\sigma}^T - \boldsymbol{b} = \boldsymbol{0} \tag{1.32}$$

In component form we have three equations

$$\begin{aligned}
\frac{\partial \sigma_{xx}}{\partial x} + \frac{\partial \sigma_{yx}}{\partial y} + \frac{\partial \sigma_{zx}}{\partial z} - b_x &= 0 \\
\frac{\partial \sigma_{xy}}{\partial x} + \frac{\partial \sigma_{yy}}{\partial y} + \frac{\partial \sigma_{zy}}{\partial z} - b_y &= 0 \\
\frac{\partial \sigma_{xz}}{\partial x} + \frac{\partial \sigma_{yz}}{\partial y} + \frac{\partial \sigma_{zz}}{\partial z} - b_z &= 0
\end{aligned} \tag{1.33}$$

These three equations, or equivalently the vector equation (1.32), must hold at all points within the body. If we were concerned with dynamical problems, the zeros on the right-hand side of (1.33) would be replaced by the mass density multiplying the appropriate component of the acceleration of the body. That extra step will not be necessary in the applications we intend to pursue. Note too that the negative signs preceding the body force terms in (1.32) and (1.33) are a result of our convention that compressive stresses are positive. In solid mechanics positive signs would appear there.

As with deformation and strain, the equilibrium equations change somewhat in non-rectangular coordinate systems. Appendix A describes how to change from one coordinate system to another and summarises the component forms for cylindrical polar coordinate systems.

We are now drawing near to the close of this chapter. We are also drawing nearer the point where simple problems can be formulated and solved. In

* If we were to make *tensile* stress positive, the negative sign in (1.32) would become positive. This is one of the rare occasions where the usual equations of solid mechanics differ from ours.

general, the solution of problems in elasticity and plasticity involves finding 15 variables. We need to determine the three components of the displacement vector $\boldsymbol{u}$, the six independent components of the strain matrix $\boldsymbol{\varepsilon}$ and the six independent components of the stress matrix $\boldsymbol{\sigma}$. If we know $\boldsymbol{u}$ as a function of position throughout the body we can use the strain–displacement equations (1.4) to find $\boldsymbol{\varepsilon}$. Conversely, if we know $\boldsymbol{\varepsilon}$, then we can integrate (1.4) to find $\boldsymbol{u}$ provided the compatibility relations (1.9) hold. We also know that the stress components must obey the equilibrium equations (1.33), but there are only three equations for the six independent components of stress. So we do not yet have enough equations even to attempt to find a solution. The missing link is the relationship between stress and strain. In the next chapter we will explore that link in the context of elasticity. In the chapters to follow we consider the same link in the context of plastic behaviour.

Further reading

The complete reference to *EG* is:

R.O. Davis and A.P.S. Selvadurai, *Elasticity and Geomechanics*, Cambridge University Press, New York, 1996.

Two useful books on the basics of continuum mechanics including elements of elasticity and plasticity are:

Y.C. Fung, *Foundations of Solid Mechanics*, Prentice-Hall, New Jersey, 1965.

L.E. Malvern, *Introduction to the Mechanics of a Continuous Medium*, Prentice-Hall, New Jersey, 1969.

Original references to works cited in this chapter:

Barré de Saint Venant, Éstablissment élementaire des formules et équations générales de la théorie de l'élasticité des corps solides, Appendix in: Résumé des leçons des Ponts et Chaussées sur l'Application de la Mécanique première partie, première section, *De la Résistance des Corps Solides*, by C.-L.M.H. Navier, 3rd edn, Paris, 1864.

A.L. Cauchy, Recherches sur l'équilibre et le mouvement intérieur des corps solides ou fluides, élastiques ou non élastiques, *Bull. Soc. Philomath*, **2**, 300–304 (1823).

O. Mohr, *Zivilingenieur*, W. Ernst und Sohn, Berlin, 1882.

K. Terzaghi, *Erdbaumechanik auf bodenphysikalischer Grundlage*, Franz Deuticke, Vienna, 1925.

Exercises

1.1 Find the strain matrix as a function of x, y and z associated with each displacement field:

(a) $u_x = \alpha - \beta x, u_y = u_z = 0$

(b) $u_x = \xi y, u_y = u_z = 0$

(c) $u_x = ax^2 - bxy + cy^2, u_y = u_z = 0.$

1.2 Consider the elastic halfspace in Figure 1.13. Under the action of gravity the components of stress have the form

$$\sigma_{xx} = \alpha x, \quad \sigma_{yy} = \sigma_{zz} = \beta x, \quad \sigma_{xy} = \sigma_{yx} = \sigma_{zy} = \sigma_{yz} = \sigma_{xz} = \sigma_{zx} = 0$$

where α and β are constants. Use Cauchy's relationship (1.11) to determine the Cartesian components of the traction vector $\boldsymbol{T}$ as a function of r and θ on the cylindrical surface that aligns with the z-axis shown in the diagram.

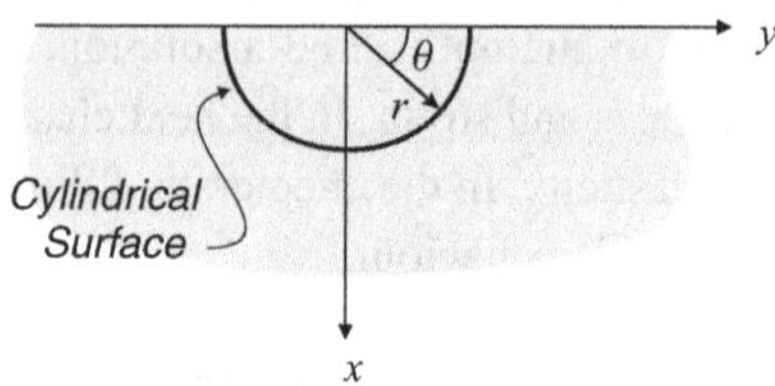

Figure 1.13.

1.3 Construct the Mohr stress circle for an arbitrary point on the cylindrical surface in Exercise 1.2. Use the circle to find the components of the traction vector $\boldsymbol{T}$ that act normal and tangential to the surface.

1.4 Given the general stress matrix shown in equation (1.12), form a new matrix $\mathbf{S} = \boldsymbol{\sigma} - \frac{1}{3}tr(\boldsymbol{\sigma})$. Find the invariants of the matrix $\mathbf{S}$ using the form given in (1.18). That is, find $S_1 = tr(\mathbf{S})$, $S_2 = \frac{1}{2}[(tr\mathbf{S})^2 - tr(\mathbf{S}^2)]$ and $S_3 = \det(\mathbf{S})$.

1.5 What is the smallest quantity of numbers required to specify fully the state of stress at a point in a continuum? Explain why the total of 12 numbers involved in $\sigma_1, \sigma_2, \sigma_3$ and $\hat{\boldsymbol{n}}_1, \hat{\boldsymbol{n}}_2, \hat{\boldsymbol{n}}_3$ are not all needed to provide a specification of the stress state.

2
Elastic and inelastic material behaviour

2.1 Introduction

Geotechnical engineers have made good use of the theory of elasticity for a number of decades. It became clear near the end of the nineteenth century that a variety of problems involving an elastic halfspace could be solved using techniques developed by the French mathematician Joseph Boussinesq. Boussinesq solved the problem of a point load resting on the surface of a homogeneous isotropic linearly elastic halfspace. He also developed the solution for a rigid circular footing resting on the halfspace surface. His work inspired others to investigate related problems with the result that by the middle of the twentieth century a wide range of problems involving both homogeneous and layered halfspaces with isotropic and anisotropic elastic materials had been solved for a variety of loading conditions. Solutions continue to appear in the geotechnical literature as well as in other disciplines. There are also coupled solutions in which porous materials saturated with pore fluid are modelled incorporating both elastic deformation and pore fluid flow.

In this chapter we will outline the basic elements of behaviour of elastic materials. The stress–strain relations for isotropic materials are given in a variety of forms and relationships between the elastic constants are derived. We will note the bounds imposed on the elastic constants by thermodynamic requirements and we discuss some special classes of problems such as plane strain problems and problems involving incompressible materials. Much of this material is also presented in *EG*, often in more detail. Readers familiar with the basic elements of elasticity, particularly with regard to geomechanics, may wish to skip portions of this chapter.

The relevance of an elastic solution for any problem in geomechanics will depend upon two things. First, there is a requirement that the stresses within the soil mass should be reasonable in the sense that the average stress state is well below the yield stress. In many problems there may be small regions of high stress such as near the edge of a rigid footing, but the average stress will often be less than half the yield value and elasticity can be applied profitably to estimate the settlement. The second condition for relevance is the need to select appropriate values for the elastic constants. This point is discussed in more detail in *EG*. The choice of appropriate parameter values will depend upon the quantity and quality of test data available and on the experience and good judgement of the engineer involved.

Elastic solutions are obviously of limited value when the stress level in a soil mass becomes too severe. At this point the response of the soil becomes much more difficult to model. Deformations grow in magnitude and the accompanying strains may display unexpected attributes. Shearing strains may continue to grow despite decreasing stress levels. The volume of the soil may expand despite an overall compressive stress regime. Good-quality test data becomes more difficult to obtain. If we could look inside the soil at a microscopic scale we would expect to see individual soil particles undergoing fracture and crushing, and relatively large-scale rearrangement of the interparticle structure or fabric of the soil. These are irreversible effects that we classify as an inelastic response and that we attempt to approximate by recourse to the theories of plasticity. The second part of this chapter discusses some general aspects of inelastic behaviour.

2.2 Hooke's law

The foundation stone of elasticity was fashioned by Robert Hooke in a lecture to the Royal Society in 1660 when he postulated a linear relationship between the applied tension and the elongation of a spring. That first step was followed by a series of developments leading to the fundamental idea that stresses and strains should be related linearly. For an isotropic elastic material it was found that two elastic constants were required to relate stresses to strains fully. The resulting stress–strain relationships can be written in a variety of ways, but one of the most convenient is the matrix form,

$$\boldsymbol{\sigma} = \Lambda e \mathbf{I} + 2G\boldsymbol{\varepsilon} \tag{2.1}$$

Here $\boldsymbol{\sigma}$ and $\boldsymbol{\varepsilon}$ are the stress and strain matrices, $e = tr\,\boldsymbol{\varepsilon}$ is the volumetric strain, Λ is a material constant called the Lamé constant* and G is also a material constant called the shear modulus. Since there are six independent components

* Often the Lamé constant is represented by the lowercase λ. We will use λ extensively later for other purposes.

of both $\boldsymbol{\sigma}$ and $\boldsymbol{\varepsilon}$, equation (2.1) actually represents six separate equations:

$$\begin{aligned} \sigma_{xx} &= \Lambda e + 2G\varepsilon_{xx} & \sigma_{xy} &= 2G\varepsilon_{xy} \\ \sigma_{yy} &= \Lambda e + 2G\varepsilon_{yy} & \sigma_{yz} &= 2G\varepsilon_{yz} \\ \sigma_{zz} &= \Lambda e + 2G\varepsilon_{zz} & \sigma_{zx} &= 2G\varepsilon_{zx} \end{aligned} \tag{2.2}$$

The Lamé constant Λ and the shear modulus G are the only elastic constants needed to characterise the stress–strain behaviour of an isotropic linearly elastic material completely.

While (2.1) may be a convenient way to write Hooke's law it may not be the most familiar way. For many people the most familiar elastic constants are Young's modulus E and Poisson's ratio ν. They appear if we invert equations (2.2) to obtain the components of the strain matrix:

$$\begin{aligned} \varepsilon_{xx} &= \frac{1}{E}[\sigma_{xx} - \nu(\sigma_{yy} + \sigma_{zz})] & \varepsilon_{xy} &= \frac{1}{2G}\sigma_{xy} \\ \varepsilon_{yy} &= \frac{1}{E}[\sigma_{yy} - \nu(\sigma_{zz} + \sigma_{xx})] & \varepsilon_{yz} &= \frac{1}{2G}\sigma_{yz} \\ \varepsilon_{zz} &= \frac{1}{E}[\sigma_{zz} - \nu(\sigma_{xx} + \sigma_{yy})] & \varepsilon_{zx} &= \frac{1}{2G}\sigma_{zx} \end{aligned} \tag{2.3}$$

These six equations form the most familiar set of expressions for Hooke's law, but there is no fundamental difference between (2.2) and (2.3). One gives stresses in terms of strains and the other gives strains in terms of stresses.

Note that (2.3) contains three constants E, ν and G. Only two of these can be independent. When we invert (2.2) to arrive at (2.3) we find the following relations between elastic constants:

$$\begin{aligned} E &= \frac{G(3\Lambda + 2G)}{\Lambda + G} \\ \nu &= \frac{\Lambda}{2(\Lambda + G)} \end{aligned} \tag{2.4}$$

Thus knowing Λ and G implies knowledge of E and ν. We can invert these relations to obtain

$$\begin{aligned} \Lambda &= \frac{\nu E}{(1+\nu)(1-2\nu)} \\ G &= \frac{E}{2(1+\nu)} \end{aligned} \tag{2.5}$$

It is possible to derive other relationships as well. For example,

$$\nu = \frac{E - 2G}{2G} \tag{2.6}$$

There is one other commonly used elastic constant. It is called the bulk modulus and is usually represented by K. It relates the mean stress to the volumetric strain,

$$p = Ke \tag{2.7}$$

To find how K is related to the other elastic constants we take the trace of (2.1) and use the definitions of p and e, which gives

$$K = \Lambda + \frac{2}{3}G \tag{2.8}$$

It is possible to derive further relationships between these five elastic constants. All of the 30 possibilities are tabulated in *EG*.

In some applications it may be convenient to treat the six independent components of stress and strain as components of six-dimensional vector fields. We can write

$$\begin{bmatrix} \sigma_{xx} \\ \sigma_{yy} \\ \sigma_{zz} \\ \sigma_{xy} \\ \sigma_{yz} \\ \sigma_{zx} \end{bmatrix} = \begin{bmatrix} \Lambda + 2G & \Lambda & \Lambda & 0 & 0 & 0 \\ \Lambda & \Lambda + 2G & \Lambda & 0 & 0 & 0 \\ \Lambda & \Lambda & \Lambda + 2G & 0 & 0 & 0 \\ 0 & 0 & 0 & 2G & 0 & 0 \\ 0 & 0 & 0 & 0 & 2G & 0 \\ 0 & 0 & 0 & 0 & 0 & 2G \end{bmatrix} \begin{bmatrix} \varepsilon_{xx} \\ \varepsilon_{yy} \\ \varepsilon_{zz} \\ \varepsilon_{xy} \\ \varepsilon_{yz} \\ \varepsilon_{zx} \end{bmatrix} \tag{2.9}$$

The square matrix containing the elastic constants is often called the *elasticity matrix*. Later we will identify it by $\mathbf{M}^e$.

Yet another way to write Hooke's law involves introducing the *deviatoric* stress and strain matrices. These are representations of stress and strain that characterise shearing rather than isotropic response. They are defined as follows:

$$\begin{aligned} \boldsymbol{\sigma}^d &= \boldsymbol{\sigma} - p\mathbf{I} \\ \boldsymbol{\varepsilon}^d &= \boldsymbol{\varepsilon} - \frac{1}{3}e\mathbf{I} \end{aligned} \tag{2.10}$$

Here $\mathbf{I}$ is the identity matrix. Note that both equations have the same form. In each case one-third of the trace of the matrix is subtracted from each of the diagonal matrix components. The resulting matrices, sometimes called the stress and strain *deviator matrices*, represent that part of the stress or strain state that deals with shearing. Note that both $\boldsymbol{\sigma}^d$ and $\boldsymbol{\varepsilon}^d$ are traceless. If we use (2.1) in (2.10) together with the definitions for mean stress and volumetric strain we find that $\boldsymbol{\sigma}^d$ and $\boldsymbol{\varepsilon}^d$ are related by

$$\boldsymbol{\sigma}^d = 2G\boldsymbol{\varepsilon}^d \tag{2.11}$$

Six equations are represented by (2.11), but only five are independent since $tr\,\sigma^d \equiv 0$. We can use (2.11) together with (2.7) to obtain six independent equations and thus have another way to write Hooke's law.

2.3 Values for elastic constants

Whenever we are willing to model a soil as an isotropic elastic material, values for two of the elastic constants are required. It makes no difference which two since, if we know the values for two constants, we can use the relationships between constants to determine the remaining three. There are theoretical restrictions on the values of the constants that result from the principles of thermodynamics. To understand these restrictions we first need to introduce the idea of an elastic *stored energy function*, W_e. The stored energy is exactly what its name implies. It represents the amount of work per unit volume done on the material by the applied stresses as a function of position in the body. The work is stored in the body and can be represented as

$$W_e = \frac{1}{2} tr(\sigma \varepsilon) \tag{2.12}$$

In the context of elasticity this energy will be recoverable in the sense that, during unloading, the body itself can do exactly the same amount of work against the applied forces as was done by those forces when the body was initially loaded. Basically (2.12) represents the area under the stress–strain curve for the material, although it is a bit more general in the sense that we are considering all the components of stress and strain.

The thermodynamic restrictions on the elastic constants follow from the requirement that the stored energy function for an elastic material should be *positive definite*. Positive definite means that $W_e \geq 0$ at all times, and $W_e = 0$ *only* when the stresses and strains are equal to zero. If we accept the premise of positive definiteness, then it can be shown* that the following bounds must apply to the elastic constants:

$$E > 0, \quad G > 0, \quad K > 0, \quad \Lambda + \frac{2}{3}G > 0, \quad -1 < \nu \leq 0.50 \tag{2.13}$$

These bounds seem innocuous but are of some importance. Particularly interesting is the restriction on Poisson's ratio, especially the upper limit. All common engineering materials have Poisson ratio values that are positive, so the lower limit is only of academic interest. Note that the upper limit may be less than *or equal* unlike all of the other bounds. Let us consider what happens when Poisson's ratio takes on a value of 0.50.

* See Chapter 2 in *EG* for a derivation of these results.

Figure 2.1. Prismatic bar in simple tension.

Recall the definition of ν. It represents the negative of the ratio of the lateral strain to the longitudinal strain experienced by an elastic bar subjected to simple tension such as the one shown in Figure 2.1. Suppose the axis of the bar is aligned with the x-axis and a uniform tension σ_{xx} is applied at the ends of the bar. The axial strain is given by $\varepsilon_{xx} = \sigma_{xx}/E$, where E represents Young's modulus for the bar. As the bar is stretched it also grows more slender and the lateral strains are given by

$$\varepsilon_{yy} = \varepsilon_{zz} = -\nu\varepsilon_{xx} \tag{2.14}$$

These are all familiar concepts. We go through them here because we want to find what the volumetric strain in the bar is. Recalling (1.8) we see that the volumetric strain is given by

$$e = \varepsilon_{xx} + \varepsilon_{yy} + \varepsilon_{zz} = (1 - 2\nu)\varepsilon_{xx} = (1 - 2\nu)\frac{\sigma_{xx}}{E} \tag{2.15}$$

Evidently, if ν takes the value of the upper limit in (2.13), there will be no volumetric strain. In that case we would say that the material is *incompressible.*

Incompressible materials form an interesting subgroup of all elastic materials. Of course no material is entirely incompressible, in the sense that no volume change occurs regardless of what stresses are applied, but many materials are nearly incompressible. Rubber is a good example. It is compressible but its shear modulus is far less than its bulk modulus, which means that changes of shape will dominate the deformation of rubber in comparison with changes of volume. Poisson's ratio for rubber is very nearly 0.50 and for many purposes we can assume it to be exactly 0.50. Thus, if we consider a rubber band we can safely assume that its volume remains unchanged regardless of how far we may wish to stretch it.

Note that for incompressible materials *only one* elastic constant is needed since Poisson's ratio is known. The stress–strain relationships may be written more simply and the relationships between elastic constants are also simplified. For incompressible materials we have

$$E = 3G, \qquad K = \Lambda = \infty \tag{2.16}$$

These simplifications often have dramatic effects when the time comes to begin solving actual problems. Because of this a large volume of research devoted to problems in incompressible elasticity has developed over many years.

There is one situation in which soils may behave as incompressible materials. That is the case of fully saturated soils deforming in *undrained* conditions. Of course, neither the soil particles themselves nor the pore water is incompressible, but, if no drainage is permitted and the pores are filled with water, then the shear modulus of the soil will be far smaller than the bulk modulus and Poisson's ratio will effectively be equal to 0.50. The requirement for undrained conditions is important and cannot be overlooked, but in the case of many clay soils drainage may occur so slowly that years are required before any significant volume change occurs. In that case it may be very useful to assume that the soil is undrained and let $\nu = 0.50$.

The range of shear modulus values encountered in natural soils is quite large. Very soft soils may have shear modulii as small as 10 MPa. At the other extreme, soft rock modulii may be of the order of 2000 MPa. If one assumes a typical value for Poisson's ratio of 0.30, the corresponding range of values for Young's modulus is roughly 25–5000 MPa. Typically the density of most soils will lie somewhere in the range of 1.5 t/m^3 for very soft, loose soils to 2.5 t/m^3 for soft rocks. For elastic materials the velocity of shear waves is given by $\sqrt{G/\rho}$, so we would expect the speed of shear waves to range between 50 and 1000 m/s. The velocity of compression waves or P-waves would be of the order of two to three times higher than the shear wave velocities. With such a broad range of possible values for soil modulii, it becomes clear that the choice of appropriate values for particular problems becomes extremely important. Techniques for estimating specific values for elastic constants are discussed in Chapter 2 of *EG*.

2.4 Solution of problems in elasticity

If values for two elastic constants are available we are in a position, at least theoretically, to solve problems. Some problems may prove too difficult to solve, but a surprising number of interesting problems have yielded solutions and no doubt more will do so in the future. There are also a great number of approximate solutions, in the sense that not all the requirements for an exact solution are met precisely, but the solutions, nonetheless, have useful attributes.

The complete solution of any problem involving a linearly elastic body will require knowledge of six components of stress, six components of strain and three components of displacement throughout the body. The strain–displacement relations (six equations), the equilibrium equations (three equations) and Hooke's law (six equations) must all be satisfied simultaneously. In addition, the stress and displacement components may be required to satisfy *boundary conditions*. Boundary conditions make the solution specific to the particular problem. We can categorise different types of boundary conditions according to whether stresses or displacements or both are involved. *Traction boundaries* are those portions of the boundary of the body for which all three components of traction are fully specified. *Displacement boundaries* are those portions of the boundary of the body where all three components of displacement are specified. *Mixed boundaries* are boundaries where some components of traction and some components of displacement are specified. Taken together, the three types of boundaries must comprise the complete boundary of the body.

An example of a traction boundary is the surface of a beam that supports prescribed loads. A surface that is free of stress such as the lower surface of a beam is also a traction boundary since zero tractions are prescribed there. An example of a displacement boundary is the built-in end of a cantilever beam. On that surface zero displacements are specified. A rigid smooth punch indenting an elastic plate gives an example of a mixed boundary. The part of the plate in contact with the punch may be subject to a specified displacement in the direction normal to the surface, and a zero traction condition (provided the face of the punch is smooth) in the tangential direction.

2.5 Plane elasticity

An important subset of problems in elasticity falls under the heading of *plane stress* and *plane strain* problems. Together they are jointly referred to as *plane elasticity*. In most geotechnical applications only plane strain conditions will be of interest, and we will focus primarily on plane strain here. Plane stress conditions are described briefly, but the equations are left as an exercise at the end of this chapter.

Both plane stress and plane strain conditions refer to situations where we may be justified in assuming that certain components of the stress or strain matrix are identically zero. Plane stress conditions correspond to problems where all stresses associated with one coordinate direction are assumed to vanish. If the coordinate system is arranged so that the particular direction of zero stress is the z-direction, then the stress matrix will look like this

$$\boldsymbol{\sigma} = \begin{bmatrix} \sigma_{xx} & \sigma_{xy} & 0 \\ \sigma_{yx} & \sigma_{yy} & 0 \\ 0 & 0 & 0 \end{bmatrix} \tag{2.17}$$

This situation may arise where a relatively thin elastic plate is subjected to edge loading with the faces of the plate (arranged normal to the z-direction) being free from stress. A typical situation is illustrated in Figure 2.2.

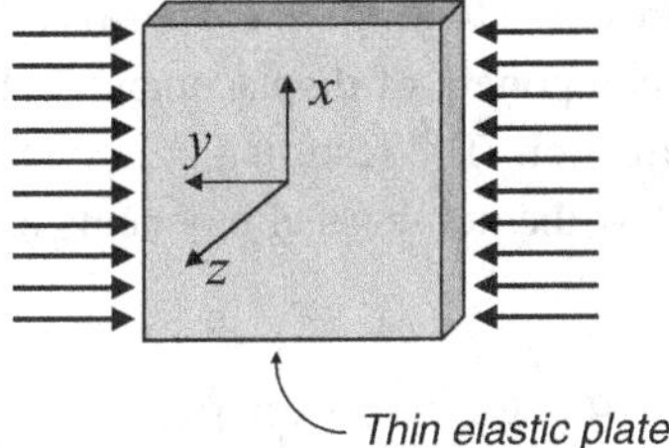

Figure 2.2. Example of a plane stress problem.

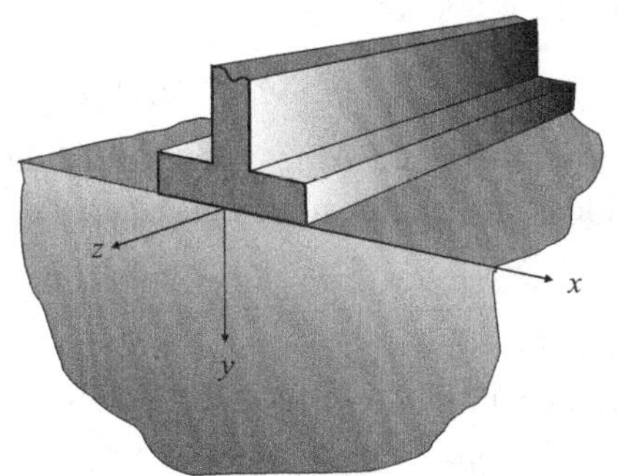

Figure 2.3. Example of a plane strain problem.

In the figure the only loads supported by the plate lie in the (x, y)-plane. If the plate is not too thick we are reasonably well justified in assuming there will be no stresses associated with the z-direction and the stress matrix in (2.17) should closely approximate the stress field in the plate. Moreover, we may also assume that the non-zero components of stress will not be functions of z.

In contrast, plane strain conditions refer to situations where we assume that all components of strain associated with one coordinate direction are identically zero. Choosing the z-direction to be the appropriate direction, the strain matrix for plane strain conditions may be written as

$$\boldsymbol{\varepsilon} = \begin{bmatrix} \varepsilon_{xx} & \varepsilon_{xy} & 0 \\ \varepsilon_{yx} & \varepsilon_{yy} & 0 \\ 0 & 0 & 0 \end{bmatrix} \tag{2.18}$$

This condition might arise if we were to consider a body of considerable length in the z-direction with loadings that are not functions of the z-direction. An example is a long-strip footing resting on an elastic halfspace, such as that shown in Figure 2.3.

In the figure the footing extends in both the positive and negative z-directions for a distance that is large compared with the footing width. If we consider conditions near the centre of the footing we are reasonably justified in assuming that there will be no z-component of displacement and that all quantities are independent of the z-direction. The resulting strain field will be that given in (2.18). Using Hooke's law the non-zero components of strain may be written as

$$\varepsilon_{xx} = \frac{1}{2G}[\sigma_{xx} - \nu(\sigma_{xx} + \sigma_{yy})], \quad \varepsilon_{yy} = \frac{1}{2G}[\sigma_{yy} - \nu(\sigma_{xx} + \sigma_{yy})], \quad \varepsilon_{xy} = \frac{\sigma_{xy}}{2G} \tag{2.19}$$

In addition, note that σ_{zz} is not zero.

$$\sigma_{zz} = \Lambda e = \nu(\sigma_{xx} + \sigma_{yy}) \tag{2.20}$$

Physically σ_{zz} is the stress required to maintain the condition $\varepsilon_{zz} = 0$. Also the compatibility conditions (1.9) reduce to a single equation

$$\frac{\partial^2 \varepsilon_{xx}}{\partial y^2} + \frac{\partial^2 \varepsilon_{yy}}{\partial x^2} = 2\frac{\partial^2 \varepsilon_{xy}}{\partial x \partial y} \tag{2.21}$$

and there are only two equilibrium equations (see (1.33)).

$$\begin{aligned} \frac{\partial \sigma_{xx}}{\partial x} + \frac{\partial \sigma_{yx}}{\partial y} - b_x = 0 \\ \frac{\partial \sigma_{xy}}{\partial x} + \frac{\partial \sigma_{yy}}{\partial y} - b_y = 0 \end{aligned} \tag{2.22}$$

Together (2.19), (2.21) and (2.22) give six equations for the three unknown components of strain $\varepsilon_{xx}, \varepsilon_{yy}, \varepsilon_{xy}$, and the three unknown stresses $\sigma_{xx}, \sigma_{yy}, \sigma_{xy}$.

Solutions to many plane strain problems may be found in the elasticity literature. One of the simplest is the case of an elastic halfspace loaded only by gravity. Consider the halfspace in Figure 2.3 (without the footing). The body force components for this situation become

$$b_x = 0, \qquad b_y = \rho g \tag{2.23}$$

Since $b_x = 0$, we can assume that there will be no dependence on x for any of the problem variables. Then the second equation of (2.22) gives the familiar result

$$\sigma_{yy}(y) = \int_0^y b_y \, dy = \int_0^y \rho g \, dy \tag{2.24}$$

where we have set the constant of integration equal to zero because of the zero-traction boundary condition at $y = 0$. The first equation of (2.22) now shows that σ_{xy} is independent of the depth y, and the zero-traction boundary then requires that σ_{xy} be zero everywhere in the halfspace. It follows from Hooke's law that $\varepsilon_{xy} = 0$ everywhere as well. If there is no x dependence, then the displacement u_x can at most be a function of y, showing that $\varepsilon_{xx} = \partial u_x/\partial x$ must also be zero. Using this in (2.19) gives

$$\sigma_{xx} = \frac{\nu}{1-\nu}\sigma_{yy} \tag{2.25}$$

and

$$\varepsilon_{yy} = \frac{1}{2G}\left[\sigma_{yy} - \nu\left(\sigma_{yy} + \frac{\nu}{1-\nu}\sigma_{yy}\right)\right] = \frac{\sigma_{yy}}{2G}\left(\frac{1-2\nu}{1-\nu}\right) \tag{2.26}$$

Finally, note that (2.20) gives

$$\sigma_{zz} = \nu\left(\frac{\nu}{1-\nu}\sigma_{yy} + \sigma_{yy}\right) = \frac{\nu}{1-\nu}\sigma_{yy} = \sigma_{xx} \tag{2.27}$$

This result is expected since the problem is invariant to rotation of the coordinates about the y-axis and therefore σ_{xx} and σ_{zz} must be the same.

A question arises here regarding total and effective stress. How should we interpret the stresses in equations (2.24)–(2.27)? Are they strictly total stresses, or may we say that the same equations apply to effective stresses? It happens that we can apply the equations to effective stresses provided the soil is saturated and we are careful. First, in (2.24) we replace the density ρ with the *submerged* or *buoyant density* ρ_b defined by

$$\rho_b = \rho - \rho_w \tag{2.28}$$

where ρ_w is the mass density of the pore fluid. We have made a tacit assumption that the soil is fully saturated; that is, the ground water table lies at the ground surface. Other circumstances are slightly more complicated but no real difficulties arise if they are considered. Next consider (2.25). How is Poisson's ratio affected by the presence of the pore fluid? We have discussed this topic earlier in the context of incompressible materials. If no drainage is permitted, the appropriate value for ν is 0.50. This would apply to both the total and the effective stress in undrained situations. If there is drainage, then ν takes on some value smaller than 0.50 and this would be the appropriate value for both the total and the effective stress. Finally, consider the strain ε_{yy} given by (2.26). The shear modulus G is independent of the pore fluid, so, provided drainage is permitted, the strain given by that equation is also correct.*

* Students of geomechanics will be aware that if drainage happens slowly, for example in the case of a clay soil, then the strain in (2.26) will also develop slowly. The theory of one-dimensional consolidation would be applicable in that case.

In a natural soil deposit with horizontal surface and zero strain in the horizontal plane, the ratio of the horizontal effective stress to the vertical effective stress is often referred to as the coefficient of lateral earth pressure at rest. It is usually identified by K_0. If we use the coordinate frame in Figure 2.3,

$$K_0 = \frac{\sigma'_{xx}}{\sigma'_{yy}} \tag{2.29}$$

Sometimes field tests are used to measure approximately the value of K_0. We see from (2.25) that, if the soil behaves as an elastic material, the value of K_0 should be $\nu/(1 - \nu)$. The restriction that ν should be less than or equal to 0.50 shows that K_0 should be no greater than 1.0. In fact, values of K_0 significantly greater than 1.0 have often been observed. Observations of $K_0 > 1$ are an indication that a soil has failed to respond elastically at some point in its history. Inelastic behaviour of this form as well as other forms is quite common. The remainder of this chapter will be used to discuss inelastic response and to put in place a general framework of ideas that will be useful for the development and application of plasticity theory given in subsequent chapters.

2.6 Indications of inelastic behaviour

There are many categories of inelastic behaviour that may be exhibited by a variety of materials. Brittle materials such as glass and ceramics may fracture, and ductile materials such as mild steel will flow. Fatigue is a common problem in many applications including pavement engineering. All of these things can happen in soils and rocks given that appropriate circumstances exist. If we focus on soils, we expect that inelastic behaviour will result when the particle structure or fabric is disrupted. This may occur because the particles are rearranged, or it may result from individual particles fracturing or crushing. Intuitively we would anticipate a combination of fracturing, crushing and rearrangement to occur simultaneously in most circumstances. For the remainder of this chapter we will consider some common ways in which inelastic behaviour may be generated. The discussion is qualitative in the sense that very few equations are used. Our purpose is to set the stage for the development of plasticity ideas that follow.

To begin, think about the soil deposit mentioned earlier for which the value of K_0 is greater than 1.0. Suppose the soil is a silt so that the particles are relatively small. If we were to look back in time to the moment when the soil came into being it could be that the particles were initially sediment at the bottom of a lake or sea, or they may have been deposited by the wind. Whatever the circumstances, it is likely that the particle structure was initially quite loose in the sense that relatively large pore spaces could be found between nearly all particles. As time passed more soil may have accumulated above our deposit and the vertical stress component would grow according to (2.24). There would be a corresponding growth of the horizontal stress (both total stress and effective stress) and, at this early stage in the life of the soil, an elastic, or at least nearly elastic, response would be expected. The particles themselves would deform and if there is rearrangement of particles it would happen on a very minor or limited scale. Equation (2.25) would be appropriate to relate the effective stress components. As the overburden accumulates both σ'_{xx} and σ'_{yy} grow, but σ'_{yy} grows faster since $\nu/(1-\nu)$ is less than 1.0. If the soil surface is horizontal it is likely that all horizontal surfaces will be principal surfaces. All vertical surfaces will also be principal surfaces since there is invariance under rotation about the y-axis. So both σ'_{xx} and σ'_{yy} are principal stresses. The difference $(\sigma'_{yy} - \sigma'_{xx})$ is the diameter of the Mohr circle and is also twice the greatest shear stress in our soil deposit. As more overburden accumulates the size of the Mohr stress circle is evidently growing and so are the shear stresses on non-principal surfaces, but the strength of the soil is also increasing owing to the increasing confining stress and failure due to excessive shear stress will not occur.

If we could look inside the deposit at this stage we would see an assemblage of particles still in a relatively loose configuration, but now supporting the weight of soil above. Individual particles may support different portions of the load depending on their particular situation. Some particles may be quite heavily loaded while others are not. There may be chains of heavily loaded particles and between these chains there may be groups of particles that are isolated from the overall loading regime. On average the stresses are described by (2.24) and (2.25), but on a microscopic scale a far more complex and heterogeneous process is evolving. At some point there may be sufficient overburden so that the particle structure can no longer respond elastically. Perhaps one particle in one heavily loaded chain fractures. The new smaller particles that result now rearrange themselves into some new configuration. The load supported by the broken particle must be redistributed to surrounding particles. There are many lightly loaded particles at this stage and it may be relatively easy for those particles to take up the load shed by the fractured particle, but there has been an irreversible, inelastic event.

Further build-up of overburden will lead to more and more events like that just described. With each event, more and more particles become fragmented and fractured, and greater and greater amounts of particle rearrangement occur. Each fracture and rearrangement is an irreversible event. A thermodynamicist would point out that the entropy of the soil is growing. The soil is now becoming more densely packed and the relative amounts of volume taken up by particles and by pore space have changed. Fragmenting particles produce smaller offspring that are more suited to filling the gaps between the larger particles that remain. The remaining pores are much smaller. The chains of loaded particles are not so distinct and the load is more uniformly spread among all particles. The soil would now be described as dense or tightly packed. All the deformations that have occurred because of particle fracture and rearrangement are referred to as inelastic deformation.

We can digress for a moment and ask whether, at this point, after innumerable fractures and rearrangements, the value of K_0 will be larger than 1.0. The answer is no. The vertical normal effective stress will still be greater than the horizontal stress. So how do larger values of K_0 come about? The answer lies in unloading. Perhaps at some point in the history of our soil conditions change and the overburden is no longer accumulating. Instead, erosional processes may begin to occur and previously deposited overburden is now removed. This results in a decrease in the vertical effective stress σ'_{yy}. Does the horizontal stress also decrease? The answer is yes, but, crucially, it will not decrease as rapidly. The boundary conditions for the two stress components are different. The vertical stress simply responds to the removal of overburden by decreasing an appropriate amount. The horizontal stress will sense the decrease in vertical stress, but its response is governed by the fact there is *no horizontal displacement* in the soil. So the removal of overburden affects σ'_{yy} directly, but only affects σ'_{xx} indirectly. The horizontal stress is, to some extent, *locked-in* by the requirement for zero horizontal displacement. If sufficient overburden is removed, the vertical stress may decrease to a value that is smaller than the horizontal stress and, consequently, K_0 will be greater than 1.0. The entire process of overburden accumulation (loading) and erosion (unloading) is shown schematically in Figure 2.4.

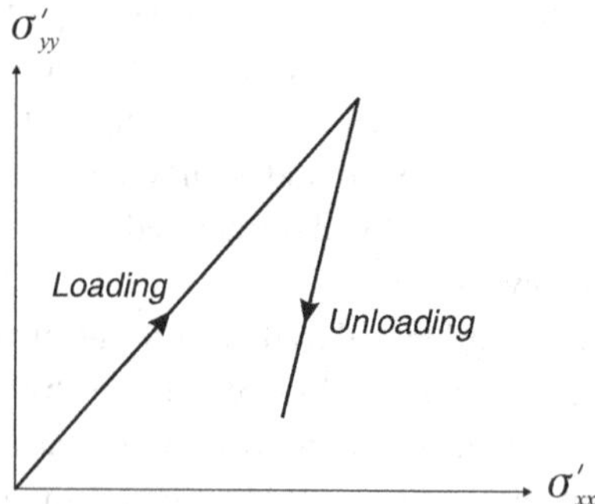

Figure 2.4. Increasing and decreasing stress caused by accumulation and subsequent erosion of overburden.

There are many physical situations in which 'locked-in' stresses may arise. A common example occurs in shrink fitting a gear or washer on to a circular shaft. The shaft has a certain outside diameter and the gear or washer has a hole of smaller diameter. By heating the gear uniformly, it is possible to expand the size of the hole significantly. The shaft is then easily inserted into the expanded hole. Upon cooling, the mass around the hole shrinks back to its original size and frictional forces create a highly effective joint. The amount of locked-in stress can be calculated from the known dimensions of the gear and shaft.

The locked-in stress fixing a shrink-fitted gear to its shaft results from heating, while the locked-in horizontal stress in an unloading soil deposit results from particle breakage and rearrangement. The former is, at least theoretically, reversible, but the latter clearly is not. Literally any particle rearrangement within a soil mass cannot be reversed by any natural process and an irreversible response results whenever particle fracture and rearrangement occur. Irreversible processes are by definition inelastic. The locked-in horizontal stress in our soil mass is an artefact of the inelastic response, but the inelastic process itself is *particle fracture and rearrangement*. Broadly speaking, there are two categories of loading that will generate inelastic behaviour in soil. One category is related to the increase of normal stresses and the other to the increase of shear stresses. Of course, shear and normal stresses are closely interrelated but, in the qualitative analysis we pursue here, it is enough to think of the two types of stress independently. The situation in which a natural soil deposit is subjected to an increasing overburden stress falls under the heading of inelastic effects due to normal stress increase. In contrast, slope failures are an example of excessive shear stresses. There are many other situations in which both inelastic normal stress and inelastic shear stress effects occur. Soon we will consider two particular laboratory tests in which the two categories of response are emphasised. Before describing those tests however, it is useful to discuss shear stress effects in more detail.

The classic stability problems of geomechanics involve foundations, retaining walls and slopes. All three problems arise due to the development of zones or bands of intense shearing deformation. These intense shear strains occur because of large shear stresses. It is useful here to think about how a shearing failure might develop at both macroscopic and microscopic scales.

We begin by assuming that at some point within a soil mass we can identify a surface for which the inelastic response due to high shear stress is imminent. We are naturally interested in the normal stress acting on that surface as well as the shear stress because of the frictional nature of the interparticle contacts. If the combined stress state reaches a critical point, the soil particle structure will be disrupted in two important ways. First, at macro-scales, the formation of a

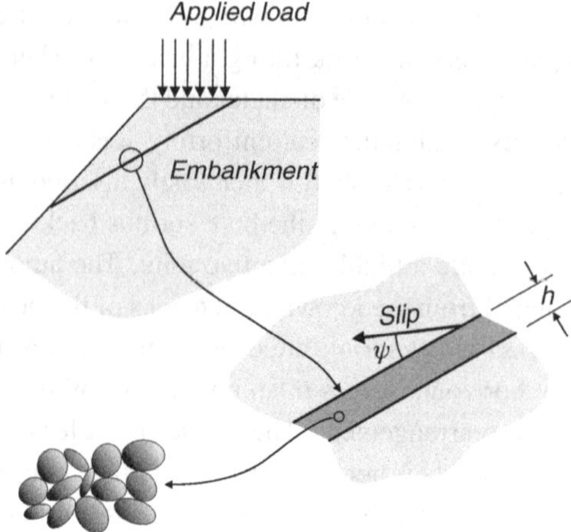

Figure 2.5. Failure of an embankment due to excessive shear stress.

band of shearing will divide the soil mass into two segments that may displace relative to each other. Second, at micro-scales, there will be fragmentation and rearrangement of particles within the shear band.

Consider the typical problem sketched in Figure 2.5. A load of some form has caused failure to occur in an embankment. We will assume that the embankment is composed of compacted sand. The failure region is a planar shear band. The mass of soil above the shear band is sliding downhill in relation to the stationary mass of soil below. A simple large-scale description of the situation tells us that the average shear stress acting on the shear surface has exceeded the soil strength. Following Coulomb, the soil strength is composed of two elements: cohesion and friction. Since we are dealing with sand, the cohesion will be zero and the frictional strength is therefore given by the product of the average effective normal stress acting on the surface multiplied by the coefficient of friction for the soil. In geomechanics we represent the coefficient of friction by the tangent of an angle φ called the *angle of internal friction*. All of these elements are familiar to students of geomechanics.

Now consider what may happen on a micro-scale. The shear band will have a thickness h that is of the order of a few particle diameters in size. The deformation is said to have *localised* within this band. If we could observe carefully inside the band as the applied load is increased toward failure, we would expect to see a random arrangement of particles with an extremely complex system of interparticle contact forces that, on a macro-scale, result in the average values of shear and normal stress required for equilibrium. Typically, the interparticle contact surfaces are oriented in a random way with a few contacts parallel to the direction of slip, but with most oriented acutely to it. In Figure 2.5 the particles are depicted as smooth ellipsoids, but in fact we are aware of a vast range of possible particle shapes, both smooth and rough.

It is immediately apparent that, for large shearing deformation to occur within the shear band, some particles must either fracture conveniently on surfaces parallel to the direction of slip or else move out of the way so that other particles may slide past. Some fracturing will occur as the average shear stress grows, but it will most probably involve the breaking off of asperities from the particle surfaces rather than a major fracture along the slip direction. The greatest part of the shearing deformation will occur because some particles will be forced to move perpendicular to the slip surface, thereby permitting other particles to move past. As a result the soil mass above the shear surface *moves upward*, normal to the surface, as well as slipping tangentially downhill. The property of deformation normal to the shearing surface is called *dilatancy*. It is a characteristic associated with all granular materials, especially densely packed materials. If the sand in our embankment were *very* loosely packed we might not observe dilation, but that would be an unlikely occurrence in a constructed embankment.

Assume the embankment in Figure 2.5 is about to fail. The soil mass above the shear surface will begin to move and its displacement will have components both parallel and normal to the shear surface. The direction of motion will lie at an angle ψ, called the *angle of dilatancy*, above the shear surface as shown in Figure 2.5. In general, the angle of dilatancy will be smaller than the angle of internal friction. If we recall that the soil shear strength is given by the product of the angle of internal friction and the normal effective stress on the slip surface, we can now see that the friction angle actually represents *two* sources of strength. One is the frictional resistance caused by particles grinding past one another and the second is *interlocking*. Interlocking of particles causes dilation and may contribute significantly to the overall soil strength.

To summarise, we have described two different aspects of inelastic behaviour. The first involved natural compression of soil due to accumulating overburden and the second resulted from an excessive ratio of shear stress to normal stress. In the former we observed crushing and fragmentation of the soil particles leading to fewer large particles, more smaller particles and more dense particle packing. In the latter we find some particle fracturing but more rearrangement, inducing an increase in soil volume within the shearing zone that we call dilatancy. Both of these aspects of soil behaviour can be accommodated in modern theories of soil plasticity as will be seen in Chapter 7.

To bring this chapter to a close we will now discuss the laboratory tests mentioned earlier. One test is associated with inelastic compression due to increasing normal stress. It is the *oedometer test*. The second test is associated with development of inelastic shearing deformation due to increasing shear stress. It is called the *triaxial test*. Both are commonly used tests in nearly all geomechanics laboratories.

2.7 The oedometer test

All of the processes involved in loading and unloading a natural soil deposit can be simulated in the laboratory. An oedometer test subjects a cylindrical sample of soil to conditions similar to the deposit of silt described in the preceding section. Basically, the oedometer consists of a very stiff steel ring enclosing the soil sample as depicted in Figure 2.6. The sample is loaded vertically through the loading cap by the applied load. The rigid ring prohibits any horizontal deformation just as occurs in the ground in our discussion above.

If we are dealing with a saturated sample, drainage may be arranged at the upper and lower surfaces of the soil. The applied load is increased in increments and then decreased again to simulate the processes of overburden accumulation and removal that occurred in the natural soil deposit. In most oedometer tests we would not attempt to measure the horizontal stress applied to the sample by the rigid ring, but it would be possible to do so if we wished and we could determine K_0 directly.

Results from oedometer tests are usually presented in terms of the vertical stress and the *void ratio* of the soil. In most geomechanics texts the void ratio is denoted by e, but we have used e earlier to represent volumetric strain. Therefore we will employ $\tilde{e}$ for the void ratio. The definition of $\tilde{e}$ is

$$\tilde{e} = \frac{\text{volume of voids}}{\text{volume of solid particles}} \tag{2.30}$$

where the two volumes are measured in the same total soil volume, which is

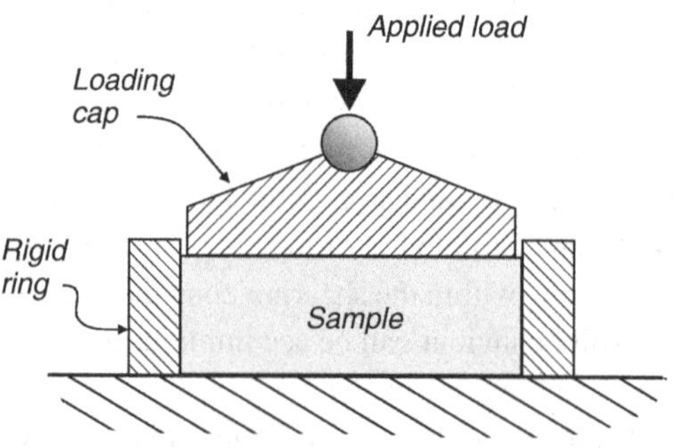

Figure 2.6. Oedometer test apparatus.

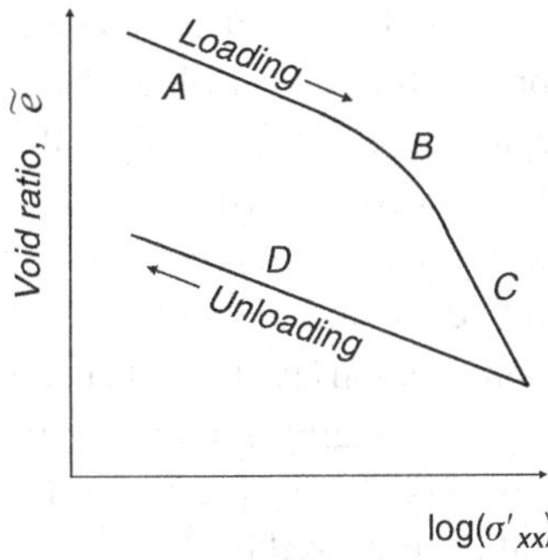

Figure 2.7. Typical response found in an oedometer test.

assumed to be of a representative size. If we let $\tilde{e}_0$ denote the initial void ratio at the beginning of the test, then it is easy to see that the void ratio and the volumetric strain are related by

$$e = -\frac{\tilde{e} - \tilde{e}_0}{1 + \tilde{e}_0} \tag{2.31}$$

where the negative sign ensures that the compressive volumetric strain is positive. If we know the void ratio we can find the volumetric strain and vice versa.

Typical data from an oedometer test are illustrated in Figure 2.7. The graph shows how the void ratio changes as the applied load increases. The load is represented by the logarithm of the vertical effective stress. It is usually assumed that a uniform vertical stress is applied to the sample, although this may not be exactly true. A more correct assessment of the problem would suggest that we should use a mixed boundary condition on all the sample surfaces. The normal component of displacement would be specified together with a frictional condition on the tangential component of traction. While this would be possible theoretically, the additional complication is assumed to be unwarranted. The uniform stress approximation will be sufficiently accurate in an average sense. There are three segments of more or less linear response* in Figure 2.7. In the early stages of loading the graph is linear and this portion may be interpreted as primarily an elastic response (region A). In this region the soil particle structure is basically unchanged from its original state and little if any fragmentation and rearrangement have occurred. The void ratio is decreasing, but this is primarily

* Of course the logarithmic scale means that void ratio and effective stress are *not* linearly related, and this tells us that any assumption we might make concerning the use of the *linear theory* of elasticity may be slightly flawed. A *non-linear elastic* relation of the form $de/dp = \kappa/p$, where κ is constant, is sometimes used to replace equation (2.7) and thereby more accurately represent this part of the soil response.

because of elastic deformation of the particles themselves. The first important particle fracture corresponds to the point at which the graph becomes non-linear (region B). The graph begins to steepen smoothly at this point as more fractures and particle rearrangement occur. The void ratio decreases more quickly as these irreversible processes take place. A second region of linear response then appears (region C). In this region the soil structure continues to evolve in the sense that fracture and fragmentation lead to further particle rearrangement. The distribution of particle sizes is changing as more and more particles are broken. There are theoretical arguments that suggest why a linear response may occur in this region and they will be discussed in Chapter 7. Finally, the applied load is reduced and unloading commences. Once again we find an approximately linear response, which we now identify as elastic rebound of the compressed soil skeleton (region D). If the applied load is reduced to zero the void ratio is permanently reduced and some degree of volumetric strain is permanently locked into the particle structure. During the unloading stage we would expect to find that the horizontal effective stress would become greater than the vertical effective stress due to the locked-in stress effect.

We can clearly see the region of inelastic response in Figure 2.7. It begins when the first particle fractures occur and the loading curve begins to bend downward and it ends when unloading commences. One of the main challenges geomechanics poses to the theory of plasticity is how to model this type of inelastic behaviour. An interesting point about the oedometer test response or the response of natural soil deposits described earlier, is that while there are significant shear stresses developed in the soil during loading, the inelastic behaviour we observe is not directly related to shearing. In fact, we could produce a similar inelastic response in a test where the soil is subjected to a purely isotropic stress. The classical theories of plasticity relate solely to inelastic response caused by shear. Some important changes are required to develop a theory that encompasses behaviour similar to that shown in Figure 2.7.

2.8 The triaxial test

Measurement of soil shear strength in the laboratory is often performed using the *triaxial compression test*. A cylindrical sample of soil is placed in a pressure chamber called a triaxial cell and subjected to an isotropic compressive stress. A plunger passing through the top part of the pressure cell applies an additional axial load on the sample. The basic elements of the test equipment are illustrated in Figure 2.8. The pressure cell is usually filled with water and the sample is encased in a rubber membrane to isolate it from this 'cell fluid'. When the

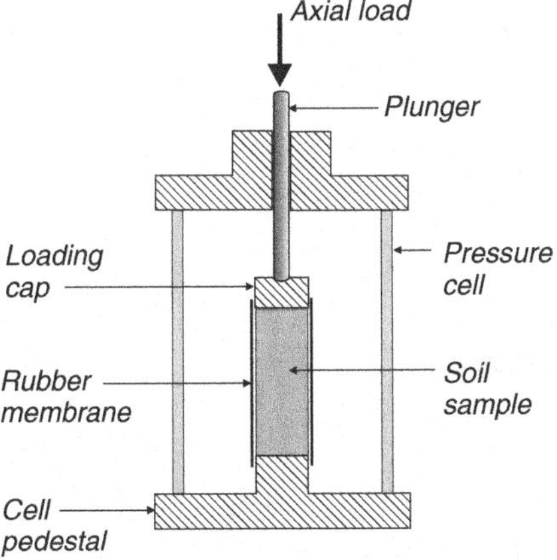

Figure 2.8. Triaxial test apparatus.

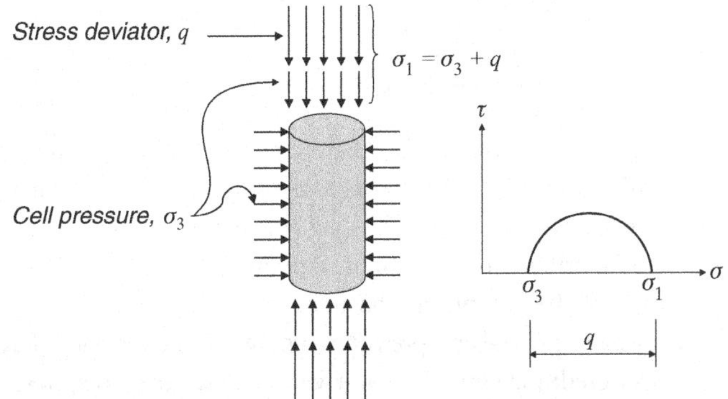

Figure 2.9. Stress state in a triaxial test.

water is pressurised, a uniform isotropic stress called the *cell pressure* acts on all surfaces of the sample. We assume that horizontal and vertical surfaces are principal surfaces and that the axial stress in the sample is given by the sum of the cell pressure plus the axial plunger load divided by the sample cross-sectional area. Since the cell pressure is a principal stress we represent it by σ_3. The axial stress is denoted by σ_1. Note that all vertical surfaces support the same stress, hence the intermediate principal stress σ_2 is equal to σ_3.

The stress state in the sample is assumed to be homogeneous and has the form illustrated in Figure 2.9. We use the symbol q, called the *stress deviator*, to represent the principal stress difference $(\sigma_1 - \sigma_3)$. Note that q is equal to the diameter of the Mohr stress circle and is a general measure of the shear stress supported by the sample. In performing the test the cell pressure is usually

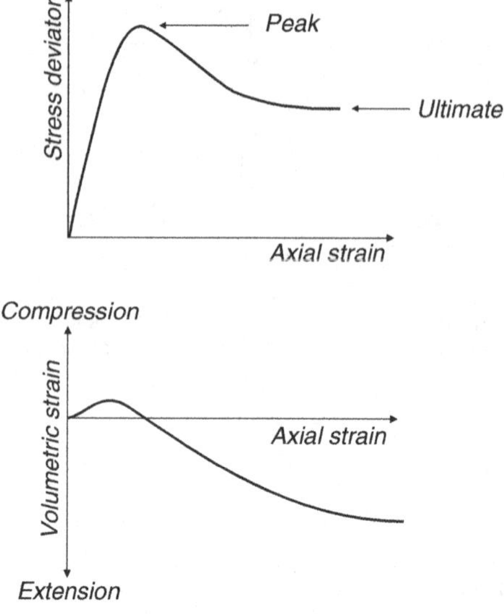

Figure 2.10. Typical response found in a triaxial test.

increased to a specified value and then held constant while the stress deviator is increased until the soil fails. Failure is a result of shear on a planar surface or surfaces that makes an angle of $(45^\circ + \varphi/2)$ with the horizontal. Students of geotechnical engineering will be familiar with this basic test.

The behaviour of the sample as the axial strain is increased is the most interesting feature of a triaxial compression test. Two aspects of the soil response are usually considered: the stress deviator versus axial strain response and the volumetric strain versus axial strain response. Figure 2.10 illustrates typical results for a compacted or dense sand sample. Note how our measure of shear stress, the stress deviator q, reaches a peak and then decreases, finally reaching an ultimate state. At the same time the volumetric strain exhibits a small amount of compression, but then changes to negative values consistent with dilation of the sample.

The stress–strain response shown in Figure 2.10 is interpreted as follows. For small values of q the sample responds elastically and the stress–strain curve is essentially linear. A further increase in axial strain leads first to a stress peak and then to a decreasing stress level and ultimately to a constant value of q. Tests on samples with different degrees of compaction show that the magnitude of the peak q value increases as the sample becomes denser. Because of this the stress peak is attributed to the effects of interlocking – the greater the interlocking, the higher the peak stress value is. The ultimate stress level represents the strength of the soil after all interlocking mechanisms have been broken down and large amounts of deformation have occurred. Dilatancy accompanies the breakdown of interlocking. Initially the volumetric strains are positive, indicating compression. The onset of shear localisation leads to a reversal in the trend of the volumetric strain. Shortly after this the volumetric strain becomes negative and the volume of the soil actually increases. This is highly inelastic behaviour. Among other things, the value of Poisson's ratio would need to be greater than 0.50 in order to accommodate a volume increase under these conditions, even though that would violate the restrictions imposed on the elastic response discussed above.

In Figure 2.10, after the peak stress occurs, the portion of the stress–strain curve where q is decreasing is referred to as *strain softening*. The sample seems to be spontaneously losing strength once the effects of interlocking are overcome. This type of response poses significant challenges for the theory of plasticity, especially from a computational standpoint. Questions also arise as to which level of strength, peak or ultimate, is appropriate for use in design. Conservative engineers will generally adopt the ultimate strength for applications, being aware of course that the peak strength is a reserve element in the overall safety of a project. Serious problems can arise, however, if the peak strength is ever exceeded since the abrupt loss of strength that follows is highly unstable and very large deformations may ensue.

The main point we take away from this general discussion of inelastic effects is that all inelastic responses are associated with particle fracture and rearrangement. Furthermore, there exist two categories of failure, one associated with normal compressive stress and one associated with shear stress. The importance of interlocking and dilatancy in the development of shear strength has also been emphasised. A very wide range of inelastic behaviour is possible in any soil. As we will see later in this book it is possible to represent nearly all of these possibilities mathematically, at least in a conceptual way. In the next chapter we begin to investigate mathematical models for the onset of plastic response.

Further reading

The early works cited near the beginning of this chapter may be found in:

J. Boussinesq, 'Équilibre d'élasticité d'un solide isotrope sans pesanteur, supportant différents poids', *C. Rendus Acad. Sci. Paris*, **86**, 1260–1263 (1878).

R. Hooke, *The Posthumous Works, Containing his Cutlerian Lectures, and other Discourses, Read at the Meetings of the Illustrious Royal Society*; *London*, Samuel Smith and Benjamin Walford for Richard Waller, London, 1705.

There are many good reference works on the elements of elasticity.

R.J. Aitken and N. Fox, *An Introduction to the Theory of Elasticity*, Longman, London, 1980.
J.R. Barber, *Elasticity*, Kluwer Academic, Dordrecht, The Netherlands, 1992.
R.W. Little, *Elasticity*, Prentice Hall, New Jersey, 1965.
A.P.S. Selvadurai, *Partial Differential Equations in Mechanics 2: The Biharmonic Equation, Poisson's Equation*, Vol. 2, Springer-Verlag, New York, 2000.

Useful discussions on the shear strength of soils may be found in many geotechnical text books including:

M.D. Bolton, *A Guide to Soil Mechanics*, Macmillan, London, 439 pp., 1979.
T.W. Lambe and R.V. Whitman, *Soil Mechanics*, John Wiley and Sons, New York, 553 pp., 1979.
R.F. Scott, *Principles of Soil Mechanics*, Addison-Wesley, Reading, MA, 550 pp., 1963.

Exercises

2.1 Use equations (2.2) and (2.3) to prove the relationships given in (2.4).

2.2 Invert equations (2.4) to obtain (2.5).

2.3 For the triaxial compression test the stress matrix has this form

$$\boldsymbol{\sigma} = \begin{bmatrix} \sigma_3 + q & 0 & 0 \\ 0 & \sigma_3 & 0 \\ 0 & 0 & \sigma_3 \end{bmatrix}$$

Derive the corresponding stress deviator matrix. The invariants of the stress deviator are sometimes represented by J_1, J_2 and J_3. Using the definitions of stress invariants from equations (1.19), show that

$$J_1 = 0, \qquad J_2 = -\frac{1}{3}q^2, \qquad J_3 = \frac{2}{9}q^3$$

2.4 Consider a traction vector $\boldsymbol{T}$ acting on a surface with unit normal vector $\hat{\boldsymbol{n}}$. The angle between the two vectors is often referred to as the *angle of obliquity*. Show that the maximum value for the angle of obliquity for any surface in a triaxial compression test sample is given by $\sin^{-1}[(\sigma_1 - \sigma_3)/(\sigma_1 + \sigma_3)]$.

2.5 A deep deposit of homogeneous dry sand has bulk density $\rho = 2.0$ t/m^3. Assume that the deposit has a horizontal surface, is elastic, and that Poisson's ratio has a value of 0.25. Calculate the shear and normal stresses at a depth of 5.0 m for surfaces with unit normal vectors oriented at angles of 30°, 45° and 60° from the horizontal.

2.6 For plane stress problems such as that illustrated in Figure 2.2, the stress matrix has the form shown in equation (2.17). Assuming there is no z dependence for any of the stress components, derive the following relationships for the extensional strains:

$$\varepsilon_{xx} = \frac{1}{E}(\sigma_{xx} - \nu\sigma_{yy}), \quad \varepsilon_{yy} = \frac{1}{E}(\sigma_{yy} - \nu\sigma_{xx}), \quad \varepsilon_{zz} = \frac{\nu}{1-\nu}(\varepsilon_{xx} + \varepsilon_{yy})$$

3
Yield

3.1 Introduction

The term *yield* refers to the onset of inelastic behaviour such as described in the preceding chapter. In this chapter we will try to make a precise description of yielding. In particular, we will try to establish a set of mathematical conditions for yielding that will be referred to as the *yield criterion.*

There have been many different yield criteria suggested by different researchers and engineers. Coulomb set down the first useful yield criterion in 1773. It forms one of the cornerstones of our understanding of the way soils behave and it will be considered in detail later in this chapter. First, however, we will investigate some of the yield criteria suggested for ductile metals. Metals are a bit simpler than geomaterials, and many of the basic ideas can be developed in a simpler context.

A yield criterion can be visualised as a mathematical function. We will represent it by f. The arguments of f might be almost anything to do with the state of the body at the onset of plastic behaviour, but the most obvious candidates for arguments would be the components of stress or strain or both. Modern developments in plasticity accept that the most appropriate arguments are the individual components of the stress matrix. Realising that there are only six independent stress components, we can write our prototype yield criterion as follows:

$$f(\sigma_{xx}, \sigma_{yy}, \sigma_{zz}, \sigma_{xy}, \sigma_{yx}, \sigma_{zx}) = k \tag{3.1}$$

where k represents a constant. It may be zero, but in many cases it will be convenient to have a non-zero constant. As for the left-hand side of (3.1) we will think of this as simply some function of the stress components which is, as yet, undetermined. Yielding is signalled when this function becomes equal to the constant k. Also, there may be functional arguments other than stress that we might wish to add later, but stress will be the central criterion for yielding in most modern plasticity models.

Next, we are aware from our study of Mohr's circle that it may not be necessary to know all the components of stress. We can recreate the stress components in any coordinate frame using the principal stresses, $\sigma_1, \sigma_2, \sigma_3$, provided we know the respective orientations of the principal directions. This would suggest that the six stress components in (3.1) could be replaced by the three principal stresses, plus some information concerning the orientation of the principal directions. At this point, the developments that follow will benefit from the introduction of an important material characteristic that removes all dependence on orientation of the principal directions. The material characteristic is *isotropy*. For an isotropic material, there can be no dependence of material response on a specified direction. Thus, for isotropic bodies, (3.1) can be rewritten, without any loss of generality, in terms of just principal stresses:

$$f(\sigma_1, \sigma_2, \sigma_3) = k \tag{3.2}$$

Here k is still a constant, although its form may differ from that in (3.1).

Alternatives to the principal stresses are the principal invariants I_1, I_2, I_3. Equation (1.19) gave the relations between the invariants and the principal stresses. Obviously we could recast (3.2) as

$$f(I_1, I_2, I_3) = k \tag{3.3}$$

In some circumstances this form for the yield criterion may be more convenient than (3.2), but in most cases we will find (3.2) to be the more useful description.

Much of what follows in this chapter will be directed towards visualising various yield criteria. In this effort, the form (3.2) will be most useful. Because of this it will be convenient to digress for a moment and introduce a new three-dimensional space with coordinates proportional to the values of the principal stresses themselves.

3.2 Principal stress space

If we know the values of the principal stresses at some position within a deforming body, we can plot the values in a Cartesian space with axes having dimensions of stress. Figure 3.1 shows such a space. We call it the *principal stress space*. A point in this space then represents the state of stress in the body. It is called the *stress point*.

As the stress state in the body changes, the stress point will change its position in principal stress space. We might intuitively expect that there will be regions within principal stress space corresponding to elastic response, and other regions

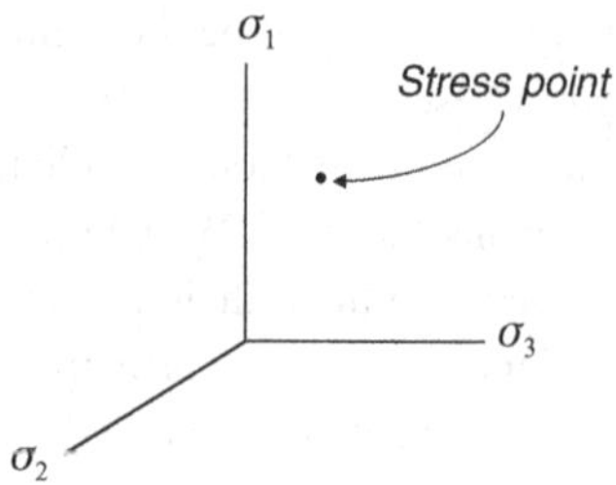

Figure 3.1. Principal stress space.

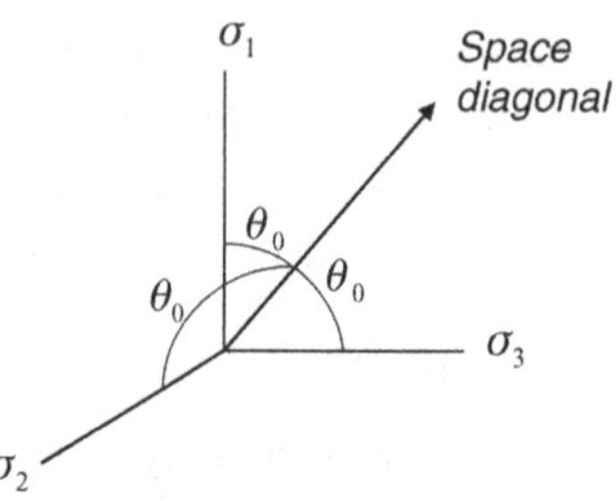

Figure 3.2. The space diagonal in principal stress space.

for plastic response. In fact, the yield criterion f will enable us to describe exactly stress states for which the material will exhibit either elastic or plastic response.

Referring back to (3.2), we see that in general $f = k$ defines a geometric surface in principal stress space. The terms *yield surface* and *yield locus* are both used to describe the surface $f = k$. Plastic response will be confined to the surface. Also, so long as we do not add any more arguments to the function f, the surface will be fixed in space. Later in our development, another argument will be added alongside the principal stresses in (3.2) and its aim will be to permit the surface to move in certain ways.

In principal stress space, a line of special significance is that making equal angles with the three principal stress axes. We call this line the *space diagonal*. It is illustrated in Figure 3.2. The angle between the space diagonal and any of the coordinate axes is denoted by θ_0. It is easy to see that

$$\cos\theta_0 = \frac{1}{\sqrt{3}} \tag{3.4}$$

Note that whenever the stress point lies on the space diagonal, the values of all three principal stresses are the same and we have a state of isotropic stress.

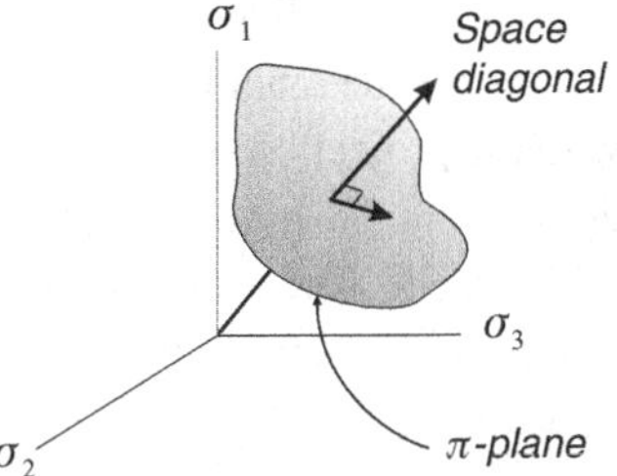

Figure 3.3. Schematic view of the π-plane.

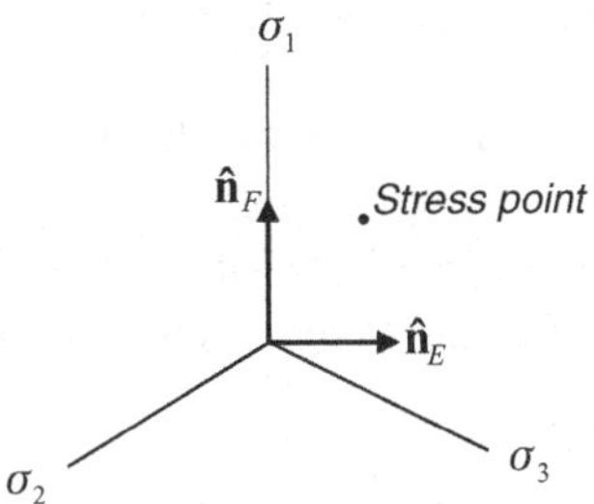

Figure 3.4. Looking down the space diagonal.

Next, it will be helpful to construct a planar surface perpendicular to the space diagonal. This surface is usually called the *π-plane*.* A sketch of the π-plane, placed at an arbitrary point along the space diagonal, is shown in Figure 3.3. Only some irregular segment of the plane is shown but, in fact, it extends in all directions perpendicular to the space diagonal. If we now look down along the space diagonal towards the origin of principal stress space, we obtain a view such as that shown in Figure 3.4. The three principal stress axes are shown, but we realise that these are simply images of the axes projected on to the π-plane. The stress point can also be seen on this sketch.

When choosing to describe the position of a point in space we can select from a number of orthogonal coordinate systems. We usually choose a system that provides the best advantage for the purpose we have in mind. With certain coordinate systems the description may be neater and more compact. For example, the equation of a sphere of radius a in rectangular Cartesian coordinates is $x^2 + y^2 + z^2 = a^2$, whereas, in spherical polar coordinates, it is simply $R = a$. Keeping this in mind, it will be convenient for us to construct a local coordinate system with axes along and perpendicular to the space diagonal. Even

* There have been other names associated with the π-plane. These include Pi-plane, octahedral plane, deviatoric plane and Haigh–Westergaard plane.

specifying one axis along the space diagonal we can still choose from an infinite number of orientations for the other two axes. A convenient choice will be to align one of these with the projection of σ_1 on the π-plane. We then define three orthogonal unit vectors denoted by $\hat{\mathbf{n}}_D, \hat{\mathbf{n}}_E, \hat{\mathbf{n}}_F$. Two of these lie in the π-plane and are shown in Figure 3.4. The third, $\hat{\mathbf{n}}_D$, is the unit vector coinciding with the space diagonal. These vectors have the following component descriptions in principal stress space:

$$\hat{\mathbf{n}}_D = \frac{1}{\sqrt{3}}\begin{bmatrix}1\\1\\1\end{bmatrix}, \quad \hat{\mathbf{n}}_E = \frac{1}{\sqrt{2}}\begin{bmatrix}0\\-1\\1\end{bmatrix}, \quad \hat{\mathbf{n}}_F = \frac{1}{\sqrt{6}}\begin{bmatrix}2\\-1\\-1\end{bmatrix} \tag{3.5}$$

Note that each is a unit vector and observe that they are mutually orthogonal. Now let $\boldsymbol{\sigma} = [\sigma_1 \ \sigma_2 \ \sigma_3]$ denote the stress vector joining the origin and the stress point in principal stress space. Then the components of $\boldsymbol{\sigma}$ in the D, E and F directions are given by the inner products of $\boldsymbol{\sigma}$ with the three unit vectors:

$$\begin{aligned}
\sigma_D &= \frac{1}{\sqrt{3}}(\sigma_1 + \sigma_2 + \sigma_3) = \sqrt{3}p \\
\sigma_E &= \frac{1}{\sqrt{2}}(-\sigma_2 + \sigma_3) \\
\sigma_F &= \frac{1}{\sqrt{6}}(2\sigma_1 - \sigma_2 - \sigma_3)
\end{aligned} \tag{3.6}$$

Note that the component σ_D, along the space diagonal, is proportional to the mean stress $p\,(= I_1/3)$. The components of $\boldsymbol{\sigma}$ are illustrated schematically in Figure 3.5. Note that they are mutually orthogonal.

In the π-plane the stress components σ_E and σ_F appear as shown in Figure 3.6. Another interesting quantity is the radial distance from the space diagonal to

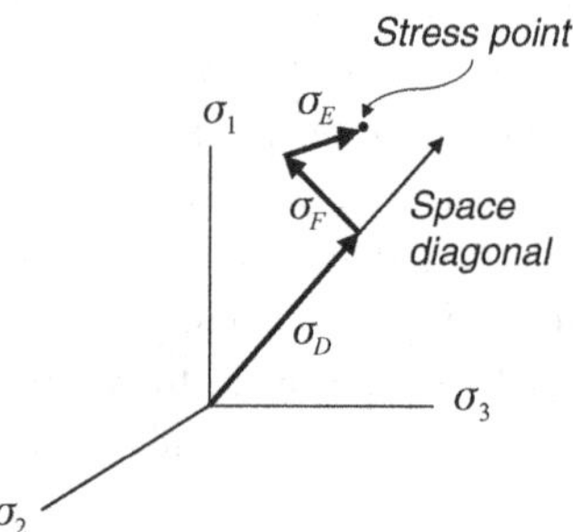

Figure 3.5. Locating the stress point in principal stress space.

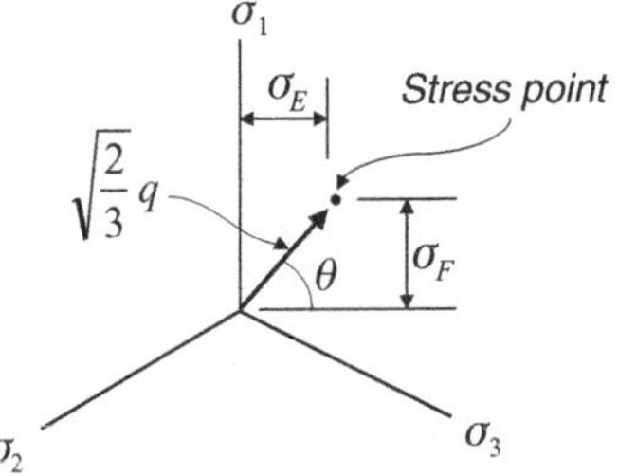

Figure 3.6. Alternative methods for locating the stress point in principal stress space.

the stress point. We can find it easily using the Pythagorean theorem,

$$\left(\sigma_E^2 + \sigma_F^2\right)^{1/2} = \sqrt{\frac{2}{3}}\left(\sigma_1^2 + \sigma_2^2 + \sigma_3^2 - \sigma_1\sigma_2 - \sigma_2\sigma_3 - \sigma_3\sigma_1\right)^{1/2} \quad (3.7a)$$

$$= \sqrt{\frac{2}{3}}q \quad (3.7b)$$

In the second equation above, q is called the *deviatoric stress*. It can be readily shown that

$$\begin{aligned} q &= \left(\sigma_1^2 + \sigma_2^2 + \sigma_3^2 - \sigma_1\sigma_2 - \sigma_2\sigma_3 - \sigma_3\sigma_1\right)^{1/2} \\ &= \frac{1}{\sqrt{2}}[(\sigma_1 - \sigma_2)^2 + (\sigma_2 - \sigma_3)^2 + (\sigma_3 - \sigma_1)^2]^{1/2} \\ &= \left(I_1^2 - 3I_2\right)^{1/2} \end{aligned} \quad (3.8)$$

where I_1 and I_2 are, respectively, the first and second principal stress invariants. Note also that in a triaxial test, where σ_2 and σ_3 are equal, q reduces to $(\sigma_1 - \sigma_3)$, which is identical to the stress deviator introduced in Chapter 2.

A final quantity of interest, also shown in Figure 3.6, is the angle between the horizontal (E) direction and the radial dimension to the stress point. We see that

$$\tan\theta = \frac{\sigma_F}{\sigma_E} = \frac{2\sigma_1 - \sigma_2 - \sigma_3}{\sqrt{3}(\sigma_3 - \sigma_2)} \quad (3.9)$$

The angle θ is called the *Lode angle* after the German engineer W. Lode who first used it in 1926.

To summarise, we see that, in principal stress space, the stress point may be described in any of three ways: by the principal stresses $\sigma_1, \sigma_2, \sigma_3$, by the stress components $\sigma_D, \sigma_E, \sigma_F$, or by the quantities p, q, θ that virtually create a separate polar coordinate system. In some cases one description may be preferable to another, but all three describe exactly the same state of stress in the body.

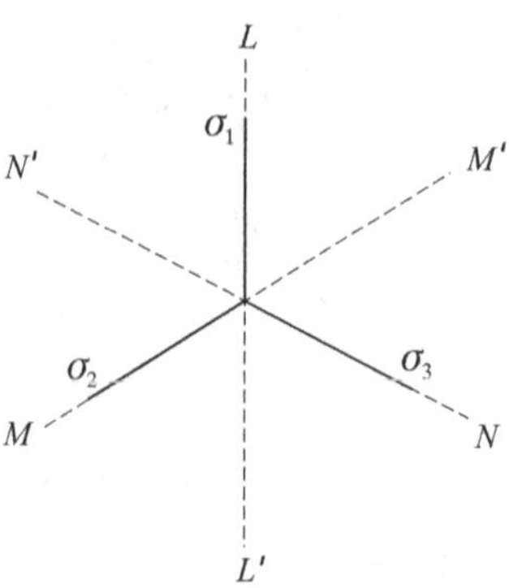

Figure 3.7. Symmetries in principal stress space.

Finally, we can remark that there are certain symmetries that must hold for any yield surface in the π-plane. To see this, note that if the principal stresses $(\sigma_1, \sigma_2, \sigma_3)$ result in plastic behaviour, then so also must $(\sigma_1, \sigma_3, \sigma_2)$ because of isotropy. Referring to Figure 3.7, it is clear that the yield surface must be symmetric with respect to the line LL'. Similar arguments lead to symmetries about lines MM' and NN' as well. These symmetries divide the π-plane into six 60° segments that must possess similar properties.

3.3 Yield surfaces for metals

The first yield criterion for metal was suggested by the French engineer H. Tresca in 1864. His experiments suggested that plastic behaviour would commence when the maximum shear stress reached a critical value. Recalling Mohr's circle, we see that the maximum shear stress will always be half the difference between the major and minor principal stresses. One can easily discover the critical stress by performing a simple tension test on a bar of the metal. If we denote the tensile stress at failure (i.e. the onset of plastic behaviour) by σ_T, then the maximum shear stress is exactly half σ_T. Therefore if we consider the case where $\sigma_1 \geq \sigma_2 \geq \sigma_3$, the yield function f in (3.2) becomes $(\sigma_1 - \sigma_3)/2$ and the constant k is $\sigma_T/2$. The yield criterion can be written as

$$\sigma_1 - \sigma_3 = \sigma_T \tag{3.10}$$

Note that the surface defined by (3.10) does not depend on the mean stress p. The function f depends only on the diameter of the Mohr circle. This implies that the yield surface image in the π-plane will be independent of the position on the space diagonal. We can investigate the yield surface by considering the values of σ_E and σ_F that correspond with (3.10). If we eliminate σ_2 from the

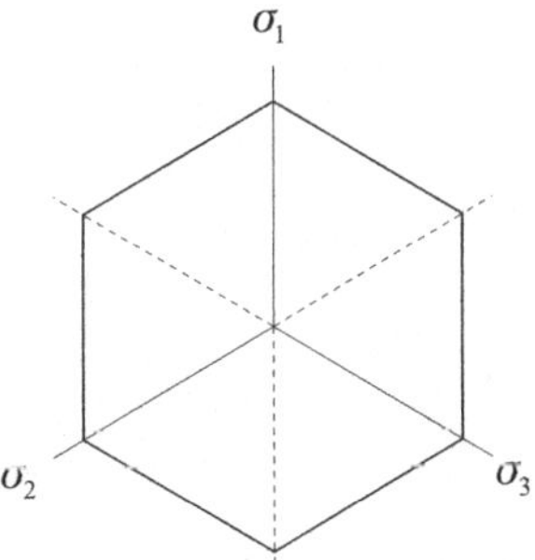

Figure 3.8. The Tresca yield surface.

second two equations of (3.6) we find

$$-\sigma_E + \sqrt{3}\,\sigma_F = \sqrt{2}(\sigma_1 - \sigma_3) = \sqrt{2}\,\sigma_T \tag{3.11}$$

which shows the yield surface is a straight line in the π-plane. In fact, because of symmetry, the line can apply only over one of the 60° segments shown in Figure 3.7. To map the entire surface, one must consider the other possibilities, i.e. $\sigma_1 \geq \sigma_3 \geq \sigma_2$, $\sigma_2 \geq \sigma_1 \geq \sigma_3$, and so on. The complete yield surface has the shape of a regular hexagon. Its intersection with the π-plane is shown in Figure 3.8.

It is possible to rewrite Tresca's yield criterion in terms of p, q, θ or the stress invariants I_1, I_2, I_3, but the equations would be much more complex than the simple form given in (3.11). The more important issue here is the *visualisation* of the yield surface. In principal stress space we see an infinitely long prism. Its cross-section is a hexagon and its central axis is the space diagonal. The volume enclosed, by definition, represents the set of all stress states for which the material will be elastic. If the stress point touches the surface, then yielding will occur.

The second yield criterion of general interest for metals was suggested by R. von Mises in 1913. He suggested that yield will occur when the value of the deviatoric stress q reaches a critical value. We write the von Mises yield condition as

$$q = k \tag{3.12}$$

Recalling equation (3.7b) we see that yield will occur when, in the π-plane, the radial distance from the origin to the stress point reaches the value $\sqrt{2/3}\,k$. As with Tresca's criterion, we can determine the value of k from a simple tension test. If we set $\sigma_1 = \sigma_T$, the tensile yield stress, and we let $\sigma_2 = \sigma_3 = 0$, then we find that q is exactly σ_T and therefore so is k. An alternative explanation for

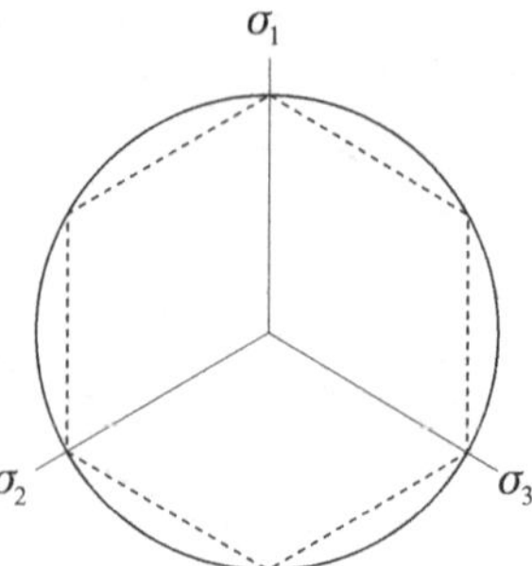

Figure 3.9. The von Mises yield surface.

Mises' criterion was supplied by the German engineer H. Hencky. In 1924 he pointed out that (3.12) is equivalent to requiring that the elastic stored energy owing to distortion (shearing) must equal a critical value. The intersection of the von Mises surface with the π-plane is a circle passing through the vertices of the Tresca hexagon (Figure 3.9). The complete surface is an infinitely long cylinder whose central axis coincides with the space diagonal.

When a ductile metal yields we see, on a microscopic level, displacements occurring between the atoms that make up the crystal lattice. These are called *dislocations*. A dislocation can move through the lattice, displacing one atom after another producing a small, irrecoverable deformation. Very large numbers of dislocations may occur as the applied stress reaches the yield criterion, and they will be manifest on a macroscopic level as a plastic deformation. This is not exactly the situation one envisions in a soil as it approaches failure, but some similarities may exist. On a macroscopic level both a ductile metal and a soft clay may appear to flow when the stresses become severe. Both metals and soils often exhibit localisation of deformation within relatively narrow regions or bands when failure is imminent. On a microscopic level, dislocations in the atomic lattice of a metal bear at least a vague similarity to the fracture and rearrangement of particles in a yielding soil. Workers in geotechnical engineering have often attempted to adapt aspects of metal plasticity theories for use in soil mechanics. The reverse, however, is also true since the very first practical yield criterion was derived specifically for soil. It was the work of the great French engineer Charles Augustus Coulomb.

3.4 The Coulomb yield criterion

Coulomb wrote his first scientific paper in 1773. In it he considered a number of problems involving the strength of building materials prevalent in his day,

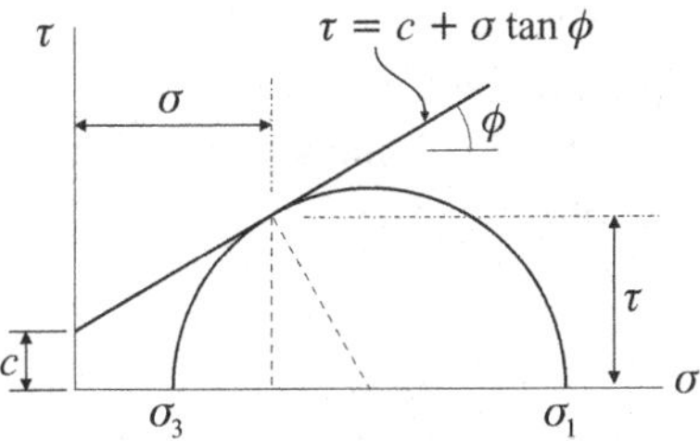

Figure 3.10. The Coulomb failure criterion.

namely wood, stone, masonry and soil. His interest in soil stemmed from the design of retaining walls. As a military engineer he had been involved in the construction of several large earth-retaining structures. He began by observing that all the materials derived strength from two sources: cohesion and friction. His observations of real soils suggested that failure will usually be associated with a surface of rupture within the soil mass. Restricting attention to this surface he wrote his failure criterion as

$$\tau = c + \sigma \tan\phi \tag{3.13}$$

where τ and σ represent the shearing stress and normal stress on the physical plane through which material failure occurs. The constant c is called the cohesion. It has dimensions of stress. The quantity $\tan\phi$ is similar to a coefficient of friction. The angle ϕ is referred to as the angle of internal friction. Coulomb did not write the criterion exactly as we have done here, but his words clearly expressed the meaning we associate with the equation today.

We can recast (3.13) in the form of (3.2) by referring to Mohr's circle. The graph of (3.13) is a straight line on the Mohr diagram as shown in Figure 3.10.

If failure is to occur for a combination of principal stresses $\sigma_1 \geq \sigma_2 \geq \sigma_3$, the critical Mohr stress circle, derived from σ_1 and σ_3 must be a tangent to this line. Therefore, the values of τ and σ can be related to the principal stresses σ_1 and σ_3 by considering the geometry of the dashed triangle

$$\begin{aligned} \tau &= \frac{1}{2}(\sigma_1 - \sigma_3)\cos\phi \\ \sigma &= \frac{1}{2}(\sigma_1 + \sigma_3) - \frac{1}{2}(\sigma_1 - \sigma_3)\sin\phi \end{aligned} \tag{3.14}$$

Using these relations in (3.13) we find

$$\sigma_1(1 - \sin\phi) - \sigma_3(1 + \sin\phi) = 2c\cos\phi \tag{3.15}$$

Relating this expression to (3.2) we see the yield function f on the left-hand side. On the right-hand side is the constant k. Note that f does not depend upon the intermediate principal stress σ_2.

To discover the form of the Coulomb surface in the π-plane, we can begin by inverting equations (3.6). This gives

$$\begin{aligned}\sigma_1 &= \frac{1}{\sqrt{3}}\sigma_D + \sqrt{\frac{2}{3}}\sigma_F \\ \sigma_2 &= \frac{1}{\sqrt{3}}\sigma_D - \frac{1}{\sqrt{2}}\sigma_E - \frac{1}{\sqrt{6}}\sigma_F \\ \sigma_3 &= \frac{1}{\sqrt{3}}\sigma_D + \frac{1}{\sqrt{2}}\sigma_E - \frac{1}{\sqrt{6}}\sigma_F\end{aligned} \qquad (3.16)$$

Substituting the first and last of these equations into (3.15) gives

$$-\sqrt{3}\,\sigma_E(1+\sin\phi) + \sigma_F(3-\sin\phi) = 2\sqrt{6}\,c\,\cos\phi + 2\sqrt{2}\,\sigma_D\,\sin\phi \qquad (3.17)$$

We have moved σ_D to the right-hand side of this equation since it will be constant in the π-plane. Equation (3.17) shows that σ_E and σ_F are linearly related, and therefore the intersection of the yield surface with the π-plane will be a straight line. Of course the straight line will only apply over one of the 60° segments, exactly the same as for the Tresca yield surface. Here, however, there are two important differences with respect to the Tresca surface. The first is that the relative slopes of the surface in the various 60° segments are different. We will see this more clearly in a moment. The second and more important difference is this: the size of the surface depends upon σ_D and hence upon the mean stress p. Graphing the yield surface for all six of the 60° segments results in the irregular hexagonal shape shown in Figure 3.11. For the purposes of constructing this figure we have taken ϕ to be 30°.

Each of the vertices of the hexagon has a particular physical meaning. All vertices occur on the lines of symmetry where two of the principal stresses are equal. The uppermost vertex corresponds to the condition where $\sigma_1 > \sigma_2 = \sigma_3$.

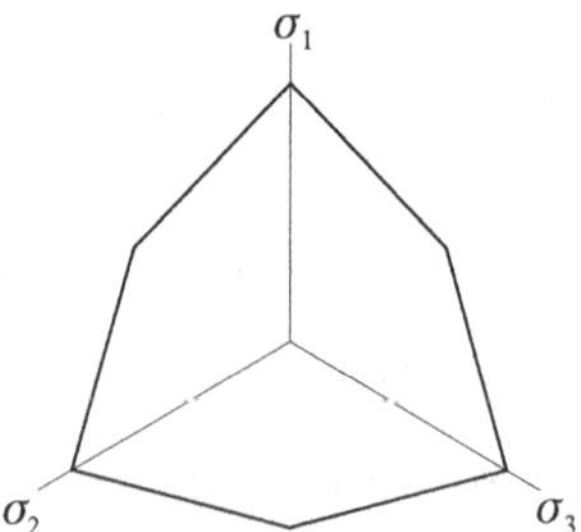

Figure 3.11. Cross-section through the Coulomb yield surface.

The lowermost vertex is directly opposite and it corresponds to $\sigma_1 < \sigma_2 = \sigma_3$. Each of the other vertices is obtained by permuting the indices in these two expressions. We can relate each vertex physically to a type of triaxial test in the laboratory. The major vertices such as the uppermost one represent conventional compression triaxial tests. The minor vertices, such as the lowermost one, represent triaxial *extension* tests. The triaxial extension test is performed in exactly the same way as a compression test, but, after application of the cell pressure, the axial stress is reduced rather than increased. The principal stresses then have the form $\sigma_1 < \sigma_2 = \sigma_3$.

The resemblance between the Coulomb and Tresca surfaces is more than a passing one. Note that if we set $\phi = 0$ in (3.13), Coulomb's criterion is essentially the same as Tresca's, namely that failure occurs when the greatest shear stress reaches a critical value. If we set $\phi = 0$ in (3.17), we obtain (3.11) provided we set $2c = \sigma_T$. The difference between the yield surface shapes in the π-plane stems solely from ϕ. But this is not the most important difference. That distinction belongs to the dependence of Coulomb's criterion on the mean stress p. Note how σ_D (and hence p) enters the right-hand side of (3.17). Because of this, the size of the yield surface grows as the mean stress increases. Whereas Tresca's surface was an infinitely long uniform hexagonal prism, Coulomb's surface has an expanding pyramid shape as shown in Figure 3.12.

We can see a less complicated picture if we consider the intersection of the yield surface with the (σ_D, σ_F)-plane. This is shown in Figure 3.13. The yield surface is now just the two straight lines shown in the figure. The upper line corresponds to (3.17) with σ_E set to zero. The lower line corresponds to the lowermost vertex of the Coulomb hexagon. To find its equation we consider the

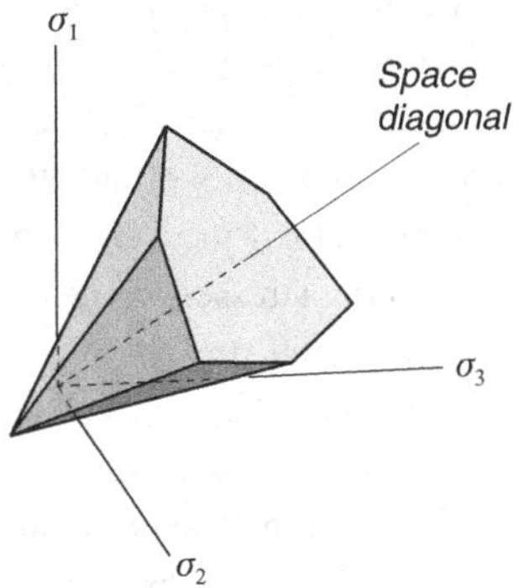

Figure 3.12. Perspective view of the Coulomb yield surface.

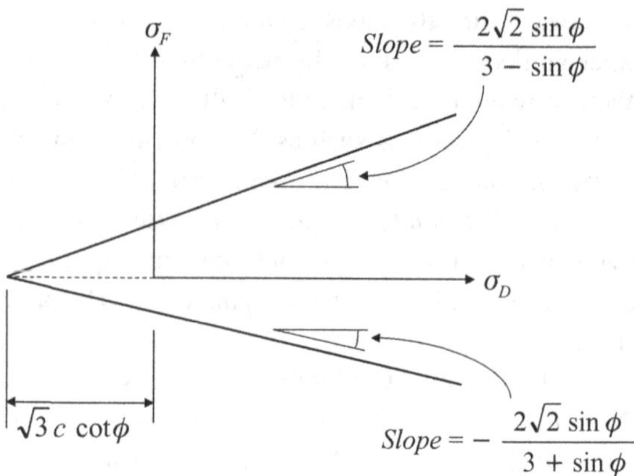

Figure 3.13. Long section through the Coulomb yield surface.

case where $\sigma_1 < \sigma_2 = \sigma_3$. For this situation, (3.15) is replaced by

$$\sigma_3(1 - \sin\phi) - \sigma_1(1 + \sin\phi) = 2c\cos\phi$$

Then using (3.16) to replace σ_1 and σ_3 we find

$$\sqrt{3}\,\sigma_E(1 - \sin\phi) - \sigma_F(3 + \sin\phi) = 2\sqrt{6}\,c\cos\phi + 2\sqrt{2}\,\sigma_D\sin\phi \quad (3.18)$$

and the equation of the lower line is obtained by setting $\sigma_E = 0$.

In this figure, note how the magnitudes of the slopes of the failure lines differ. The extension test leads naturally to a weaker condition than does the compression test. Also, note how the cohesion controls the extent of the negative values for σ_D. There are obvious parallels between this diagram and the more common Mohr diagram such as that in Figure 3.10, but the two diagrams are, in fact, complementary. Figure 3.13 is a slice through the general three-dimensional yield surface shown schematically in Figure 3.12.

We must also note here that nothing has been said concerning effective or intergranular stress associated with particulate materials such as soils. In fact, all of the normal stresses in these equations may be effective stresses. Specifically, we assume that the stress σ in (3.13), as well as all of the principal stresses $\sigma_1, \sigma_2, \sigma_3$, and the stress component σ_D measured on the space diagonal, are all effective stresses. The remaining stresses, τ, σ_E, σ_F, are shear stresses and by definition are unrelated to pore fluid stresses.

At this point one might ask whether the Coulomb hexagonal yield surface is in fact realistic. Only a small number of real soils have been sufficiently well tested to give an answer to this question. The problem lies in creating test conditions in which all three principal stresses can be varied independently. The conventional triaxial test, despite its name, does not permit this. A much more specialised test configuration is needed, the so-called *cubical true triaxial test*. As its name suggests, the test employs a cubical soil sample. Normal stresses are applied to each face and these are the principal stresses. They may be varied independently as the sample is loaded to the point of failure. One interesting thing to do with such a device is to carry out a sequence of tests for which, at the point of failure, the principal stresses are all different, but the mean stress p remains the same. In this way we may plot actual points corresponding to failure on the π-plane. For the small number of soils that have been tested, it appears the Coulomb hexagon is probably a reasonably good approximation for the true locus of yielding in the π-plane. The data suggest that the true surface is similar to but somewhat smoother than the theoretical hexagon. This is perhaps not surprising since the sharp vertices of Coulomb's surface would be very demanding for nature to reproduce. Those sharp vertices are also demanding from a computational viewpoint. Because of this, some researchers have suggested slight modifications to Coulomb's criterion that produce smoother surfaces.

3.5 Modifications to Coulomb's criterion

There are three modified forms of the Coulomb criterion that we will consider. The first was proposed in 1952 by two of the most prominent researchers from the field of both metal and soil plasticity: D.C. Drucker and W. Prager. They suggested that the von Mises yield criterion could be modified easily by introducing a dependence on the mean stress p,

$$q - \xi p = k \tag{3.19}$$

Here the additional term ξp will change the von Mises yield surface from an infinitely long cylinder to a cone. We can select the values of the constants ξ and k in such a way that the cone will agree with the Coulomb surface at the major vertices. First, recall that

$$\sqrt{\frac{2}{3}}\,q = \sqrt{\sigma_E^2 + \sigma_F^2}, \quad \sqrt{3}\,p = \sigma_D \tag{3.20}$$

Then for the case where $\sigma_1 > \sigma_2 = \sigma_3$, we have $\sigma_E = 0$ and

$$\sqrt{\frac{3}{2}}\,\sigma_F - \frac{\xi}{\sqrt{3}}\,\sigma_D = k \tag{3.21}$$

Comparing this with (3.17) (after setting $\sigma_E = 0$) we conclude

$$k = \frac{6c\cos\phi}{3 - \sin\phi}, \quad \xi = \frac{6\sin\phi}{3 - \sin\phi} \tag{3.22}$$

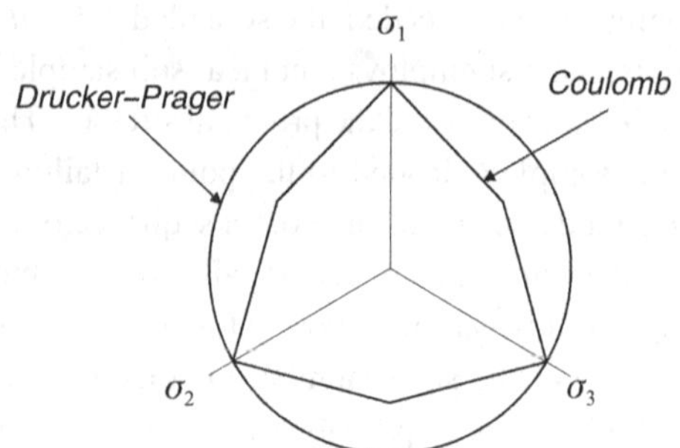

Figure 3.14. The Drucker–Prager and Coulomb yield surfaces.

The graph of Drucker and Prager's yield surface in the π-plane is a circle that touches the Coulomb hexagon as shown in Figure 3.14.

Thinking again about real soil response, tests show that the Drucker–Prager surface is not as accurate a representation as the Coulomb hexagon. Even the relatively common triaxial extension test gives results that lie far closer to the minor vertex of the hexagon than to the circle. Nevertheless, the Drucker–Prager criterion possesses the significant virtue of simplicity, and because of this it is an important addition to the repertoire of the soil plastician.

Other versions of the Drucker–Prager surface have been put forward. For example, a smaller cone which intersects the Coulomb hexagon at the minor vertices may be constructed by replacing (3.22) by

$$k = \frac{6c \cos\phi}{3 + \sin\phi}, \quad \xi = \frac{6 \sin\phi}{3 + \sin\phi}$$

The second modified form of Coulomb's surface was developed in 1975 by P.V. Lade and J.M. Duncan. Their yield criterion was proposed expressly for cohesionless soils. It can be written in the form

$$\sigma_1\sigma_2\sigma_3 = \kappa p^3 \tag{3.23}$$

where κ is a constant. On the left-hand side of this equation we see the product of all three principal stresses, which we know to be the third principal stress invariant. In the form shown above, the criterion looks deceptively simple. If we recast it using σ_D, σ_E and σ_F, it appears a bit more formidable.

$$3\sigma_D\left(\sigma_E^2 + \sigma_F^2\right) + \sqrt{2}\,\sigma_F\left(3\sigma_E^2 + \sigma_F^2\right) = 2(1 - \kappa)\sigma_D^3 \tag{3.24}$$

This form is more useful for visualising the yield surface. First, we need to select an appropriate value for the constant κ. This is usually done by requiring (3.23) and (3.24) to agree with the Coulomb hexagon at its major vertices. Then taking $\sigma_1 > \sigma_2 = \sigma_3$, we can set both $\sigma_E = 0$ and $c = 0$ in (3.17) to obtain

$$\sigma_F = \frac{2\sqrt{2}\,\sigma_D \sin\phi}{3 - \sin\phi} \tag{3.25}$$

Note that we have set $c = 0$ in this equation since the Lade–Duncan criterion applies to cohesionless soils. Using this, together with $\sigma_E = 0$ in (3.24) we find

$$\kappa = 1 - \frac{12\,\sin^2\phi}{(3 - \sin\phi)^2} + \frac{16\,\sin^3\phi}{(3 - \sin\phi)^3} \tag{3.26}$$

If $\phi = 30^\circ$ for example, we find $\kappa = 0.648$. Next, we wish to graph the shape of the yield surface in the π-plane. To do so, we can solve (3.24) for σ_E

$$\sigma_E = \pm\left[\frac{(1-\kappa)\sigma_D^3 + \sqrt{2}\,\sigma_F^3 - 3\sigma_D\sigma_F^2}{3\sigma_D + 3\sqrt{2}\,\sigma_F}\right]^{1/2} \tag{3.27}$$

which shows immediately that the Lade–Duncan surface will be symmetric about the σ_F axis in the π-plane. We now hold σ_D constant and graph σ_E for a range of values of σ_F. This gives a graph such as that shown in the right half of Figure 3.15.

Also shown in Figure 3.15 is the third modification of the Coulomb criterion. This was derived by H. Matsuoka and T. Nakai in 1974. Their yield equation can be written as

$$\sigma_1\sigma_2\sigma_3 = \xi\ p(\sigma_1\sigma_2 + \sigma_2\sigma_3 + \sigma_3\sigma_1) \tag{3.28}$$

where ξ is a constant. This equation is also restricted to cohesionless soils. Remarkably, it agrees with the Coulomb hexagon at both major and minor vertices, provided the appropriate value for ξ is used. If we recast (3.28) in terms of $\sigma_D, \sigma_E, \sigma_F$, we find

$$2\sigma_F^3 - 6\sigma_E^2\sigma_F + 2\sqrt{2}(1 - 3\xi)\sigma_D^3 - 3\sqrt{2}(1 - \xi)\sigma_D\left(\sigma_E^2 + \sigma_F^2\right) = 0 \tag{3.29}$$

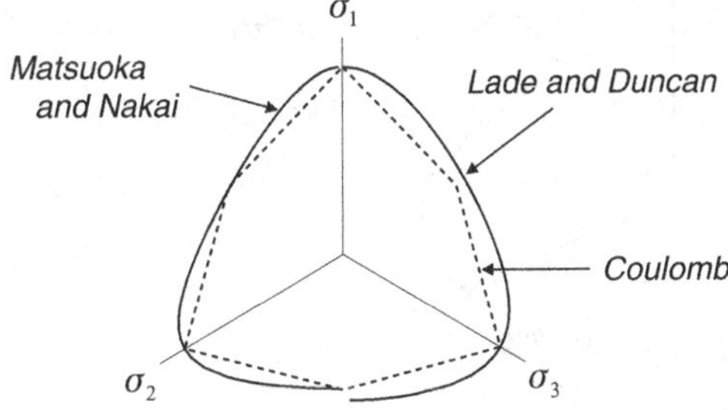

Figure 3.15. Lade–Duncan and Matsuoka–Nakai yield surfaces compared with the Coulomb yield surface.

Using this together with (3.17) we find the expression for ξ necessary to force the yield locus to agree with the Coulomb hexagon,

$$\xi = \frac{3(1 - \sin\phi - \sin^2\phi + \sin^3\phi)}{9 - 9\sin\phi - \sin^2\phi + \sin^3\phi} \tag{3.30}$$

The graph of the yield locus in the π-plane can be constructed in the same way as the Lade–Duncan surface was. Both the Lade–Duncan surface and the Matsuoka–Nakai surface provide good agreement with the available cubical triaxial test data.

At this point, one might be excused for thinking the last word on yield surfaces for soil has been written, but that would be a mistake. There remains a question to which none of these criteria can provide adequate answers. The question is: what happens if we simply increase the mean stress without applying any shearing? The answer is discussed in the next section.

3.6 The Cambridge models

Recall the void ratio versus the logarithm of effective stress response from an oedometer test discussed in Chapter 2. We realise the void ratio is directly related to the volumetric strain, e, and we can, if we wish, replace the applied stress σ_{xx} by the mean stress p. Thus the usual curve of the void ratio versus the logarithm of the effective stress can be regarded as being analogous to the e–log p response, or vice versa. We are aware that typical undisturbed soil samples produce loading curves that display a distinctive shape similar to that in Figure 3.16. We see two (more or less) straight-line segments joined by a smooth bend or 'knee'. The level of stress at the knee is called the *preconsolidation pressure*. It is interpreted as being the greatest vertical effective stress the soil

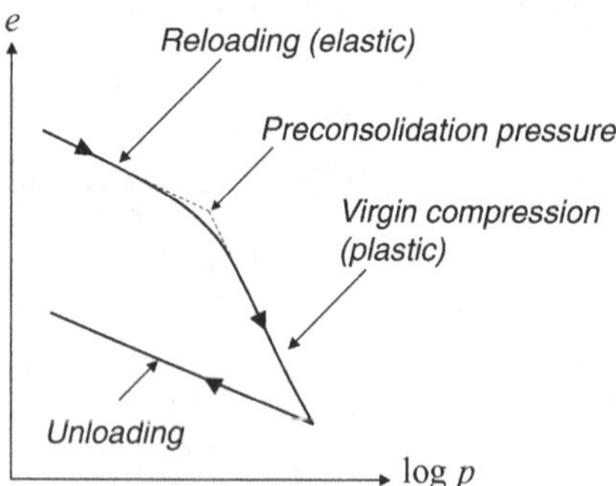

Figure 3.16. Typical one-dimensional compression response for an undisturbed soil sample.

has experienced in its past history. The upper straight portion is interpreted as reloading, since presumably the sample was unloaded when it was removed from the ground. The reloading continues until the preconsolidation pressure is reached. At that point the soil skeleton enters a new loading regime. At no time in its past history has the matrix of particles been required to support the intensity of stress that now exists. We refer to this as *virgin* or *normal compression*. The e–log p response continues as a new straight line but with a steeper slope. If the soil is then unloaded, the unloading response is found to parallel the original loading response.

Clearly this cycle of loading and unloading has produced permanent deformation of the soil skeleton. That alone would suggest plastic behaviour; but, in addition, we observe that reloading again will follow the unloading curve until the latest peak pressure is reached, whereupon the response continues down the virgin compression curve. Thus the straight-line segment with a flatter slope appears to represent an elastic response, while the steeper virgin compression curve represents plastic yielding. The somewhat vaguely defined preconsolidation pressure represents the yield stress.

If we are to represent these effects, it appears that the yield surface must somehow close on the space diagonal. Sometimes this is termed a 'capped' yield surface. For smaller values of p, the surface can expand, but if p approaches the preconsolidation pressure the surface must contract and eventually close on itself as illustrated in Figure 3.17. The notion of a completely closed yield surface is a radical departure from the basic ideas of metal plasticity, and a significant change to the 'Coulomb-type' yield surfaces discussed above. There are also obvious questions concerning what happens to the surface when yielding does occur. If the stress point simply follows the space diagonal until it reaches the closed surface, how can we arrange to let the surface grow under the influence of further increase in p? The answer to that question is a phenomenon called *hardening*; it will be deferred until Chapter 7.

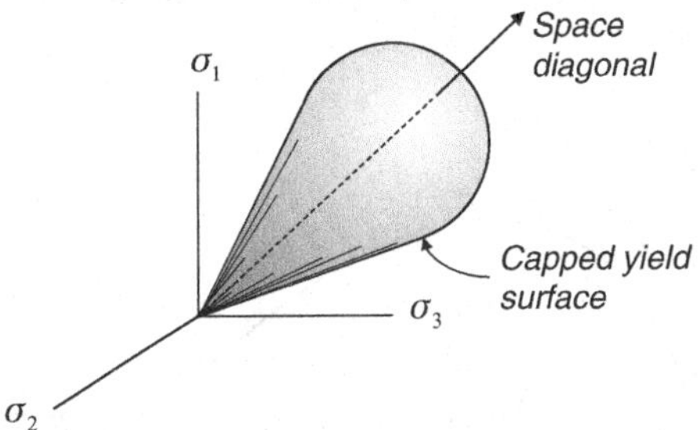

Figure 3.17. A closed or capped yield surface in principal stress space.

For the time being we only want to consider two special cases of closed surfaces.

The first researchers to propose the concept of a closed yield surface were from the Cambridge University soil mechanics group. The ideas they formulated have given rise to a host of new soil plasticity models; too many to discuss in a basic text such as this. We will confine our attention to two of the earliest yield criteria named Cam Clay and Modified Cam Clay.

The first complete plasticity theory for soil incorporating a closed yield surface was devised by K.H. Roscoe together with several co-workers. It was called the Cam Clay model (named after the river Cam, which flows behind the Cambridge Engineering Department laboratories). Their yield criterion can be written in terms of the invariants q and p as follows:

$$q + Mp\left(\ln \frac{p}{p_c} - 1\right) = 0 \tag{3.31}$$

Here both M and p_c are material parameters, while q and p are the deviatoric and the mean stress, respectively. There are certain similarities between this equation and the Drucker–Prager yield surface in (3.19). In the π-plane the Cam Clay surface will be circular just as the Drucker–Prager surface was. The major difference lies in the term in parentheses. This term causes the surface to close on the space diagonal. If we plot (3.31) we find the situation depicted in Figure 3.18.

We can visualise the full yield surface as being the curved line in Figure 3.18 revolved about the mean stress axis as shown in Figure 3.19. In principal stress space we see a pointed, bullet-shaped surface aligned with the space diagonal. The parameter p_c is called the *critical state pressure*. It will gain more significance in Chapter 7, but for the time being it is simply the ordinate of the maximum deviatoric stress. We see from (3.31) that when $p = p_c$, the deviatoric stress is given by $q = Mp = Mp_c$. If we wish this surface to agree with the major vertices of the Coulomb hexagon when $p = p_c$, then we make $M = 6 \sin \phi/(3 - \sin \phi)$, the same as the constant ξ in (3.22).

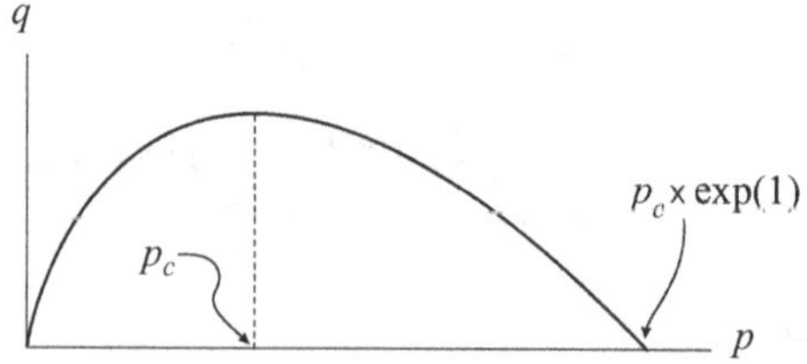

Figure 3.18. Cam Clay yield surface in (q, p)-space.

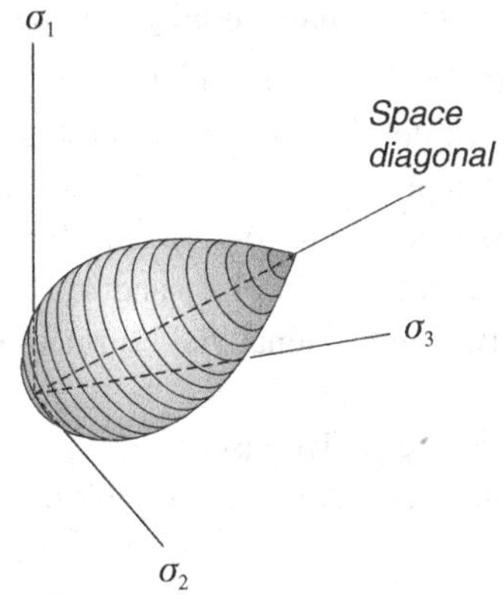

Figure 3.19. Perspective view of the Cam Clay surface in principal stress space.

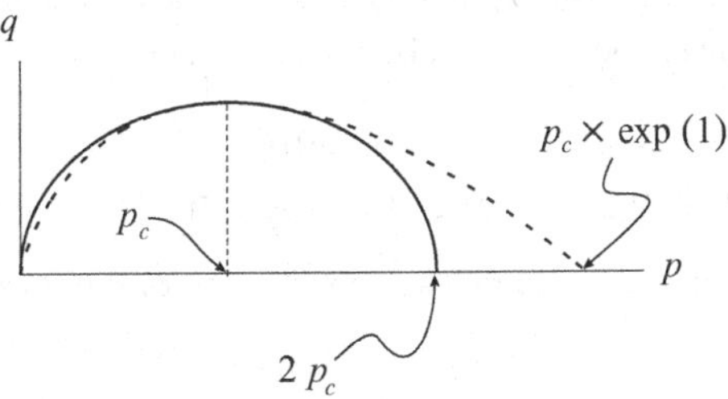

Figure 3.20. Modified Cam Clay yield surface.

The pointed vertex at the tip of the Cam Clay surface was viewed by some researchers as being a weakness of the initial Cambridge model. The Modified Cam Clay model eliminated the point and introduced an elliptical surface with the form

$$q^2 = M^2 p(2p_c - p) \tag{3.32}$$

where M and p_c play the same roles as in (3.31). The shape of the yield surface is now as shown in Figure 3.20. The original Cam Clay surface is also shown as a dashed line. The two surfaces agree exactly when $p = p_c$, but the modified surface closes on the mean stress axis at a value of $2p_c$ rather than at $2.718p_c$ for the original model.

Both Cam Clay and Modified Cam Clay have the property of permitting plastic behaviour in response to an isotropic stress increase; the motivation for closing the surface on the mean stress axis mentioned at the beginning of this section. Later we shall see that they possess additional virtues. They offer the ability to predict more rational estimates for strains than do any of the other yield surfaces discussed in this chapter.

3.7 Two-dimensional yield loci

Sometimes it is useful to consider just two stress components in visualising the yield locus. This will often be the case in problems dealing with plane stress or plane strain. For example, if we wished to consider plane strain where the extensional strain associated with σ_2 is zero, then it may be convenient to plot the yield surface as a function of σ_1 and σ_3. Our objective here is to investigate the resulting two-dimensional yield surfaces that correspond to this condition.

We can begin by considering the Tresca criterion. If we assume $\sigma_1 \geq \sigma_2 \geq \sigma_3$ then (3.10) results. But there are five other possibilities, one example is $\sigma_2 \geq \sigma_1 \geq \sigma_3$. Altogether the six possible arrangements of the principal stresses give the six sides of the Tresca hexagon shown in Figure 3.8. Let us suppose that all the three principal stresses are initially equal to zero. Then consider the range of possible values for σ_1 and σ_3 that will satisfy the Tresca yield criterion.

With $\sigma_2 = 0$, the six possible combinations for the principal stresses become

$$\left.\begin{array}{lll} \sigma_1 \geq 0 \geq \sigma_3, & 0 \geq \sigma_3 \geq \sigma_1, & \sigma_3 \geq \sigma_1 \geq 0 \\ \sigma_3 \geq 0 \geq \sigma_1, & 0 \geq \sigma_1 \geq \sigma_3, & \sigma_1 \geq \sigma_3 \geq 0 \end{array}\right\} \tag{3.33}$$

In each case Tresca's criterion gives an expression of the form

$$\sigma_m - \sigma_n = \sigma_T \tag{3.34}$$

where m and n take distinct values from 1, 2 and 3. Thus we find six straight lines to graph in the (σ_1, σ_3)-plane. The resulting yield locus is shown in Figure 3.21. For example, the uppermost horizontal line corresponds to $m = 1$ and $n = 2$ in (3.34).

Figure 3.21 gives us a picture of the yield locus when all three principal stresses are zero. If we hold $\sigma_2 = 0$, then a line can map the (σ_1, σ_3) stress point trajectory on the plane of the figure. In fact, σ_2 may not remain equal to

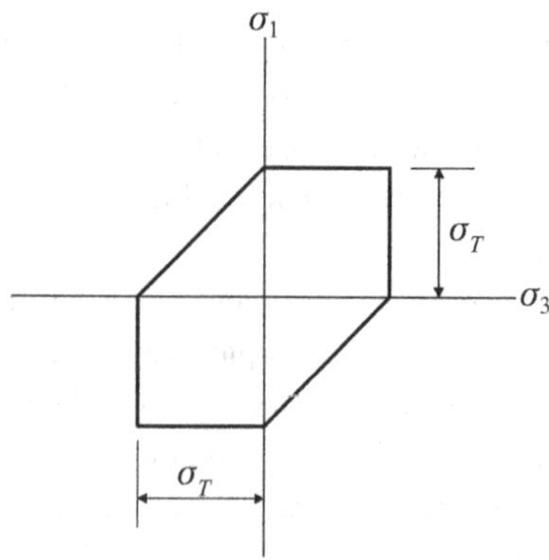

Figure 3.21. Two-dimensional Tresca yield surface.

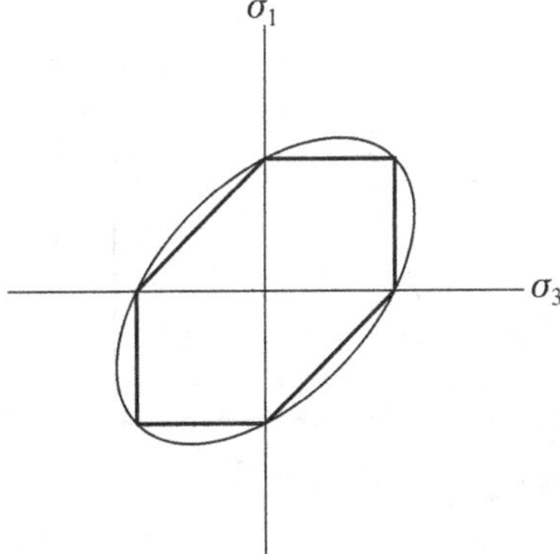

Figure 3.22. Tresca and von Mises yield surfaces in two dimensions.

zero. For example, in plane strain elasticity, Hooke's law immediately shows that $\sigma_2 = \nu(\sigma_1 + \sigma_3)$. Therefore only stress paths with $\sigma_1 = -\sigma_3$ will ensure $\sigma_2 = 0$. Nevertheless, even if σ_2 does change, the only effect on Figure 3.21 would be to shift the origin. Yield will occur whenever the two-dimensional stress point touches the yield locus.

The von Mises criterion was given by (3.12). If we set $\sigma_2 = 0$ in (3.8) we find

$$q = \left(\sigma_1^2 + \sigma_3^2 - \sigma_1\sigma_3\right)^{1/2}$$

If we use this in (3.12), and replace k by σ_T, the resulting graph of σ_1 versus σ_3 is shown in Figure 3.22. The Tresca condition is also shown. As we might expect from our experience with the π-plane, the von Mises condition passes through the 'corners' of the Tresca polygon.

Physically, we can think of Figures 3.21 and 3.22 as being the intersection of the three-dimensional yield surfaces with the plane surface defined by $\sigma_2 = 0$. In the case of Tresca, the three-dimensional surface is a hexagonal prism. Its intersection with the $\sigma_2 = 0$ surface gives the polygon shown in Figure 3.21. Similarly, the von Mises cylinder gives the ellipse shown in Figure 3.22.

Now we can move on to the Coulomb yield surface. We expect to find a similar result here as for the Tresca criterion. Setting $\sigma_2 = 0$ and considering each of the six possible combinations of principal stresses shown in (3.33) we find the two-dimensional yield locus shown in Figure 3.23. Each straight-line segment corresponds to an equation of the form

$$\sigma_m(1 - \sin\phi) - \sigma_n(1 + \sin\phi) = 2c\cos\phi \tag{3.35}$$

where once again m and n take on distinct values between 1 and 3. As an example, the uppermost horizontal line in Figure 3.23 corresponds to $m = 1$ and $n = 2$. For this figure we have taken ϕ to be 30°. Use of a different value

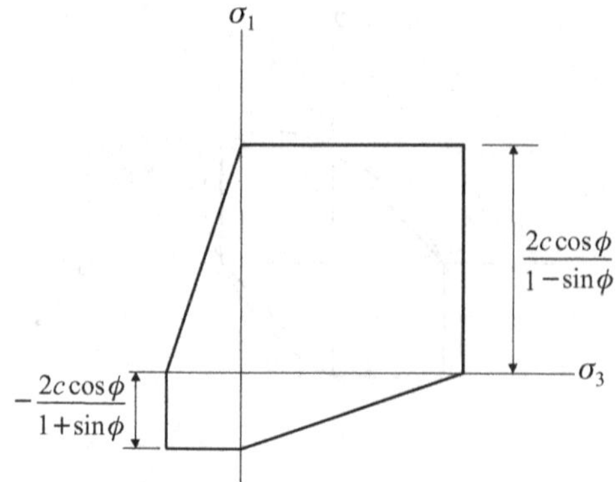

Figure 3.23. Two-dimensional Coulomb yield surface.

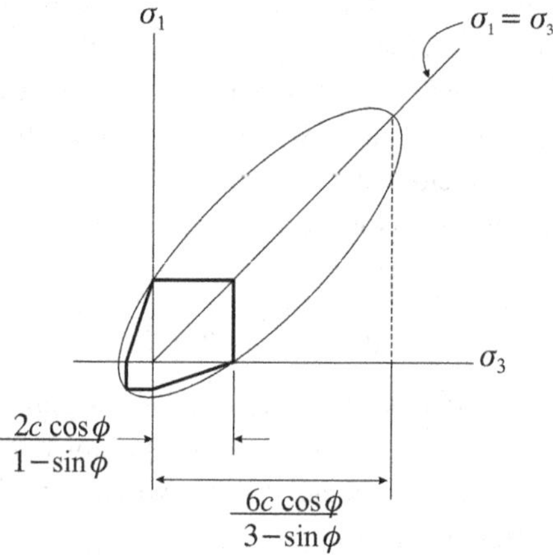

Figure 3.24. Two-dimensional Drucker–Prager and Coulomb yield surfaces.

of ϕ would alter the shape of the figure. If we were to set ϕ equal to zero, the Tresca polygon would re-emerge.

Finally, we can investigate the two-dimensional form of the Drucker–Prager condition (3.19). Recalling that this condition was represented by a cone-shaped surface in three dimensions, we expect to find a conic section in the $\sigma_2 = 0$ plane. Moreover, looking at Figure 3.14, we would anticipate that the conic section will agree with the Coulomb polygon at three vertices. Setting $\sigma_2 = 0$ in both q and p, and using the result in (3.19) together with (3.22) we find

$$\left(\sigma_1^2 + \sigma_3^2 - \sigma_1\sigma_3\right)^{1/2} - \frac{2\sin\phi}{3-\sin\phi}(\sigma_1 + \sigma_3) = \frac{6c\cos\phi}{3-\sin\phi} \tag{3.36}$$

The graph of this equation is shown in Figure 3.24. The corresponding Coulomb yield locus is also shown. The value of ϕ is assumed to be 30°.

The surprising observation in Figure 3.24 is the degree by which the Drucker–Prager failure condition deviates from the Coulomb polygon in the upper right-hand quadrant. This quadrant corresponds to the two conditions where $\sigma_1 \geq \sigma_3 \geq 0$ and $\sigma_3 \geq \sigma_1 \geq 0$. For each case one of σ_1 or σ_3 is the intermediate principal stress while the other is the major principal stress. We can easily investigate the situation where σ_1 and σ_3 are equal. The Coulomb condition gives

$$\sigma_1 = \sigma_3 = \frac{2c \cos\phi}{1 - \sin\phi} \tag{3.37}$$

This follows directly from the Coulomb yield criterion when $\sigma_2 = 0$ is the minor principal stress. The corresponding expression for the Drucker–Prager condition is

$$\sigma_1 = \sigma_3 = \frac{6c \cos\phi}{3 - 5\sin\phi} \tag{3.38}$$

This follows from (3.19) together with (3.36) and $\sigma_1 = \sigma_3$. Clearly (3.38) gives a significantly greater range for the yield stress for these conditions than does (3.37). For $\phi = 30^\circ$, the ratio of the Drucker–Prager yield stress to that for Coulomb is exactly 3. Alternative forms of the Drucker–Prager criterion, which do not pass through the major vertices of the Coulomb hexagon, will naturally give a better result in the two-dimensional case.

It is possible to graph the Lade–Duncan and Matsuoka–Nakai criteria for two-dimensional stress conditions, but this will be left as an exercise for the reader. As one might expect, both criteria will agree more closely with the Coulomb polygon than does the Drucker–Prager condition.

3.8 Example – plane strain

Even though two-dimensional yield representations are quite useful for visualisation of stress conditions, there are some potential difficulties involved. The difficulties arise because of the dependence of the Coulomb yield criterion on the mean stress (cf. (3.17)). In many problems the mean stress will change and this will cause the position and the size of the two-dimensional yield locus to change as well.

We can illustrate the difficulties that may occur by considering a simple example. Suppose we have a block of cohesionless sand loaded in plane strain conditions as shown in Figure 3.25. In the figure the coordinate directions are principal directions, and the z-direction is the direction of zero extensional strain. The stress matrix is diagonal and the three normal stresses are principal stresses. We will assume that the block is initially unstressed, then the stresses σ_{xx} and σ_{yy} are increased, equally, to a value of p_0. We also assume isotropic

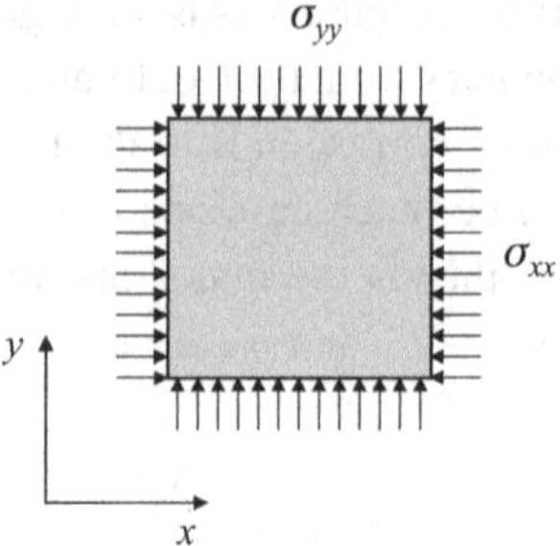

Figure 3.25. Example problem – a block of sand in plane strain with applied normal tractions.

elastic behaviour for the sand during the application of p_0. Thus the three components of stress are

$$\sigma_{xx} = \sigma_{yy} = p_0, \quad \sigma_{zz} = 2\nu\, p_0 \tag{3.39}$$

where ν is Poisson's ratio and we note that $0 \le \nu \le \frac{1}{2}$. Also note that $\sigma_{zz} = \nu\,(\sigma_{xx} + \sigma_{yy})$ for plane strain conditions. We will begin by finding the two-dimensional form of the Coulomb yield locus for this condition.

It should be carefully noted that the two-dimensional yield locus we seek will not necessarily be the yield locus at later stages of the problem. Altering any stress component may alter the mean stress and hence alter the size and position of the yield locus. For the stress state given in (3.39), we can locate the two-dimensional yield locus as follows. Since we want the yield locus in the $(\sigma_{xx}, \sigma_{yy})$-plane, we hold σ_{zz} constant. Then the greatest possible values of σ_{xx} and σ_{yy} can be obtained from the Coulomb criterion with σ_{zz} as the minor principal stress. Since the material is cohesionless, we have $c = 0$, and the Coulomb criterion (3.15) gives

$$\max \sigma_{xx} = \max \sigma_{yy} = N\sigma_{zz} = 2\,\nu\, N\, p_0 \tag{3.40}$$

where

$$N = \frac{1 + \sin\phi}{1 - \sin\phi} \tag{3.41}$$

The smallest possible values for σ_{xx} and σ_{yy} are found by taking σ_{zz} to be the major principal stress. This gives

$$\min\,\sigma_{xx} = \min\,\sigma_{yy} = \frac{1}{N}\sigma_{zz} = \frac{2\nu}{N}\, p_0 \tag{3.42}$$

These minimum and maximum values for σ_{xx} and σ_{yy} define the horizontal and vertical straight-line segments of the two-dimensional yield surface; they

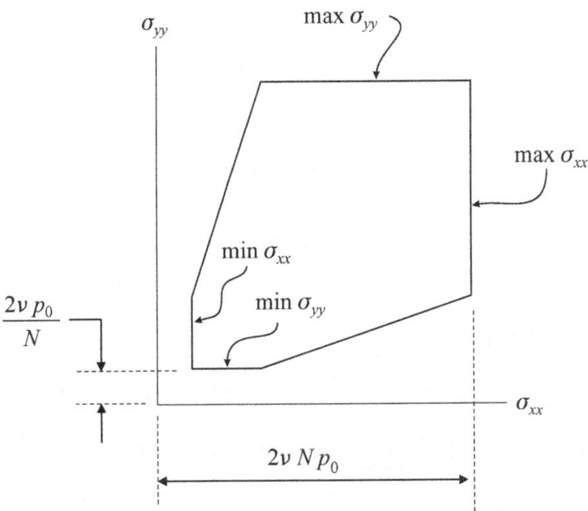

Figure 3.26. Two-dimensional Coulomb yield surface.

are shown in Figure 3.26. The remaining two sloping sides of the yield locus correspond to the condition where σ_{zz} is the intermediate principal stress. If σ_{xx} is the minor principal stress, we can set it equal to its minimum value and solve for σ_{yy} from the yield criterion,

$$\sigma_{yy} = N\sigma_{xx} = N\left(\frac{2\nu\, p_0}{N}\right) = 2\nu\, p_0 \tag{3.43}$$

This value, $2\nu\, p_0$, is the value of σ_{zz} and it identifies the four remaining vertices of the yield locus shown in Figure 3.27.

We can think physically of the yield locus in Figure 3.27 as a slice through the three-dimensional Coulomb hexagon of Figure 3.12. It is the intersection of the three-dimensional locus with the plane $\sigma_{zz} = 2\nu\, p_0$. The initial stress point $\sigma_{xx} = \sigma_{yy} = p_0$ for the initial conditions (3.39) is also shown in Figure 3.27. The main point we wish to make is this; the yield locus shown in Figure 3.27 will change for almost any change in the stresses σ_{xx} and σ_{yy}. This occurs because of the plane strain condition $\sigma_{zz} = \nu(\sigma_{xx} + \sigma_{yy})$. The only way one can alter σ_{xx} and σ_{yy} without changing σ_{zz} is to require $\sigma_{xx} = -\sigma_{yy}$.

Suppose we hold σ_{yy} constant while reducing σ_{xx}; that is, let

$$\sigma_{xx} = p_0 - p_*, \quad \sigma_{yy} = p_0 = \text{constant} \tag{3.44}$$

The stress trajectory on the two-dimensional yield locus is a horizontal line moving away from the initial stress point as shown in Figure 3.27. As σ_{xx} decreases, so will σ_{zz}. We have $\sigma_{zz} = 2\nu\, p_0 - \nu p_*$, and this will cause the

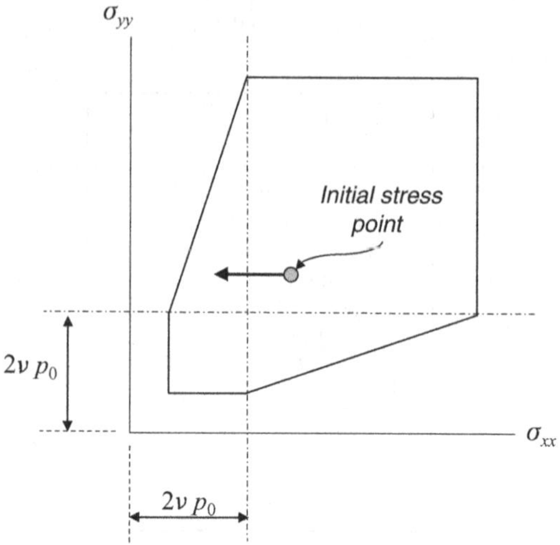

Figure 3.27. Stress point trajectory for example problem.

yield locus to shrink. At some point the value of p_* will be sufficient so that σ_{xx} will equal σ_{zz}. That condition occurs when

$$p_* = p_0 \left(\frac{1 - 2\nu}{1 - \nu} \right) \tag{3.45}$$

Up to this point, σ_{zz} has been the minor principal stress. If σ_{xx} is now further reduced, it will take the role of the minor principal stress and σ_{zz} will be the intermediate principal stress. We can then reduce σ_{xx} further until the Coulomb yield criterion is satisfied when $\sigma_{xx} = \sigma_{yy}/N = p_0/N$. The corresponding values for p_* and for σ_{zz} are

$$p_* = p_0 \left(1 - \frac{1}{N} \right), \quad \sigma_{zz} = \nu p_0 \left(1 - \frac{1}{N} \right) \tag{3.46}$$

This sequence of stress change is shown in Figure 3.28 for the case where $\nu = \frac{1}{3}$ and $N = 3$. The stress points are marked as 1, 2 and 3, respectively, for the initial state, the point where $\sigma_{xx} = \sigma_{zz}$, and the point where yield occurs. The corresponding yield loci are also shown. Note the extent to which the yield locus shrinks as σ_{xx} decreases.

The situation illustrated in Figure 3.28 is only one possibility. Different values for the parameters ν and N may result in different outcomes. For example, the yield locus may shrink more rapidly (due to smaller ν or smaller N) and it may fall upon the horizontal stress trajectory before point 2 is reached. The criterion

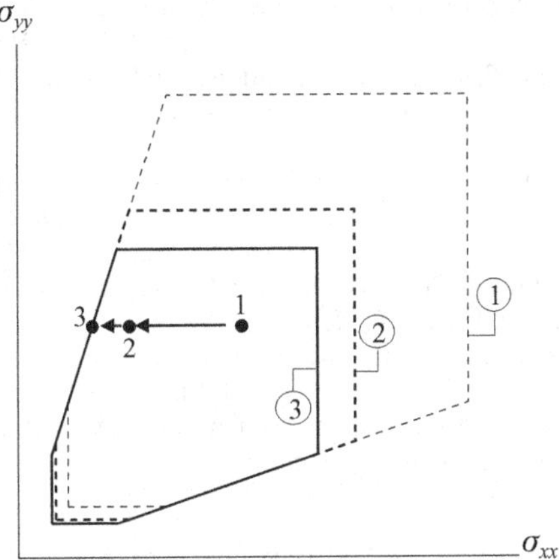

Figure 3.28. Successive Coulomb yield surfaces for example problem.

for this condition can be shown to be

$$N\left(\frac{\nu}{1-\nu}\right) < 1 \tag{3.47}$$

For instance, if $\nu = \frac{1}{3}$, any value of N smaller than 2 will result in this alternative scenario for yielding.

Finally, we can generalise our problem to take cohesive soils into account. If c is not zero, the yield criterion can be written as (3.15) with the major and minor principal stresses interpreted appropriatly. We can then rearrange (3.15) to become

$$\sigma_1 + \frac{c}{\tan\phi} = N\left(\sigma_3 + \frac{c}{\tan\phi}\right) \tag{3.48}$$

It is evident from this equation that we need only add the constant $c/\tan\phi$ to all stresses in order to account for cohesion. That is, we can define a new stress matrix

$$\boldsymbol{\sigma}^* = \boldsymbol{\sigma} + (c\cot\phi)\mathbf{I} \tag{3.49}$$

Then the stresses $\boldsymbol{\sigma}^*$ will obey a cohesionless yield criterion.

Further Reading

Original references for the various yield surfaces in the order introduced above, with the exception of the Cambridge models, may be found in:

H. Tresca, Sur l'ecoulement des corps solids soumis à de fortes pression, *Comptes Rendus Acad. Sci. Paris,* **59**, 754 (1864).

R. von Mises, Mechanik der festen Koerper im plastisch-deformablen Zustand, *Nachr. d. K. Ges. d. Wiss Göttingen, Math.-Phys. Kl.*, 582–592 (1913).

H. Hencky, Zur Theorie plastisches Deformationes und des hierdurch in Material hervorgerufenen Nachspannungen, *Zeit. Angew. Math. Mech.*, **4**, 323–334 (1924).

C.A. Coulomb, Essai sur une application des règles des maximis et minimis à quelques problemes de statique relatifs à l'architecture, *Mem. Acad. Roy. Pres. divers Sav.*, **5**, 7 (1776).

D.C. Drucker and W. Prager, Soil mechanics and plastic analysis or limit design, *Quart. Appl. Maths.*, **10**, 157 (1952).

P.V. Lade and J.M. Duncan, Elastoplastic stress strain theory for cohesionless soil, *J. Geotech. Eng. Div. ASCE*, **101**, 1037 (1975).

T. Matsuoka and K. Nakai, Stress-deformation and strength characteristics of soil under three different principal stresses, *Proc. Japan. Soc. Civil Engineers*, **232**, 59 (1974).

A detailed discussion of Coulomb's contribution to geomechanics together with an English translation of his memoir may be found in:

J. Heyman, *Coulomb's Memoir on Statics – an Essay in the History of Civil Engineering*, Cambridge University Press, Cambridge, 1972.

The Cambridge plasticity models were developed in the sequence of papers listed below:

K.H. Roscoe, A.N. Schofield and C.P. Wroth, On yielding of soils, *Geotechnique*, **8**, 28 (1958).

K.H. Roscoe and H.B. Poorooshasb, A theoretical and experimental study of strains in triaxial tests on normally consolidated clays, *Geotechnique*, **13**, 12 (1963).

K.H. Roscoe, A.N. Schofield and A. Thurairajah, Yielding of clays in states wetter than critical, *Geotechnique*, **13**, 211 (1963).

K.H. Roscoe and J.B. Burland, On the generalised stress–strain behaviour of 'wet' clay, in *Engineering Plasticity* (ed. J. Heyman and F.A. Leckie), Cambridge University Press, Cambridge, 1968.

The Cam Clay model was described more completely in:

A.N. Schofield and C.P. Wroth, *Critical State Soil Mechanics*, McGraw-Hill, New York, 1968.

Exercises

3.1 An undisturbed sample of silty sand taken from an ancient embankment was tested in a conventional drained triaxial compression test. The effective confining stress for the test was 70 kPa. The angle of internal friction for the sample was found to be 32°.

(a) Calculate the axial stress in the test specimen at failure.

(b) Next, a special triaxial *extension* test is performed on a similar sample of the same soil. In this test the *mean stress* p is held constant – equal to the mean stress at failure in the initial compression test. Estimate the value of the axial stress at failure for this extension test.

(c) For this soil, sketch the likely shape for the cross-section of the three-dimensional yield surface in the π-plane. Identify the failure points for the compression and extension tests.

(d) For this soil, sketch the likely shape of the complete three-dimensional yield surface in principal stress space.

(e) Suppose another sample of the same soil is tested in a drained, cubical true triaxial device where all three principal stresses can be altered independently. Given that two of the principal effective stresses are held constant at 70 and 50 kPa, respectively, estimate two possible values for the third principal effective stress at failure.

3.2 A sample of dry cohesionless sand is placed in a cubical, true triaxial device. The sample is first subjected to a hydrostatic stress $p = 100$ kPa. Next, the principal stresses are controlled in such a way that

$$\sigma_1 = p + s, \quad \sigma_2 = p - \alpha s, \quad \sigma_3 = p - (1 - \alpha)s$$

where α is a constant between 0 and 0.5, and s can be increased continuously until the sample fails. The angle of internal friction for the sand is 30° and it may be assumed the Coulomb yield condition applies.

(a) Show that the value of s at failure will always be 200 kPa$/(4 - 3\alpha)$.

(b) For values of α ranging from 0.0 to 0.50 in increments of 0.10, calculate the corresponding values at failure for s, σ_1, σ_2 and σ_3.

(c) For each of the values of α in (b), calculate σ_E and σ_F.

(d) For each of the values of α, plot the stress point in the π-plane. Note that the points lie on a straight line.

3.3 A sample of clay is tested in a special test device. In the test, all the normal stress components are zero ($\sigma_{xx} = \sigma_{yy} = \sigma_{zz} = 0$). The shear stress components are controlled in such a way that

$$\sigma_{xy} = 0, \quad \sigma_{xz} = S, \quad \sigma_{yz} = T$$

and S and T may be adjusted as the test progresses.

(a) Write out the stress matrix σ in terms of S and T.

(b) Solve the eigenvalue problem to find the values of the three principal stresses in terms of S and T.

Suppose particular values of S and T result in failure of the clay.

(c) Find the values of mean stress, p, and deviatoric stress, q, in terms of S and T.

(d) Find the corresponding values of the coordinates σ_D, σ_E and σ_F for the stress point in principal stress space.

(e) Plot the locus of all possible stress points for this test in the π-plane.
(f) Given that the cohesion c and angle of friction ϕ for the clay are 25 kPa and 10°, respectively, find the set of all possible values for S and T.

3.4 Create two graphs similar to Figure 3.24 showing the Lade–Duncan and Matsuoka–Nakai yield loci for the condition $\sigma_2 = \text{constant}$. Take the angle of internal friction to be 30°. Use any convenient positive value for the constant σ_2.

3.5 Rework the plane strain example problem at the end of this chapter for the situation where $\nu = \frac{1}{4}$ and $\phi = 15°$. What is the value of p_*/p_0 for which yield occurs?

3.6 Show that the Lode angle θ can also be expressed in the form

$$\theta = \frac{1}{3}\sin^{-1}\left(-\frac{3\sqrt{3}\,I_3}{2\,I_2^{3/2}}\right)$$

4
Plastic flow

4.1 Introduction

In Chapter 3 we formulated conditions describing when yielding may or may not occur. In this chapter we begin to explore what may happen if the stress point arrives at the yield surface. We intuitively expect that yielding will be accompanied by some form of increased deformation, over and above the elastic deformation that has gone on while the stress point has been inside the yield surface. We expect plastic behaviour to be softer than elastic behaviour, with the result that strains will accumulate more quickly. The term *plastic flow* is used to describe the deformation following yield.

One of the main differences between plastic response and elastic response is that plastic flow will be irreversible. While the material is elastic we can increase the stress with a consequent increase in strain, and then completely recover those strains by simply returning the stress state to its initial value. If yield occurs this will not be possible. Plastic deformation will not be recoverable from simple unloading. If we do reduce the stress to its initial value we will recover whatever elastic strain that has occurred in getting to the yield state, but the plastic strain will be locked within the body.

In order to describe plastic flow we might attempt to derive a constitutive relationship linking plastic strain to the current stress state. But this will immediately lead to difficulties owing precisely to the irreversibility mentioned above. There can clearly be no unique one-to-one relationship between plastic strain and stress since there may be an unknown amount of plastic deformation already locked within the body at the start of any loading episode. As a result we choose to seek a relationship between stress and *plastic strain rate*. By looking at the rate of change of plastic strain rather than the total amount we avoid the problem of irreversibility. Of course, if the plastic strain rate is known throughout some loading process, then a simple integration will give the total amount of plastic strain that has accumulated during that process.

Obviously, it may be convenient to differentiate between plastic and elastic strain. We do this by using superscripts. The superscript e denotes elastic strain while p denotes plastic strain. The total strain is the sum of the elastic and plastic parts. For example,

$$\varepsilon_{xx} = \varepsilon^{e}_{xx} + \varepsilon^{p}_{xx} \tag{4.1}$$

and similar expressions would hold for the other strain components. This decomposition of strain is central to nearly all theories of plastic material behaviour.

In summary, our goal in this chapter is to make some progress towards establishing a description for plastic flow based on the rate of change of plastic strain. We will denote the plastic strain rates using a raised dot, indicating a time derivative, for example, $\dot{\varepsilon}^{p}_{xx}$. Many textbooks prefer the notation $\delta\varepsilon^{p}_{xx}$, called the *plastic strain increment*. This quantity is basically the same as $\dot{\varepsilon}^{p}_{xx}$, but it does not involve time directly. Of course, we can view time simply as a parameter, and $\delta\varepsilon^{p}_{xx}$ then represents the amount of plastic strain occurring within some given parameter increment. The use of one nomenclature or the other makes no difference to the end result, but is primarily a matter of personal preference.

Despite the obvious fact that plastic strains may possibly become quite large, we will continue to use the small-strain definitions introduced in Chapter 1. This is done partly because of the simplicity inherent in the small-deformation theory and partly because, in many applications, it is the conditions at the onset of plastic behaviour that are of most interest, rather than the precise description of strain magnitude.

4.2 Normality

Our ultimate aim will be to formulate a functional relationship between the components of the plastic strain rate and the components of stress. It is reasonable to assume that the components of the plastic strain rate can be arranged into a square matrix exactly as for the elastic strains. We will denote this matrix by $\dot{\boldsymbol{\varepsilon}}^{p}$. We expect $\dot{\boldsymbol{\varepsilon}}^{p}$ to be symmetric and to have the principal values $\dot{\varepsilon}^{p}_{1}, \dot{\varepsilon}^{p}_{2}, \dot{\varepsilon}^{p}_{3}$.

An important assumption concerning plastic strains relates to Saint-Venant's hypothesis. This assumes that the principal directions of both the stress matrix $\boldsymbol{\sigma}$ and the plastic strain rate matrix $\dot{\boldsymbol{\varepsilon}}^{p}$ are aligned. If the material is isotropic, and if $\dot{\boldsymbol{\varepsilon}}^{p}$ depends only on $\boldsymbol{\sigma}$, then Saint-Venant's hypothesis is no longer an assumption but is required by the rules of linear algebra. In many cases we may be happy to assume that our material is isotropic, but it may be that $\dot{\boldsymbol{\varepsilon}}^{p}$ possesses a functional dependence on more than just the components of the stress matrix.

For example, we might reasonably wish to consider a material for which $\dot{\boldsymbol{\varepsilon}}^p$ is a function of both $\boldsymbol{\sigma}$ and $\dot{\boldsymbol{\sigma}}$. In that case Saint-Venant's hypothesis might not be appropriate. Regardless of these considerations we will take Saint-Venant's hypothesis as an important assumption for the developments in this chapter.

A consequence of Saint-Venant's hypothesis is that we can relate stresses and plastic strain rates spatially by plotting them on the same graph. For example, in the π-plane we can plot both the principal stresses and the principal components of the plastic strain rate matrix on the same graph. The axes for σ_1 and $\dot{\varepsilon}_1^p$ fall on the same line, and a similar result applies for the intermediate and minor principal values of both matrices. Of course, the scales of the respective axes are different and we are not directly comparing stresses with strain rates, but the ability to plot both together will be useful in visualising some parts of our development.

We now want to begin thinking about what happens when the stress point reaches the yield surface. Some important details can be described using a simple example. Consider a block of some material resting on a horizontal plane surface as sketched in Figure 4.1. Let the block have mass M and let the static coefficient of friction between the block and plane be μ. Suppose we now apply, in a quasi-static fashion, horizontal forces F_x and F_y to the faces of the block as shown in the figure. The resultant force acting on the block will be

$$F = \sqrt{F_x^2 + F_y^2}$$

The criterion for slip between the block and the plane is

$$F = Mg\mu$$

or

$$F_x^2 + F_y^2 = (Mg\mu)^2 \tag{4.2}$$

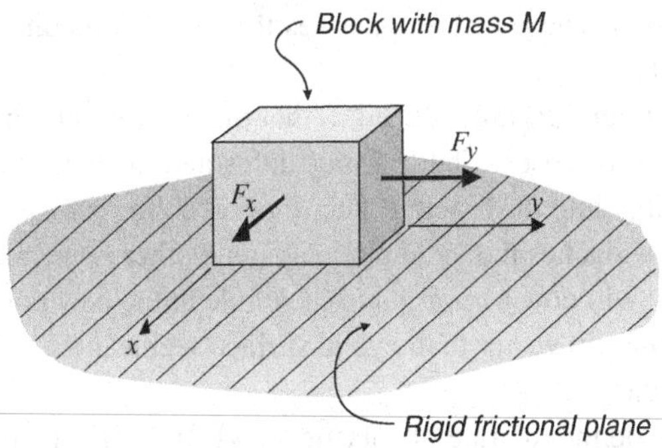

Figure 4.1. Block sliding on a rigid frictional surface.

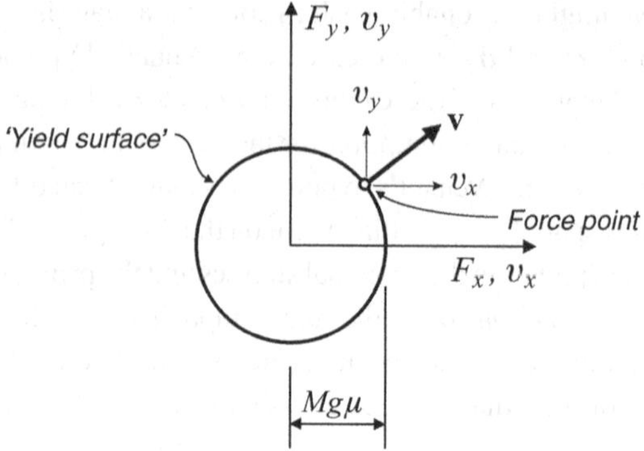

Figure 4.2. Circular 'yield surface' for block sliding on frictional surface. Note how the velocity vector is normal to the circle at the force point.

This expression is clearly analogous to a yield criterion. The forces F_x and F_y are analogues of the principal stresses. If they satisfy (4.2), slip will occur. The 'yield surface' is a circle with radius $Mg\mu$. Evidently the block can be viewed as a simple model of yielding. Plastic flow is represented by slip.

Let v_x and v_y be the rates of slip or 'plastic' deformation in the x- and y-directions. We expect that the direction of slip will be aligned with the direction of the resultant force; that is

$$\frac{v_y}{v_x} = \frac{F_y}{F_x} \tag{4.3}$$

where we assume that F_x is non-zero. Now suppose we plot the 'yield surface' (4.2) as shown in Figure 4.2. Then for any combination of forces F_x and F_y such that the 'force point' (F_x, F_y) lies on the yield surface, we can also plot the slip rate vector $\mathbf{v}$ with components v_x and v_y as shown. In the figure we have plotted $\mathbf{v}$ as if it originates at the force point. This is done solely for convenience. In fact, the vector $\mathbf{v}$ could have been plotted from any origin. We choose to plot it at the force point because that emphasises the fact that its direction is *normal* to the yield surface.

If the forces F_x and F_y are reduced so that the force point moves back inside the circle, then slip ceases. The 'plastic' deformations that have occurred are locked irreversibly into the block. If later the forces are increased and the force point once again touches the yield circle at some other point, slip begins again and the vector $\mathbf{v}$ will now be normal to the circle at the new point. This idea of 'normality' of the slip vector to the yield surface seems intuitively obvious, but is quite important.

Next, we can define the rate of 'plastic' work $\dot{W}_p$. This is the rate at which the forces F_x and F_y do work whenever the block slips. Assuming that F_x and

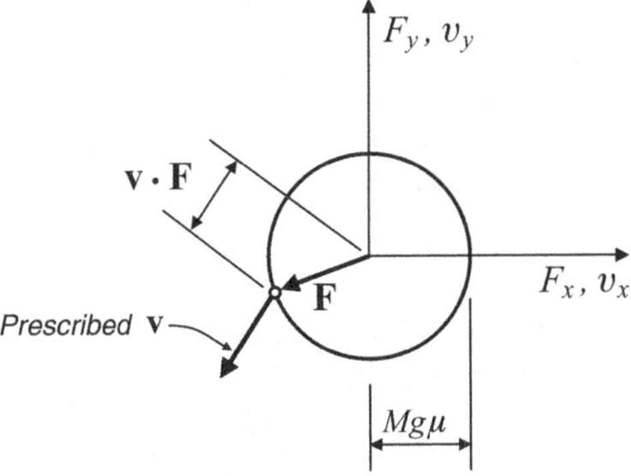

Figure 4.3. The prescribed velocity vector must be normal to the circle in order to maximise the rate of plastic work.

F_y are sufficiently large so that the force point touches the circle we have

$$\dot{W}_p = F_x v_x + F_y v_y \tag{4.4}$$

Note that $\dot{W}_p$ is the inner product of the slip rate vector $\mathbf{v}$ with the vector $\mathbf{F}$, the components of which are F_x and F_y.

Now consider a new experiment. Suppose we cause the block to slip in a certain direction; that is, we prescribe the slip rate components v_x and v_y. Then ask the question: what values of F_x and F_y were required to accomplish this slip? Obviously the answer is found by locating the force point on the circle circumference where the prescribed slip vector is normal to the circle. That particular force point defines the appropriate forces F_x and F_y. It is useful to look at things in this way since we can now see that the required values of F_x and F_y are those that *maximise* $\dot{W}_p$. Recall that $\dot{W}_p$ is the inner product $\mathbf{v} \cdot \mathbf{F}$. We can visualise the inner product as being the length of the projection of $\mathbf{F}$ on to the direction of $\mathbf{v}$. Thus, if $\mathbf{v}$ were placed at any point on the circumference of the yield circle, $\dot{W}_p = \mathbf{v} \cdot \mathbf{F}$ would be as shown in Figure 4.3. Obviously, the required forces F_x and F_y will maximise $\dot{W}_p$ only when $\mathbf{v}$ lies normal to the circle.*

The idea of maximising $\dot{W}_p$ is a convenient criterion for selecting the plastic strain rates when we return to the general problem. Before we do that however, we can generalise this simple example of sliding to something slightly more

* Note that $\dot{W}_p = \|\mathbf{v}\| \|\mathbf{F}\| \cos \Omega$, where Ω is the angle between the two vectors. The maximum value of $\cos \Omega$ for $\Omega \in (0, \pi/2)$ occurs when $\Omega \equiv 0$, that is when the vectors are aligned. Also note that if $\Omega \in (0, \pi)$, then negative work is done for $\Omega \in (\pi/2, \pi)$, which is a violation of the laws of thermodynamics. This aspect of normality was exploited by D.C. Drucker in a celebrated paper discussing the requirements of normality for non-negative work during plastic deformation; see Appendices E and F.

interesting. Suppose the friction coefficient in the x-direction is different from that in the y-direction. We might accomplish this by scoring parallel lines in the y-direction on the base of the block. That would presumably increase the roughness in the x-direction, as well as in directions intermediate between x and y. Let μ_x represent the coefficient of friction in the x-direction and μ_y the same quantity for the y-direction.

What would be the result of this change on the yield surface? We might guess (and probably be accurate in so doing) that the circle would become an ellipse. The yield condition would become

$$\frac{F_x^2}{\mu_x^2} + \frac{F_y^2}{\mu_y^2} = (Mg)^2 \tag{4.5}$$

Note that if μ_x and μ_y are the same, we recover the original condition (4.2). The new yield surface is sketched in Figure 4.4.

Now what happens if the force point touches the ellipse? Slip will occur, but in what direction? We can investigate this question by consulting the plastic work rate. Equation (4.4) still represents $\dot{W}_p$. We wish to maximise $\dot{W}_p$ subject to the constraint that F_x and F_y satisfy the yield condition (4.5). A convenient way to accomplish this is to introduce a Lagrange multiplier into (4.4). We rewrite (4.4) as

$$\dot{W}_p^* = F_x v_x + F_y v_y - \lambda\left[\frac{F_x^2}{\mu_x^2} + \frac{F_y^2}{\mu_y^2} - (Mg)^2\right] \tag{4.6}$$

where λ is the Lagrange multiplier named after the eminent French mathematician, J.-L. Lagrange. It is undetermined, but since, at yield, the quantity it multiplies is identically zero, we see that $\dot{W}_p^*$ is the same as $\dot{W}_p$. The stationary

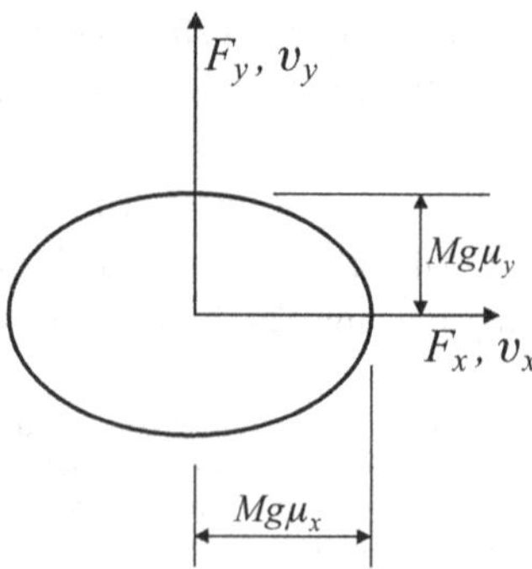

Figure 4.4. Elliptical 'yield surface' for varying friction in different directions.

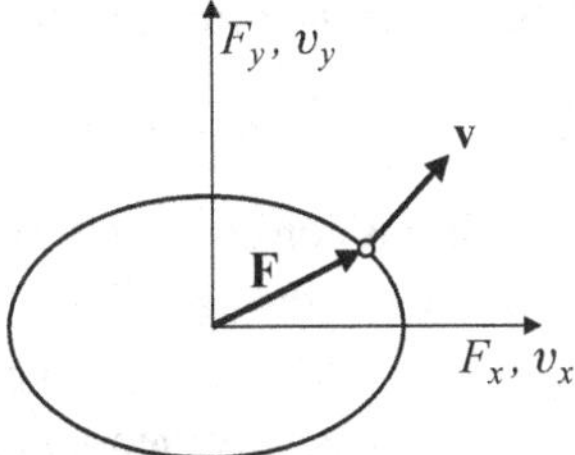

Figure 4.5. Normality of the velocity vector maximises the rate of plastic work.

values of $\dot{W}_p$ are found by setting the partial derivatives of $\dot{W}_p^*$ equal to zero,

$$\frac{\partial \dot{W}_p^*}{\partial F_x} = v_x - 2\lambda \frac{F_x}{\mu_x^2} = 0 \quad \text{and} \quad \frac{\partial \dot{W}_p^*}{\partial F_y} = v_y - 2\lambda \frac{F_y}{\mu_y^2} = 0 \tag{4.7}$$

So we conclude that the components of the slip velocity must obey

$$v_x = 2\lambda \frac{F_x}{\mu_x^2} \quad \text{and} \quad v_y = 2\lambda \frac{F_y}{\mu_y^2} \tag{4.8}$$

We still do not know what the value of λ is, but the *direction* of slip is known. We can identify the direction by this ratio

$$\frac{v_y}{v_x} = \frac{F_y/\mu_y^2}{F_x/\mu_x^2} \tag{4.9}$$

and this direction is normal to the yield ellipse at the force point as shown in Figure 4.5.

We now see that maximising $\dot{W}_p$ has led to the conclusion that the components of the slip velocity must be proportional to the components of the *gradient* of the yield surface. If we recall that the normal vector to any surface has components proportional to the gradient vector for that surface,* then it is clear that normality implies a maximum plastic work rate, and vice versa.

We can now return to the continuum problem where our material is characterised by a particular yield surface $f(\sigma_1, \sigma_2, \sigma_3)$. Plastic flow will occur if the stress point touches the yield surface anywhere. In this case the plastic work rate is given by

$$\dot{W}_p = tr(\boldsymbol{\sigma}\dot{\boldsymbol{\varepsilon}}^p) \tag{4.10}$$

* Suppose $\varphi(x, y, z) = c$, where c = constant, defines a surface in a three-dimensional Euclidean space. Then the differential of φ gives $d\varphi = \frac{\partial\varphi}{\partial x}dx + \frac{\partial\varphi}{\partial y}dy + \frac{\partial\varphi}{\partial z}dz = 0$. We can write this as the inner product of two vectors: $(\frac{\partial\varphi}{\partial x}\hat{\boldsymbol{\imath}} + \frac{\partial\varphi}{\partial y}\hat{\boldsymbol{\jmath}} + \frac{\partial\varphi}{\partial z}\hat{\mathbf{k}})\cdot(dx\,\hat{\boldsymbol{\imath}} + dy\,\hat{\boldsymbol{\jmath}} + dz\,\hat{\mathbf{k}}) = 0$. The first of these vectors is the gradient $\nabla\varphi$ and the second vector is tangential to the surface. Since the inner product is zero, we see that $\nabla\varphi$ must be perpendicular to the surface.

If our coordinate system aligns with the principal directions then this becomes

$$\dot{W}_p = \sigma_1 \dot{\varepsilon}_1^p + \sigma_2 \dot{\varepsilon}_2^p + \sigma_3 \dot{\varepsilon}_3^p \tag{4.11}$$

As one might expect from our discussion concerning the sliding block, the plastic work rate will be maximised when the direction of the vector $[\dot{\varepsilon}_1^p, \dot{\varepsilon}_2^p, \dot{\varepsilon}_3^p]$ is normal to the yield surface $f(\sigma_1, \sigma_2, \sigma_3)$, so long as we assume that the yield surface shape is convex. Most modern plasticity theories are based on this *normality condition* and non-convex yield surfaces are not permitted. The normality condition is not only a convenience, it makes possible the existence of a number of important theorems. Nevertheless, there is no fundamental reason why normality must hold. We will base much of the development here on the normality condition, but acknowledge that it is an assumption and not a physical law.

4.3 Associated flow rules

An easy way to introduce the normality condition is to define a *flow rule* of the form

$$\dot{\boldsymbol{\varepsilon}}^p = \lambda \frac{\partial f}{\partial \boldsymbol{\sigma}} \tag{4.12}$$

Here f denotes the yield condition as a general function of the components of the stress matrix $\boldsymbol{\sigma}$. The partial derivative $\partial f / \partial \boldsymbol{\sigma}$ implies the derivative with respect to any stress component from which an expression for the corresponding component of the plastic strain rate matrix $\dot{\boldsymbol{\varepsilon}}^p$ is obtained. The magnitudes of the components of the strain rate will be undetermined to within λ, which can be regarded as being similar to the Lagrange multiplier used in (4.6). The only constraint we place on λ is that it must be positive.* Equation (4.12) ensures that $\dot{\boldsymbol{\varepsilon}}^p$ will be normal to the yield surface f. If our coordinate system aligns with the principal directions of $\boldsymbol{\sigma}$, then (4.12) can be written as

$$\dot{\varepsilon}_k^p = \lambda \frac{\partial f}{\partial \sigma_k}, \quad k = 1, 2, 3 \tag{4.13}$$

where $\dot{\varepsilon}_k^p$ are the principal plastic strain rates and σ_k denote the corresponding principal stresses.

Equations (4.12) and (4.13) are called *associated* flow rules. The adjective refers to the fact the plastic strains are associated directly with the yield surface. It is possible to introduce non-associated flow rules where f in either of the equations is replaced by some other function, say g. Non-associated flow rules

* In some works λ appears with a raised dot, $\dot{\lambda}$, to emphasise the fact that the dimensions of (4.12) must be homogeneous. We merely note that our λ has dimensions of 1/time.

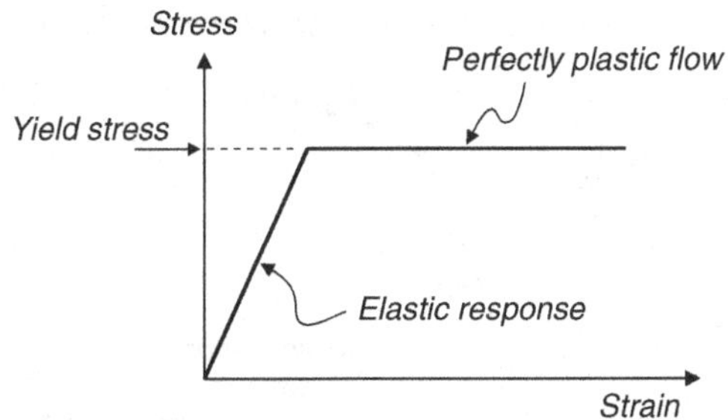

Figure 4.6. Stress–strain response for a perfectly plastic material.

will generally negate many of the advantages of the normality condition, but they may be desirable for certain types of materials or more advanced theories. We will consider non-associated flow rules later in this chapter.

Note that, because of the undetermined nature of λ, equations (4.12) and (4.13) do not specify directly the magnitude of the plastic strain rates. This is a deliberate move. In many cases the magnitude of the plastic strain rate will not be known unless more information can be supplied. In a general sense we can consider two cases.

Case 1. Perfect plasticity

We say that a material is 'perfectly plastic' if, on yielding, the plastic strains can grow without bound given that no further change in stress occurs and no outside constraints are present. The stress–strain response in simple tension for a perfectly plastic material is illustrated in Figure 4.6. We see linear elastic behaviour until the stress reaches its yield value. After yielding there is no further change in stress as plastic strains continue to accumulate. The flat response characterises perfect plasticity. The functional relationship between plastic strain and stress is multiple-valued, and knowledge of the stress does not imply that we know the magnitude of the strain. If the strain is specified, then the stress is known, but not vice versa.

Case 2. Work hardening plasticity

In contrast to perfect plasticity, 'work hardening' implies the yield surface may change in some way once initial yielding has occurred. Usually the way the yield surface changes is related to the amount of plastic strain or to the amount of plastic work that has accumulated. This introduces an extra parameter into the description of the yield surface as was mentioned in Chapter 3. The response of a work hardening material in simple tension might look something like that shown in Figure 4.7. Here the stress and strain may have a one-to-one functional relationship both before and after yield.

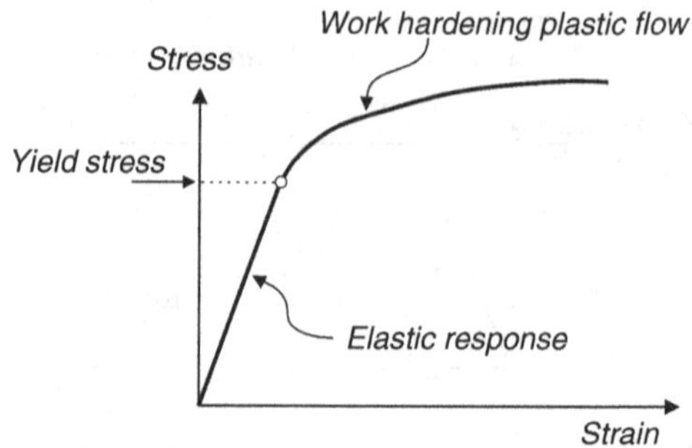

Figure 4.7. Stress–strain response for a work hardening plastic material.

In case 2 it will generally be possible to say how large the plastic strains are at any time after yield has occurred, but the same cannot be said for case 1. Often, for a perfectly plastic material, we will not be able to calculate the amount of plastic straining (although in some problems geometric constraints may permit us to do so). Nevertheless perfect plasticity may be profitably used since it permits us to take advantage of certain powerful theorems. There is a place for both perfect plasticity and work hardening plasticity in the repertoire of any geotechnical engineer. We will spend the remainder of the book dealing with one or the other. We begin with a relatively simple example using perfect plasticity.

4.4 Example – plane strain

Recall the plane strain problem of the idealised cohesionless elastic–plastic soil described in Chapter 3 (Figure 3.25). We investigated the possible yield states for this problem in detail. Suppose we reconsider the situation where σ_{xx} is reduced while σ_{yy} is held constant as shown in (3.44). The plane strain assumption implies ε_{zz} will be zero throughout the deformation. What will happen to the other two principal strains ε_{xx} and ε_{yy} given the material is perfectly plastic with an associated flow rule?

If we take the initial stress state (3.39) as our reference configuration, then the elastic strains associated with the horizontal stress trajectory shown in Figure 3.27 follow immediately from Hooke's law:

$$\begin{aligned}\varepsilon^e_{xx} &= \frac{p_0}{E}\left[1 - \nu - 2\nu^2 - (1-\nu^2)\frac{p_*}{p_0}\right] \\ \varepsilon^e_{yy} &= \frac{p_0}{E}\left[1 - \nu - 2\nu^2 + \nu(1+\nu)\frac{p_*}{p_0}\right]\end{aligned} \tag{4.14}$$

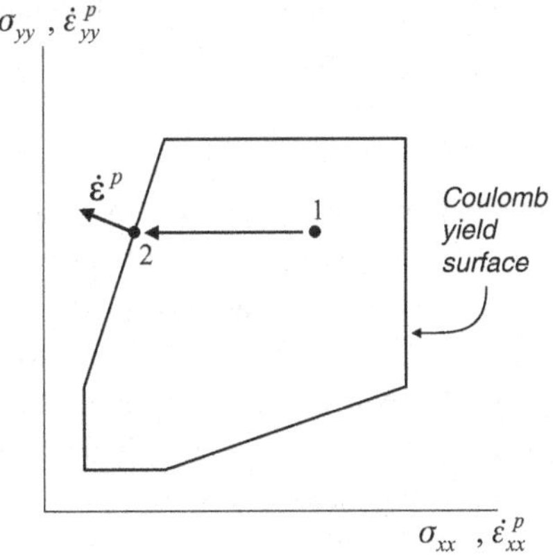

Figure 4.8. Normality of plastic strain rate vector – Coulomb yield surface.

The elastic response continues until the yield locus is reached. The value of p_* at yield was given in (3.46) as

$$\frac{p_*}{p_0} = 1 - \frac{1}{N} \tag{4.15}$$

where N was the abbreviation

$$N = \frac{1 + \sin\phi}{1 - \sin\phi} \tag{4.16}$$

At this point the elastic strains are specified by (4.14) with p_*/p_0 being given by (4.15). The situation is sketched in Figure 4.8 where the stress points marked 1 and 2 correspond to the initial and yield conditions. The yield locus shown in this figure is the same as the locus shown in Figure 3.28 for $\nu = \frac{1}{3}$ and $N = 3$.

If the normality condition applies, the plastic strain rate vector will be aligned as shown in Figure 4.8. Using the associated flow rule (4.13) with the yield function specified by

$$f = \sigma_{yy} - N\sigma_{xx} \tag{4.17}$$

we find

$$\begin{aligned} \dot{\varepsilon}^p_{xx} &= \lambda\frac{\partial f}{\partial \sigma_{xx}} = -\lambda N \\ \dot{\varepsilon}^p_{yy} &= \lambda\frac{\partial f}{\partial \sigma_{yy}} = \lambda \end{aligned} \tag{4.18}$$

where λ is our unspecified (positive) multiplier. Its dimensions are a pure rate: $(time)^{-1}$, but its value is unknown for a perfectly plastic material.

An interesting point arises here. Since $\dot{\varepsilon}_{zz} = 0$, the volumetric plastic strain rate is obtained by adding the two rates in (4.18)

$$\dot{e}^p = \lambda(1 - N) \tag{4.19}$$

This will be a negative quantity for all values of ϕ greater than zero. We conclude that the plastic flow is dilational. This will always be the case with the Coulomb yield criterion, regardless of the loading conditions, because of its mean stress dependence. The same comment applies to the Drucker–Prager, Lade–Duncan and Matsuoka–Nakai criteria as well. The normality condition implies a negative plastic volumetric strain rate whenever the shape of the yield surface expands with increasing mean stress.

4.5 Non-associated flow

If one compares the rate of dilation suggested by (4.19) with data from real soil tests, it is found that (4.19) predicts values for $\dot{e}^p$ often far in excess of values that are realistic. For many soils, shearing is accompanied by compaction rather than dilation. For other soils no volumetric strain is evident during shearing. Even for dilating soils, the rate of dilation is usually not as large as given by (4.19). We can look for a solution to this problem in two places. First, we recognise that the pressure dependence of the Coulomb criterion is partly responsible. For compacting soils we would wish the yield surface to grow smaller with increasing mean stress rather than the opposite. For soils that exhibit no volumetric strain we would want the yield surface to neither grow nor shrink. For dilating soils we require an expanding yield surface. While this may appear to be an impossible wish-list, in fact all three types of behaviour can be accommodated with the Cam Clay and Modified Cam Clay yield surfaces. The resulting theory of critical state soil mechanics will be discussed in Chapter 7. The second way to attempt to solve the problem of excessive dilation is to abandon the normality condition. We consider this possibility now.

Non-associated flow rules are mathematically similar to (4.12) and (4.13) with the essential difference being that the yield function f is replaced with another function, $g = g(\boldsymbol{\sigma})$ for example. The function g is referred to as the *plastic potential* function. In one sense we can look on (4.12) and (4.13) as being special cases of the more general flow rule

$$\dot{\boldsymbol{\varepsilon}}^p = \lambda \frac{\partial g}{\partial \boldsymbol{\sigma}} \tag{4.20}$$

with $g = f$. *Non-associated flow* occurs when g is different from f. Then the flow rule (4.20) gives plastic strain rates that will not be normal to the yield surface. There are disadvantages to dropping the normality condition, especially with regard to application of certain important theorems, but the problems of excessive plastic dilation can be rectified.

As an example, suppose we return to the plane strain problem of Figure 3.25. We can simply let g be this function (compare with (4.17))

$$g = \sigma_{yy} - M\sigma_{xx} \tag{4.21}$$

where

$$M = \frac{1 + \sin\psi}{1 - \sin\psi} \tag{4.22}$$

Here ψ is the *angle of dilatancy* for the material. Now, instead of (4.18) we have

$$\begin{aligned} \dot{\varepsilon}^p_{xx} &= \lambda \frac{\partial g}{\partial \sigma_{xx}} = -\lambda M \\ \dot{\varepsilon}^p_{yy} &= \lambda \frac{\partial g}{\partial \sigma_{yy}} = \lambda \end{aligned} \tag{4.23}$$

Again, since $\dot{\varepsilon}_{zz}$ is zero, the plastic volumetric strain rate becomes

$$\dot{e}^p = \lambda(1 - M) \tag{4.24}$$

Clearly, we can adjust the value of ψ to provide whatever plastic volumetric strain rate we require.

4.6 A loading criterion

We are aware now that the elastic response will prevail so long as the stress point lies inside the yield surface, and a plastic response occurs whenever the stress point lies on the yield surface. This is sufficient to discriminate between elastic and plastic behaviour when we are concerned only with stress. When we deal with flow and strain however, we need to think in terms of rates of change; therefore we need to investigate whether the stress point will remain on the yield surface during the next increment of flow, or whether it may move back into the elastic region inside the yield surface. The result will give us a simple criterion for whether the plastic response will continue or not.

Recall that the yield surface is described by $f(\boldsymbol{\sigma}) = k$. We can think of $\boldsymbol{\sigma}$ here as being a vector the components of which are the three principal stresses. If we take the time derivative of f we find

$$\dot{f} = \frac{\partial f}{\partial \boldsymbol{\sigma}}\dot{\boldsymbol{\sigma}} = \frac{\partial f}{\partial \sigma_1}\dot{\sigma}_1 + \frac{\partial f}{\partial \sigma_2}\dot{\sigma}_2 + \frac{\partial f}{\partial \sigma_3}\dot{\sigma}_3 \tag{4.25}$$

The only way we can continue to have a plastic response is if $\dot{f} = 0$. The other possibilities are $\dot{f} < 0$, which implies that the stress point is moving back inside the yield surface and elastic behaviour will resume, and $\dot{f} > 0$, which is impossible since the stress point cannot move outside the yield surface (although this can be relaxed for work-hardening materials, but that comes later). We can conclude then that the criterion for continued plastic loading is

$$\dot{f} = \frac{\partial f}{\partial \boldsymbol{\sigma}} \dot{\boldsymbol{\sigma}} = 0 \tag{4.26}$$

This is sometimes called the consistency condition.

The criterion can be illustrated with a simple example. For a Coulomb material $\dot{f}$ is given by

$$\dot{f} = \dot{\sigma}_1(1 - \sin\phi) - \dot{\sigma}_3(1 + \sin\phi) \tag{4.27}$$

which follows from (3.15). Let σ_{1_0} and σ_{3_0} be the present values of major and minor principal stress, and suppose the stress point lies on the yield surface so that

$$f = \sigma_{1_0}(1 - \sin\phi) - \sigma_{3_0}(1 + \sin\phi) = 2c\,\cos\phi \tag{4.28}$$

Also suppose the rate of change of σ_3 is given, for example, by $\dot{\sigma}_3 = \alpha$. Then for continued plastic loading, our criterion (4.26), together with (4.27), gives

$$\dot{\sigma}_1 = \dot{\sigma}_3 \left(\frac{1 + \sin\phi}{1 - \sin\phi}\right) = \alpha \left(\frac{1 + \sin\phi}{1 - \sin\phi}\right) = \alpha N \tag{4.29}$$

After a time interval Δt we will have

$$\begin{aligned} \sigma_1 &= \sigma_{1_0} + \dot{\sigma}_1 \Delta t = \sigma_{1_0} + \alpha N \Delta t \\ \sigma_3 &= \sigma_{3_0} + \dot{\sigma}_3 \Delta t = \sigma_{3_0} + \alpha \Delta t \end{aligned} \tag{4.30}$$

These define a new stress point at time $t + \Delta t$. To see whether this stress point remains on the yield surface we combine the stresses to form

$$\begin{aligned} &\sigma_1(1 - \sin\phi) - \sigma_3(1 + \sin\phi) \\ &\quad = (\sigma_{1_0} + \alpha N \Delta t)(1 - \sin\phi) - (\sigma_{3_0} + \alpha \Delta t)(1 + \sin\phi) \\ &\quad = \sigma_{1_0}(1 - \sin\phi) - \sigma_{3_0}(1 + \sin\phi) + \alpha \Delta t[N(1 - \sin\phi) - (1 + \sin\phi)] \\ &\quad = 2c\cos\phi + 0 \end{aligned}$$

So the new stress point also lies on the yield surface. If the equality sign in (4.26) were replaced by a less than sign, we would find the new stress point would not lie on the yield surface but would lie inside it.

Equation (4.26) gives a convenient tool, both for indicating under what conditions loading will continue, and as a constraint for problems in which loading is assured. This second point will be useful in the next section.

4.7 A complete stress–strain relationship

Recall that for perfectly plastic materials the plastic strain rates all depend upon the undetermined multiplier λ. In this section we will contrive a way to identify the value of λ in terms of the rates of change of stress and strain, provided continued plastic loading is occurring. We will eventually be able to relate directly the rate of change of stress to the rate of change of total strain. This will be done for a non-associated flow rule, but we realise that by simply setting the plastic potential function g equal to the yield function f, we will have the result for an associated flow rule as well.

To begin write down the basic decomposition of strain rate into elastic and plastic parts:

$$\dot{\boldsymbol{\varepsilon}} = \dot{\boldsymbol{\varepsilon}}^e + \dot{\boldsymbol{\varepsilon}}^p \tag{4.31}$$

Here we can think of $\dot{\boldsymbol{\varepsilon}}$ as a column vector, the components of which are the three principal strain rates. The vectors $\dot{\boldsymbol{\varepsilon}}^e$ and $\dot{\boldsymbol{\varepsilon}}^p$ have components equal to the elastic and plastic parts of each principal strain rate.

The *elastic* strain rate vector is related to the vector the components of which are the principal stress rates. We can write

$$\dot{\boldsymbol{\sigma}} = \mathbf{M}^e \dot{\boldsymbol{\varepsilon}}^e \tag{4.32}$$

where $\mathbf{M}^e$ is the elasticity matrix defined by

$$\mathbf{M}^e = \begin{bmatrix} \Lambda + 2G & \Lambda & \Lambda \\ \Lambda & \Lambda + 2G & \Lambda \\ \Lambda & \Lambda & \Lambda + 2G \end{bmatrix} \tag{4.33}$$

where $\Lambda = \nu E/(1+\nu)(1-2\nu)$ is the Lamé constant. We can use the non-associated flow rule (4.20) in (4.31) to find

$$\dot{\boldsymbol{\varepsilon}} = \dot{\boldsymbol{\varepsilon}}^e + \lambda \frac{\partial g}{\partial \boldsymbol{\sigma}} \tag{4.34}$$

Now multiply both sides of this equation, first by the elasticity matrix and then by the row vector $(\partial f/\partial \boldsymbol{\sigma})^T$

$$\left(\frac{\partial f}{\partial \boldsymbol{\sigma}}\right)^T \mathbf{M}^e \dot{\boldsymbol{\varepsilon}} = \left(\frac{\partial f}{\partial \boldsymbol{\sigma}}\right)^T \mathbf{M}^e \dot{\boldsymbol{\varepsilon}}^e + \lambda \left(\frac{\partial f}{\partial \boldsymbol{\sigma}}\right)^T \mathbf{M}^e \frac{\partial g}{\partial \boldsymbol{\sigma}} \tag{4.35}$$

If the loading criterion (4.26) is valid, then, considering (4.32), the first term on the right-hand side of this equation is zero. We can solve the remaining equation for λ to give

$$\lambda = \frac{(\partial f/\partial \boldsymbol{\sigma})^T \mathbf{M}^e \dot{\boldsymbol{\varepsilon}}}{(\partial f/\partial \boldsymbol{\sigma})^T \mathbf{M}^e \partial g/\partial \boldsymbol{\sigma}} \tag{4.36}$$

Now return to (4.34) and use the inverted form of (4.32) to find

$$\dot{\boldsymbol{\varepsilon}} = [\mathbf{M}^e]^{-1}\dot{\boldsymbol{\sigma}} + \lambda \frac{\partial g}{\partial \boldsymbol{\sigma}} \tag{4.37}$$

Replace λ in this expression with the right-hand side of (4.36) and then solve for $\dot{\boldsymbol{\sigma}}$. After some manipulation of the matrix products we find

$$\dot{\boldsymbol{\sigma}} = \mathbf{M}^e \dot{\boldsymbol{\varepsilon}} - \frac{\mathbf{M}^e \dfrac{\partial g}{\partial \boldsymbol{\sigma}} \left(\dfrac{\partial f}{\partial \boldsymbol{\sigma}}\right)^T}{\left(\dfrac{\partial f}{\partial \boldsymbol{\sigma}}\right)^T \mathbf{M}^e \dfrac{\partial g}{\partial \boldsymbol{\sigma}}} \mathbf{M}^e \dot{\boldsymbol{\varepsilon}} = \mathbf{M}^p \dot{\boldsymbol{\varepsilon}} \tag{4.38}$$

Following the second equality, $\mathbf{M}^p$ represents the collection of terms multiplying $\dot{\boldsymbol{\varepsilon}}$.

$$\mathbf{M}^p = \left[\mathbf{I} - \frac{\mathbf{M}^e \dfrac{\partial g}{\partial \boldsymbol{\sigma}} \left(\dfrac{\partial f}{\partial \boldsymbol{\sigma}}\right)^T}{\left(\dfrac{\partial f}{\partial \boldsymbol{\sigma}}\right)^T \mathbf{M}^e \dfrac{\partial g}{\partial \boldsymbol{\sigma}}} \right] \mathbf{M}^e \tag{4.39}$$

where $\mathbf{I}$ is the identity matrix.

Equation (4.38) completely specifies the rate of change of stress given the rate of change of total strain for a material obeying a non-associated flow rule, provided continued plastic loading occurs. Note how the gradients of both the yield function and the plastic potential function have been used. Note too that the quantity $(\partial f/\partial \boldsymbol{\sigma})^T \mathbf{M}^e \partial g/\partial \boldsymbol{\sigma}$ in the denominator of the fraction in (4.39) is a scalar. The numerator is not a scalar but is a square matrix. We require of course that plastic loading should continue in order for (4.38) to apply. If the loading criterion is not satisfied, then an elastic response will ensue and (4.32) will apply.

We can illustrate the use of (4.38) by considering a simple plane strain example shown in Figure 3.25. If the material obeys a non-associated flow rule then a plastic potential g given by (4.21) is appropriate. The Coulomb yield function is given by (4.17). Thus the two vectors $\partial f/\partial \boldsymbol{\sigma}$ and $\partial g/\partial \boldsymbol{\sigma}$ are given by

$$\frac{\partial f}{\partial \boldsymbol{\sigma}} = \begin{bmatrix} -N \\ 1 \end{bmatrix}, \quad \frac{\partial g}{\partial \boldsymbol{\sigma}} = \begin{bmatrix} -M \\ 1 \end{bmatrix} \tag{4.40}$$

These vectors are two-dimensional because we need not consider the intermediate principal stress σ_{zz}, and the strain ε_{zz} is zero. (In fact, things are not quite this simple since we may have zero total strain but non-zero elastic and plastic strain, but we will assume this does not occur.)

The elasticity matrix is

$$\mathbf{M}^e = \begin{bmatrix} \Lambda + 2G & \Lambda \\ \Lambda & \Lambda + 2G \end{bmatrix} \tag{4.41}$$

Using (4.40) and (4.41) in (4.39) we find that the plasticity matrix, after some manipulation, is

$$\mathbf{M}^p = \frac{2G}{(1-\nu)(1+MN) - \nu(M+N)} \begin{bmatrix} 1 & M \\ N & MN \end{bmatrix} \tag{4.42}$$

The elastic relation $\Lambda = 2G\nu/(1-2\nu)$ has been used in (4.42) to eliminate the Lamé constant Λ. We can make an interesting observation here. The determinant of $\mathbf{M}^p$ is zero. This might initially seem surprising, but in fact it is expected for the physical reason that, for a perfectly plastic material, the strains cannot be determined solely from knowledge of the stresses. If det $\mathbf{M}^p$ were non-zero, we could invert (4.38) and solve for the strain rate vector, an impossibility for perfect plasticity. We can, of course, use (4.42) in (4.38) to find the stress rates if the strain rates are specified.

Even if the stresses are specified, it may still be possible under special conditions to find the corresponding strains during plastic flow, but that can take place only when additional information is present. The next example shows how this may happen.

4.8 The pressuremeter problem

The pressuremeter is a device used for *in situ* testing of geotechnical materials. Basically it consists of a cylindrical inflatable balloon, inserted in a borehole. If the pressure inside the balloon is increased, the boundary of the borehole is forced to expand outward. It is relatively easy to measure both the balloon pressure and the change in radius of the bore. The question that arises is, how can these measurements be used to infer the properties of the soil surrounding the borehole? The problem has been analysed by many researchers for a variety of conditions and a large specialised branch of the geotechnical literature has grown up around the pressuremeter. Our aim here is not to summarise the various lines of research but instead to attempt to solve the general problem using the knowledge of perfect plasticity we now possess.

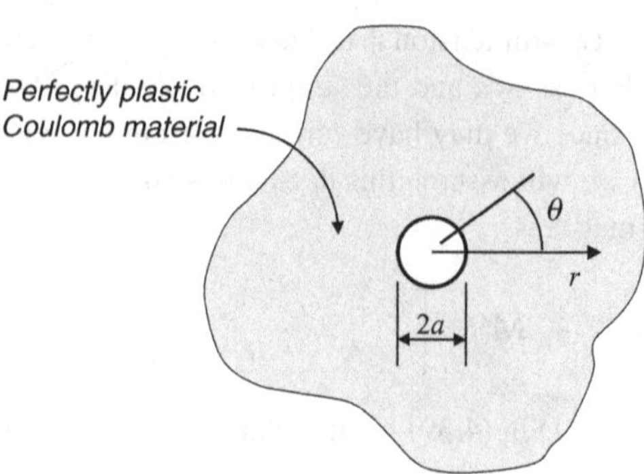

Figure 4.9. The pressuremeter problem – cylindrical cavity in an infinite body.

Usually the pressuremeter is idealised as a plane strain, cavity expansion problem. This means that we consider an infinite body containing an infinitely long cylindrical cavity of radius a as shown in Figure 4.9. We will assume that the body consists of a perfectly plastic cohesionless Coulomb material that obeys a non-associated flow rule. That is, the yield surface is specified by the Coulomb hexagonal surface and the flow rule is given by (4.20). The geometry of the cavity suggests we use cylindrical polar coordinates with the z-axis aligned with the centre of the cavity. The radial and azimuthal directions are specified by r and θ as shown in Figure 4.9. Let the cavity radius be a and suppose that the pressuremeter fills the cavity so that a uniform pressure p may be applied to the cavity wall. The condition of plane strain is implied by zero extensional strain in the z-direction.

In reality, the pressuremeter problem can be modelled in a number of ways depending on the choice of initial conditions. One possibility would be to assume that the cavity boundary is initially free from stress, such as if the borehole were completely empty before the device was inserted. This is an unlikely condition in the case of a cohesionless granular material since failure and plastic flow would no doubt occur and the bore itself would collapse. A second possible condition would be to assume that the borehole walls initially suffer no displacement such as might occur with a *self-boring* pressuremeter. This condition may be more realistic, but both displacements and stresses may have changed in the vicinity of the bore during the actual boring process. The initial condition we will use is that used by most investigators, namely that the initial state of stress in the infinite space is isotropic with all normal stresses being equal to a constant P_0 including the applied stress on the cavity boundary. This may not be a realistic initial state for the soil surrounding the bore, but it is probably not so far from reality either. Thus we will approach the pressuremeter in a mathematical sense as a simple plane strain cavity expansion problem starting from a simple isotropic initial stress state.

Because of symmetry we would expect to find only one non-zero displacement: the radial displacement u_r. The non-zero stresses will be the radial stress σ_{rr}, the hoop stress $\sigma_{\theta\theta}$ and the axial stress σ_{zz}. These will be principal stresses and initially they are all equal to the initial stress P_0. We will assume that the cavity pressure is initially P_0 but increases to a value P. Our plane strain assumption implies that $\varepsilon_{zz} = 0$.

The equations governing the problem are as follows. Since the problem is radially symmetric it follows that there is only one non-trivial equation of equilibrium. Referring to equilibrium of stress in the radial direction we have

$$\frac{d\sigma_{rr}}{dr} + \frac{\sigma_{rr} - \sigma_{\theta\theta}}{r} = 0 \tag{4.43}$$

The boundary condition at the cavity wall is

$$\sigma_{rr}(a) = P \tag{4.44}$$

while for large r we have

$$\sigma_{rr}(\infty) = \sigma_{\theta\theta}(\infty) = P_0 \tag{4.45}$$

Taking the initial state as our reference state, the strain–displacement relations are given by

$$\varepsilon_{rr} = \frac{du_r}{dr}, \qquad \varepsilon_{\theta\theta} = \frac{u_r}{r} \tag{4.46}$$

while Hooke's law gives

$$\begin{aligned} \sigma_{rr} &= \Lambda e + 2G\varepsilon_{rr} + P_0 \\ \sigma_{\theta\theta} &= \Lambda e + 2G\varepsilon_{\theta\theta} + P_0 \end{aligned} \tag{4.47}$$

where $\Lambda = \nu E/(1+\nu)(1-2\nu)$ represents the Lamé constant. The yield function for a cohesionless Coulomb material gives

$$f = \sigma_{rr} - N\sigma_{\theta\theta} = 0 \tag{4.48}$$

The plastic potential function for non-associated flow may be written as

$$g = \sigma_{rr} - M\sigma_{\theta\theta} \tag{4.49}$$

where M is given by (4.22). Finally, the stress rate–strain rate equation for plastic response is

$$\begin{bmatrix} \dot{\sigma}_{\theta\theta} \\ \dot{\sigma}_{rr} \end{bmatrix} = \frac{2G}{\zeta} \begin{bmatrix} 1 & M \\ N & MN \end{bmatrix} \begin{bmatrix} \dot{\varepsilon}_{\theta\theta} \\ \dot{\varepsilon}_{rr} \end{bmatrix} \tag{4.50}$$

which follows from (4.42), with

$$\zeta = (1-\nu)(1+MN) - \nu(M+N) \tag{4.51}$$

Note that the plane strain form for the matrix $\mathbf{M}^p$ given above works equally well in this problem as in the simpler case considered earlier.

The nine equations, (4.43)–(4.51), may seem daunting at first glance, but they are typical of the description of any well-posed plasticity problem. In fact, had we not had the plane strain assumption, the result might have looked even more complex. The first four equations, (4.43)–(4.46), are universal to the problem. They apply regardless of whether we have elastic or plastic response. Equation (4.47) describes the material when elastic response is appropriate. The remaining four equations, (4.48)–(4.51), describe plastic response. For an associated flow rule we could have eliminated one equation, (4.49), but the effect would be minor on the overall complexity of the system of equations.

We will solve the problem in a sequence of steps. First, we will determine the elastic response prior to any yielding around the cavity. This is followed by a determination of conditions for yield to first occur, and then a description of how the yield zone grows around the cavity as the applied pressure increases. Next, we can determine the stresses in the yielding soil. Finally, we will investigate the displacement in the plastic zone and relate it to the applied pressure.

To solve for the elastic response, we can combine the equilibrium equation (4.43) with Hooke's law (4.47) and the strain displacement relations (4.46) to obtain this second-order ordinary differential equation for the displacement u_r

$$\frac{d^2u_r}{dr^2} + \frac{1}{r}\frac{du_r}{dr} - \frac{u_r}{r^2} = 0 \tag{4.52}$$

Note that both elastic constants Λ and G have dropped out of this equation. The solution to (4.53) is well known.

$$u_r = C_1 r + \frac{C_2}{r} \tag{4.53}$$

where C_1 and C_2 are constants of integration. The boundary condition (4.45) cannot be satisfied unless $C_1 = 0$, hence $u_r = C_2/r$ is the most general possible solution. We can use this solution in the strain–displacement relations and use the resulting strains in Hooke's law. This provides an expression for the stress σ_{rr} that may be used in the boundary condition (4.44) to find the value of the constant C_2. The resulting elastic solution is given by

$$\begin{aligned} u_r &= -\frac{(P-P_0)a^2}{2Gr}, & \varepsilon_{rr} &= \frac{(P-P_0)a^2}{2Gr^2} = -\varepsilon_{\theta\theta}, \\ \sigma_{rr} &= \frac{(P-P_0)a^2}{r^2} + P_0, & \sigma_{\theta\theta} &= -\frac{(P-P_0)a^2}{r^2} + P_0 \end{aligned} \tag{4.54}$$

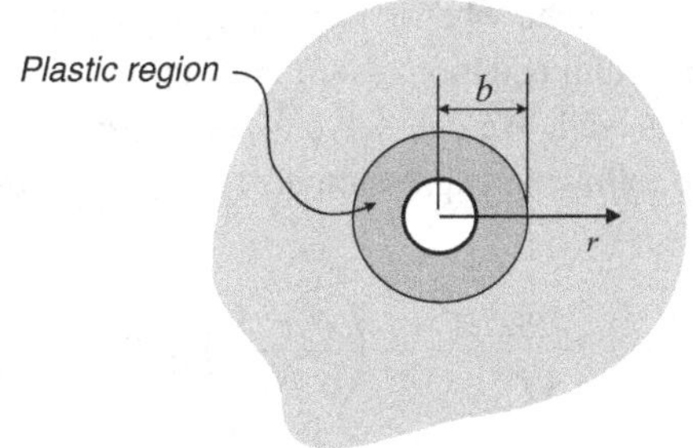

Figure 4.10. Plastic region surrounding a cavity in the pressuremeter problem.

The displacement of the cavity wall is*

$$u_r(a) = -\frac{(P - P_0)a}{2G} \tag{4.55}$$

Evidently $u_r(a)$ and P are linearly related. Since $u_r(a)$ and P can both be measured, and a is known, we can use (4.55) together with pressuremeter measurements to estimate the shear modulus G.

Yield will occur whenever (4.48) is satisfied. It is clear from (4.54) that this will first occur at the cavity wall, $r = a$. Setting r equal to a in (4.54) and using the yield condition (4.48) we find that initial yielding occurs when the applied pressure is given by

$$P = P_0\left(\frac{2N}{N+1}\right) = P_0(1 + \sin\phi) \tag{4.56}$$

What happens next? It is reasonable to assume that a plastic zone forms around the cavity as sketched in Figure 4.10. Let the radius of the zone be b. Then for $r > b$ we will have an elastic response, while for $a < r < b$ the stresses must obey (4.48). Suppose we let $\sigma_b = \sigma_{rr}(b)$, the radial stress at the elastic–plastic boundary. Then in the region $r > b$ we have an identical problem to our original elastic problem, but with a replaced by b and P replaced by σ_b. Our solution (4.54) applies with the changes noted. Moreover, at $r = b$, yielding has just occurred, showing that $\sigma_b = P_0(1 + \sin\phi)$ as we found in (4.56). Therefore in the elastic region the displacement, strains and stresses are

$$\begin{aligned} u_r &= -\frac{P_0 \sin\phi\, b^2}{2Gr}, \qquad & \varepsilon_{rr} &= \frac{P_0 \sin\phi\, b^2}{2Gr^2} = -\varepsilon_{\theta\theta}, \\ \sigma_{rr} &= \frac{P_0 \sin\phi\, b^2}{r^2} + P_0, \qquad & \sigma_{\theta\theta} &= -\frac{P_0 \sin\phi b^2}{r^2} + P_0 \end{aligned} \tag{4.57}$$

* Negative signs associated with displacements in (4.54) and (4.55) as well as some equations that follow all result from the sign convention adopted in Chapter 1.

Next, we must investigate what is happening inside the plastic region. We know that the equilibrium equation (4.43) must apply, and that the yield condition (4.48) must as well. We can solve (4.48) for $\sigma_{\theta\theta}$ and use the result in (4.43) to obtain the following expression describing equilibrium in the plastic zone:

$$\frac{d\sigma_{rr}}{dr} + \frac{\sigma_{rr}}{r}\left(\frac{N-1}{N}\right) = 0 \tag{4.58}$$

We can separate variables and integrate this equation directly. Then, using (4.57) evaluated at $r = b$, we obtain the following expressions for σ_{rr} and $\sigma_{\theta\theta}$:

$$\begin{aligned}\sigma_{rr} &= P_0(1+\sin\phi)\left(\frac{b}{r}\right)^{(N-1)/N}\\ \sigma_{\theta\theta} &= P_0(1-\sin\phi)\left(\frac{b}{r}\right)^{(N-1)/N}\end{aligned} \tag{4.59}$$

These stresses apply throughout the plastic zone $a \leq r < b$. At the cavity wall (4.59) gives the applied cavity pressure

$$P = P_0(1+\sin\phi)\left(\frac{b}{a}\right)^{(N-1)/N} \tag{4.60}$$

The only task remaining now is to determine the displacement within the plastic zone. We can accomplish this as follows. First, note that (4.50) contains only one independent equation since the determinant of the coefficient matrix is zero. Using the equation for $\dot{\sigma}_{rr}$ together with the strain–displacement relations (4.46) we find

$$M\frac{d\dot{u}_r}{dr} + \frac{\dot{u}_r}{r} = \frac{\zeta}{N}\frac{\dot{\sigma}_{rr}}{2G} \tag{4.61}$$

Noting that b is a variable, we can take the time derivative of (4.59),

$$\dot{\sigma}_{rr} = P_0(1+\sin\phi)\left(\frac{N-1}{N}\right)\left(\frac{b}{r}\right)^{(N-1)/N}\left(\frac{\dot{b}}{b}\right) \tag{4.62}$$

Use this on the right-hand side of (4.61) to obtain

$$M\frac{d\dot{u}_r}{dr} + \frac{\dot{u}_r}{r} - \frac{\zeta P_0\sin\phi}{G\;\;N}\left(\frac{b}{r}\right)^{(N-1)/N}\left(\frac{\dot{b}}{b}\right) = 0 \tag{4.63}$$

Integrating (4.63) and using the boundary condition $u_r(b) = -P_0 b\sin\phi/2G$

from (4.57) we obtain

$$u_r = -\frac{P_0 b \sin\phi}{2G}\left\{1 + \frac{BN}{N-1}\left[1 - \left(\frac{r}{b}\right)^{1/N}\right] - \frac{AM}{M+1}\left[1 - \left(\frac{b}{r}\right)^{1/M}\right]\right\} \tag{4.64}$$

where

$$B = \frac{2\zeta}{M+N}, \quad A = 1 + B \tag{4.65}$$

At this point the reader may be excused for thinking that things are becoming a little complex. There are, however, some interesting points we can note. First of all, we obtained the stresses in the plastic region (4.59) without any reference to the deformations. The stresses were determined solely from the equilibrium equations together with the yield condition. The surprising conclusion is this: in the plastic region the stresses and strains are uncoupled. This is the result of our assumption of perfect plasticity. The second important observation we can make is that, despite the material being perfectly plastic, we were able to solve for the displacement in the plastic region. (We could, of course, solve for the strains as well, but they would be of little interest.) This may seem to refute our earlier statements that the plastic strain rates arc indeterminate for a perfectly plastic material. The reason we were able to find the plastic displacements is a result of the fact that the plastic zone is surrounded or contained by the elastic zone, and we have a well-defined boundary condition at b.

One final point. We can determine the displacement at the cavity wall from (4.64) by setting $r = a$. Note that both the resulting equation for $u_r(a)$ and the applied pressure P in (4.60) are functions of b/a. Therefore we can combine the two results to determine $u_r(a)$ as a function of applied pressure. The resulting pressure–displacement relationship is plotted in dimensionless form in Figure 4.11 for $\psi = 5°$ and several values of ϕ.

This example problem illustrates the difficulties involved with detailed solution of problems in perfect plasticity. Had we not wanted the displacement in the plastic zone, things would have been considerably simpler and we could have stopped following (4.59). Fortunately, for many practical problems, the plastic deformations will not be of so much interest and simply finding the stresses for plastic response may be sufficient.

A final comment regarding the pressuremeter problem is that in certain circumstances the plastic zone may expand to an indefinitely large value and the cavity itself will expand without a further increase in pressure. This is referred to as *collapse*. There are many problems in geotechnical engineering where similar ideas apply and collapse is indicated by a failure of the soil to support

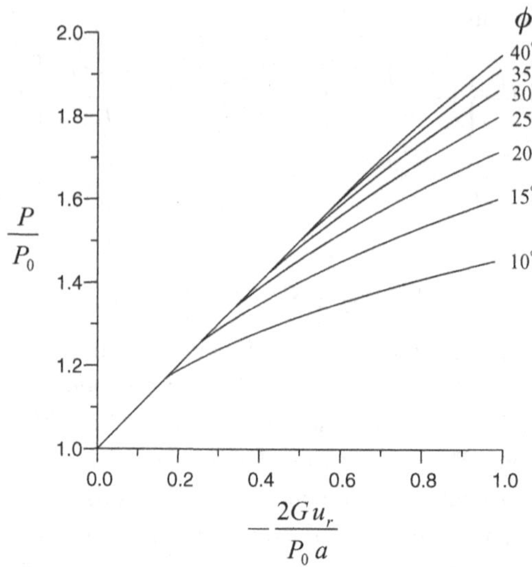

Figure 4.11. Non-dimensional response for a pressuremeter in an elastic–perfectly plastic Coulomb material.

the applied loads. The shallow foundation problem is a typical example, the *collapse load* being the ultimate foundation load the soil will support. The determination of collapse loads is an important aspect of geomechanics. We consider it in the next chapter.

Further reading

Saint-Venant originally proposed the hypothesis regarding coincidence of principal directions of stress and plastic strain in

B. de Saint-Venant, Mémoire sur l'établissement des équations différentielles des mouvements intérieurs opérés dans les corps solides ductiles au delà des limites où l'élasticitié pourrait les ramener à leur premier état, *Comptes Rendus Acad. Sci. Paris*, **70**, 473 (1870).

The ideas of normality and the associated flow rule are described in a wide variety of textbooks on metal plasticity. Three widely read books are

R. Hill, *The Mathematical Theory of Plasticity*, Clarendon Press, Oxford, 1950.

C.R. Calladine, *Engineering Plasticity*, Pergamon Press, Oxford, 1969.

C.S. Desai and H.J. Siriwardane, *Constitutive Laws for Engineering Materials*, Prentice-Hall, Englewood Cliffs, NJ, 1984.

The pressuremeter was invented in 1957 by Ménard.

L. Ménard, 'Mésures in-situ des propriétés physique des sols,' *Annales des Ponts et Chaussées*, **127**, 357–377 (1957).

A number of investigators attempted to solve the pressuremeter problem for a perfectly plastic Coulomb material. The complete solution, without additional assumptions concerning displacement in the plastic zone, was not discovered until 1986. See:

R.E. Gibson and W.F. Anderson, 'In situ measurement of soil properties with the pressuremeter', *Civil Engineering and Public Works Review*, **56**, 615–618, 1961.

J.P. Carter, J.R. Booker and S.K. Yeung, 'Cavity expansion in cohesive frictional soils', *Geotechnique*, **36**, 349–358, 1986.

Exercises

4.1 A direct shear test is performed on a sample of dry cohesionless silt with angle of internal friction equal to 20°. It may be assumed that the silt behaves as a Coulomb material and exhibits perfect plasticity. The test apparatus allows shearing deformation to be localised in an initially 1 mm thick band through the centre of the sample as shown in Figure 4.12. A normal stress equal to 100 kPa is placed on the sample and then the shearing stress is increased until yield occurs. The value of Poisson's ratio for the silt is 0.4. Assume plane strain conditions apply perpendicular to the plane of the drawing.

(a) Show that the value of the shear stress at yield is 36.4 kPa.

(b) Show that the principal stresses at yield are $\sigma_1 = 152$ kPa, $\sigma_2 = 90.6$ kPa and $\sigma_3 = 74.5$ kPa.

(c) Find the orientations of the major and minor principal stresses.

(d) Assuming the silt obeys an associated flow rule, determine the relative magnitudes of the principal plastic strain rates associated with the major and minor principal directions.

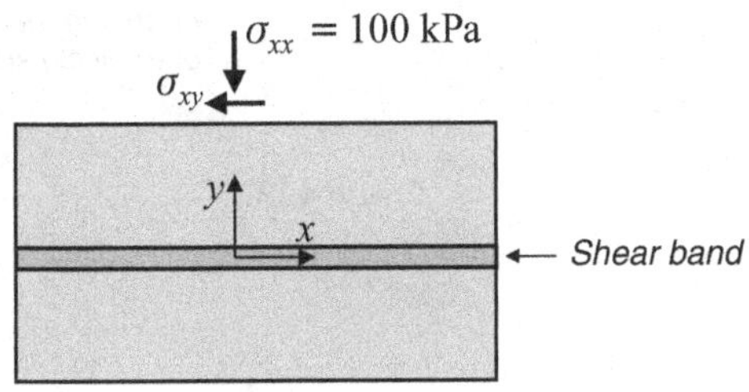

Figure 4.12.

(e) Sketch the two-dimensional yield surface for the situation described and sketch the plastic strain rate vector.

(f) Use the plastic strain rate vector from (d) to determine the direction of motion of the upper half of the sample relative to the lower half.

4.2 Formulate the plasticity matrix $\mathbf{M}^p$ defined in equation (4.39) for the case of a perfectly plastic von Mises material (i.e. $f = q =$ stress deviator) with an associated flow rule. Show that $\mathbf{M}^p$ may be written in the form

$$\mathbf{M}^p = \Lambda \begin{bmatrix} 1 & 1 & 1 \\ 1 & 1 & 1 \\ 1 & 1 & 1 \end{bmatrix} + \frac{G}{3} \begin{bmatrix} \zeta_{23}+2 & \zeta_{12}-1 & \zeta_{13}-1 \\ \zeta_{12}-1 & \zeta_{13}+2 & \zeta_{23}-1 \\ \zeta_{13}-1 & \zeta_{23}-1 & \zeta_{12}+2 \end{bmatrix}$$

where $\zeta_{mn} = 3(\sigma_m - \sigma_n)^2/\sigma_T^2$ and σ_T is the yield stress in simple tension.

4.3 Use the associated flow rule to obtain expressions for the principal plastic strain rates, $\dot{\varepsilon}_1^p$, $\dot{\varepsilon}_2^p$ and $\dot{\varepsilon}_3^p$, for plastic flow of a von Mises material obeying equation (3.12). Combine the strain rates to show that the plastic volumetric strain must always be zero. Show that the Lode angle (equation (3.9)) is related to the plastic strain rates by $\tan\theta = \sqrt{3}\,\dot{\varepsilon}_1^p/(\dot{\varepsilon}_3^p - \dot{\varepsilon}_2^p)$.

4.4 When the associated flow rule is applied to the Tresca yield surface (or, for that matter, any yield surface that is not smooth) ambiguities arise when the stress point lies at a vertex or corner such as the situation shown in Figure 4.13. The strain rate vector may take on any attitude between the limiting values shown. Discuss this ambiguity in light of the stress state that applies at the vertex. In your opinion, what is the most appropriate attitude for the strain rate vector? What implications arise when the stress point moves slightly away from the vertex?

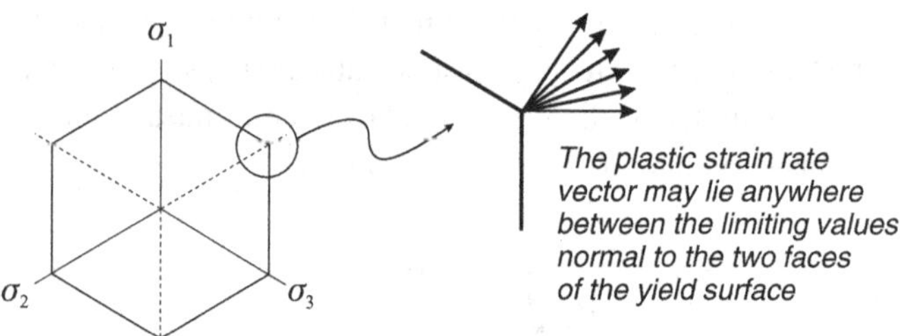

Figure 4.13.

5
Collapse load theorems

5.1 Introduction

One of the most powerful aspects of the theory of plasticity lies in its ability to easily predict approximate values for the collapse load in a very wide range of applications. This comes about through two theorems called the *upper bound theorem* and the *lower bound theorem*. As their names imply, the theorems provide bounds, or limiting values, for the collapse load. Often any usage of the theorems is referred to as *limit analysis*.

The business of predicting collapse loads is totally concerned with finding the loads* that will bring the structure or body to an imminent state of collapse. We are not concerned with what happens before or after in the sense of trying to analyse elastic strains or plastic flow. Also, we must not confuse the collapse load with the yield load. In some instances they will be the same and yield will immediately lead to collapse, but in other cases yield may happen well before collapse. As an example, yield precedes collapse by a significant margin in the shallow foundation problem where localised yielding may happen immediately near the edges of a rigid footing, well in advance of the collapse load. There are restrictions on the applicability of both theorems. A key factor in the development of limit theorems rests with the normality relationship between the yield surface and its associated plastic strain rate vector. For either rigid–perfectly plastic or elastic–perfectly plastic materials, the limit theorems can be proved rigorously (see Appendix H). In general, these conventional limit theorems do not apply to materials that obey non-associated flow rules. For such materials restricted forms of the limit theorems can be proposed. With work hardening materials, the absence of a limiting yield surface precludes the definition of a collapse load and implies the inapplicability of the conventional limit theorems.

* Here 'loads' is used in a general sense. We may be interested in boundary tractions or body forces in the form of gravity loads, or both, depending upon the particular problem being considered.

The most common problems in geotechnical engineering that fall under the heading of limit analysis are the determination of the thrust on a retaining wall and the bearing capacity of a shallow footing. Other typical problems include estimation of the capacity of shallow anchors, arching and the analysis of the stability of cuts, embankments and tunnels. In nearly all of these applications the simplest idealisation for collapse load analysis will be the plane strain assumption. While the theorems themselves apply in any general three-dimensional configuration, there are practical matters involved in their application that are usually greatly simplified if plane strain conditions apply. Because of this we will confine the developments in this chapter to a state of plane strain.

In essence, the theorems work well for perfectly plastic materials because perfect plasticity implies an uncoupling of forces and deformations. Referring back to the pressuremeter problem in Chapter 4, note that the stresses for the post-yield state could be obtained directly by combining the equilibrium equations with the yield condition. There was no need to consider strains or deformations in order to determine the stresses. This is in complete contrast with elastic behaviour where all of the field equations (strain–displacement, equilibrium and Hooke's law) are required to obtain the correct solution. The difference is emphasised by noting that to find the elastic response we were required to solve a second-order differential equation (4.52), while the post-yield stresses were obtained from a simpler first-order equation (4.61).

5.2 The theorems

Most of the development of collapse load theory took place in the 1950s. Professor Daniel C. Drucker together with co-workers established the theorems and elaborated their applications in a series of papers, first for metal plasticity and then for soil mechanics. The essence of both theorems can be encapsulated in a single sentence as follows. *An elastic–perfectly plastic body will, on the one hand, do the best it can to distribute stress in order to avoid collapse, but, on the other hand, will experience collapse if any kinematically admissable collapse mode exists*. This so-called 'anthropomorphic' explanation implies a sense of intelligent behaviour to the inanimate body that clearly cannot exist. However, it may help in applications of the theorems to view the material as if it could rationally adapt itself to whatever loads it is asked to support.

Proofs of both theorems as well as certain auxiliary theorems may be found in the appendices. We will state the theorems here and then proceed to investigate their application.

The lower bound theorem. Collapse will not occur if *any* state of stress can be found that satisfies the equations of equilibrium and the traction boundary conditions and is everywhere 'below yield'.

Note that this is a very general statement. The phrase *any state of stress* may cover a vast range of possibilities, including stress fields that may, from a physical standpoint, be completely implausible. The theorem tells us that if we can find a stress field such that equilibrium and the traction boundary conditions are satisfied, and if the stresses nowhere exceed yield values, then the body cannot collapse. Therefore the boundary tractions are safe. They provide a lower bound for the tractions that will produce collapse. In the anthropomorphic sense, we are assured that the body will possess the necessary intelligence to find a physically permissible stress field that can support the applied loads, provided any equilibrium stress field exists. Any stress field that satisfies the criteria of the lower bound theorem is referred to as a *statically admissible stress field*.

The upper bound theorem. Collapse must occur if, for *any* compatible plastic deformation, the rate of working of the external forces on the body equals or exceeds the rate of internal energy dissipation.

This is also a very general statement, again because of the word *any*. The words 'compatible plastic deformation' imply any deformation that satisfies all displacement boundary conditions and is possible kinematically. That is, 'no gaps, overlaps or separations' should occur. No reference to equilibrium is made. Naturally, there will be a great number of possible deformation mechanisms and it will be our task to investigate those that are sufficiently simple to provide useful results. The result we seek is a simple energy balance. The boundary tractions and body forces will do a certain amount of work during the deformation. If their rate of working is greater than or equals the rate at which energy is dissipated within the body, then collapse is assured. This may be a very valuable result. For example, in the field of metal forming, knowledge of the upper bound load tells us that no more than the upper bound force will be required to cause the forming process to occur. Any deformation field that satisfies the criteria of the upper bound theorem is referred to as a *kinematically admissible deformation*.

Often engineering intuition is useful in application of the theorems. An experienced practitioner can often foresee either a stress field or a compatible deformation that will provide a good result, while a less experienced person may not. There are useful guides to the process and we will consider some below. Often the most useful tools are stress and deformation fields that are discontinuous.

Application of the lower bound theorem usually proceeds like this. First, we hypothesize a statically admissible stress field. Often it will be a discontinuous field in the sense that we have a patchwork of regions of constant stress that together cover the whole soil mass. There will always be one or more particular value of stress that is not fully specified by the conditions of equilibrium. We then try to adjust these undetermined stresses so that the load on the soil is maximised but the yield condition remains unsatisfied everywhere. The resulting load becomes our lower bound estimate for the actual collapse load.

The upper bound theorem is slightly different. We must hypothesize a displacement field, and usually this will be a discontinuous patchwork of regions, each with a constant velocity. We adjust the directions of the velocity vectors of the various regions so as to ensure that there are no gaps or overlaps anywhere in the soil mass. We must then calculate two rates: the rate of working of all the external forces, including gravity forces, and the rate of energy dissipation owing to slip along the surfaces of discontinuity that separate the various regions. The rate of energy dissipation will depend upon our choice of displacement field, but it will be independent of the applied load on the soil. If we set these two rates to be equal, the resulting equation can be solved for the applied load or loads on the soil. These loads will be the upper bound estimate for the true collapse load.

In applications of geotechnical interest we will usually try to obtain both upper and lower bounds on the collapse load. Hopefully the two results will not be greatly different and the true collapse load is then closely bracketed. In some cases the two bounds are the same and hence both give the exact collapse load, but these cases are rare in practice, and occur only because the problems involved are simple and have been studied extensively. The best approach to learning how to use the theorems is by working through some simple examples. We will begin to do that now.

5.3 Discontinuities of stress and deformation

One should bear in mind that the two theorems allow wide latitude in the selection of the stress (lower bound) or deformation (upper bound) fields we may invoke. With regard to stresses, we need only satisfy the equations of equilibrium inside the body and match the applied tractions at the boundary. This allows us to consider stress fields that would not be physically reasonable under normal circumstances. For example, consider the problem illustrated in

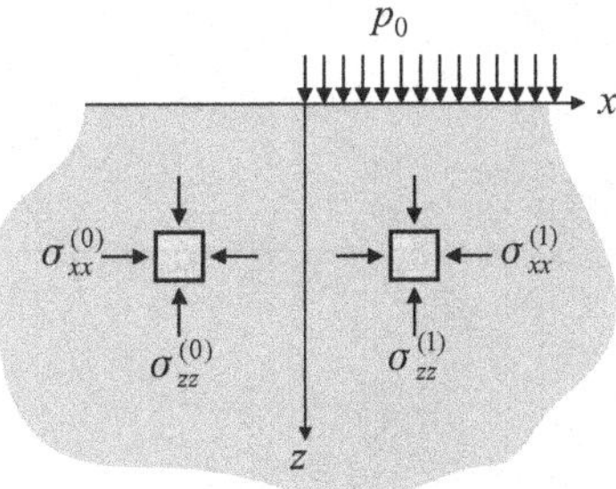

Figure 5.1. Discontinuous stresses in a halfspace with surface step-load.

Figure 5.1. The body is a homogeneous halfspace. On the boundary $z = 0$, tractions are fully specified. A normal applied stress p_0 is found for $x \geq 0$ and zero tractions apply for $x < 0$. Clearly, plane strain conditions apply. No body forces are present.

The proposed stress field inside the halfspace consists of two homogeneous stress states. Both are principal stress states with zero shear stress on surfaces perpendicular to the coordinate axes. For $x \geq 0$ constant stresses $\sigma_{xx}^{(1)}$ and $\sigma_{zz}^{(1)}$ apply everywhere. Similarly, for $x < 0$, the constant stresses are $\sigma_{xx}^{(0)}$ and $\sigma_{zz}^{(0)}$. The boundary conditions immediately tell us that $\sigma_{zz}^{(1)} = p_0$ and $\sigma_{zz}^{(0)} = 0$. Equilibrium will be satisfied provided $\sigma_{xx}^{(1)} = \sigma_{xx}^{(0)}$. Evidently these simple conditions are sufficient for the lower bound theorem provided the stresses are 'below yield'. For a Tresca material this would imply that both $|\sigma_{zz}^{(0)} - \sigma_{xx}^{(0)}| < \sigma_T$ and $|\sigma_{zz}^{(1)} - \sigma_{xx}^{(1)}| < \sigma_T$ where σ_T is the yield stress in simple tension. Obviously for other materials other conditions would apply.

Note that the proposed stress field is not continuous on the surface $x = 0$. The horizontal stress components $\sigma_{xx}^{(0)}$ and $\sigma_{xx}^{(1)}$ are continuous, but a jump occurs in the vertical stress components $\sigma_{zz}^{(0)}$ and $\sigma_{zz}^{(1)}$. From the standpoint of physical realism, our solution would be extremely implausible; yet this is a statically admissible stress field.

The Mohr circles for one possible stress system are illustrated in Figure 5.2. Since the given stresses are principal stresses, we can immediately construct the circles as shown. For a Tresca material, the stresses will satisfy all the criteria laid down by the lower bound theorem provided the diameters of both Mohr circles are smaller than σ_T. If that is the case then collapse will *not* occur under the applied stress p_0.

If we look at Figure 5.2 for a moment, it is clear that the value of p_0 shown is only one possible estimate for the lower bound. The horizontal stresses $\sigma_{xx}^{(0)}$ and $\sigma_{xx}^{(1)}$ are not specified aside from the requirement that they must be equal. Thus we can adjust $\sigma_{xx}^{(0)}$ and $\sigma_{xx}^{(1)}$ in whatever way we wish, so long as the

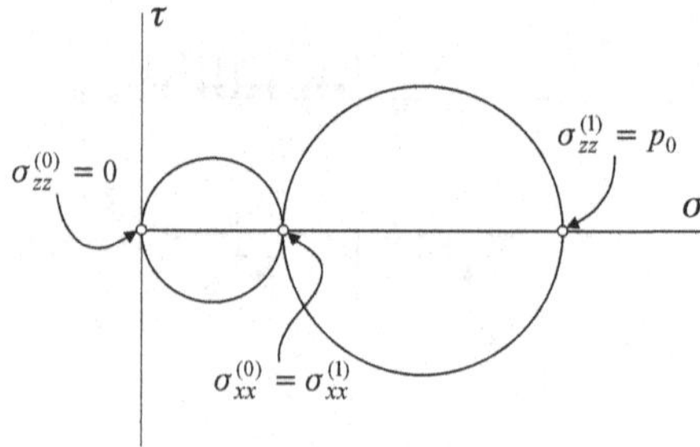

Figure 5.2. Mohr diagram for the stress field shown in Figure 5.1.

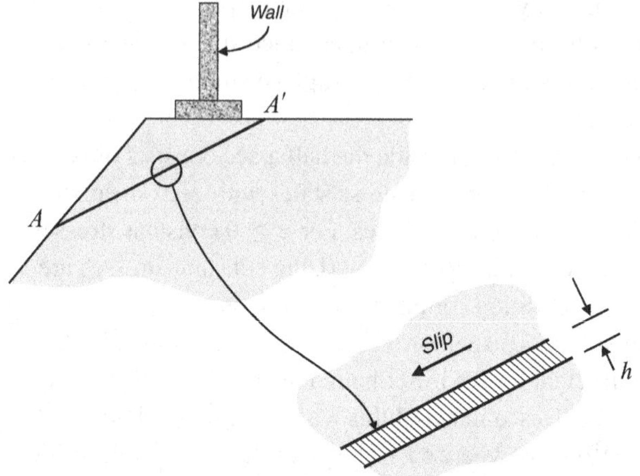

Figure 5.3. Typical collapse mechanism – slip surface with thickness h.

yield condition is not exceeded. The best lower bound estimate will be the largest possible value for p_0 and this will occur when both Mohr circles have their maximum permissible diameters; that is, when both circles have diameter σ_T. In that case we will have $\sigma_{zz}^{(0)} = 0$, $\sigma_{xx}^{(1)} = \sigma_{xx}^{(0)} = \sigma_T$ and $\sigma_{zz}^{(1)} = p_0 = 2 \times \sigma_T$. Thus our best lower bound is $2\sigma_T$. This process of optimising our lower bound estimate by taking the stress state to the most trying condition, the yield condition, will be a feature whenever the lower bound theorem is used.

Discontinuous deformation fields will be useful when applying the upper bound theorem. By discontinuous deformation we have in mind a situation where rigid blocks of material are separated by thin deforming layers. An example is the wall foundation resting near the edge of an embankment illustrated in Figure 5.3. We assume the foundation is long and plane strain conditions apply. The upper bound theorem requires a kinematically feasible deformation.

One possible deformation is slip on the planar surface AA'. The material above and below the slip surface is assumed to be rigid and all deformation is concentrated along the surface. We can think of the surface as a band of thickness h as illustrated in the enlarged segment. Within this band the material is perfectly plastic and the associated flow rule applies.

The upper bound theorem states that we must compare the rate of working of the external loads with the internal energy dissipation. For the situation shown both the wall load and gravity will do work if the block of soil slips. The internal dissipation will all take place within the slip band. It is interesting to investigate what this internal dissipation might be for the case of a Coulomb material.

By localising all slip on a single surface we have simplified the problem enormously. The Coulomb yield condition applies directly to the slip surface,

$$\tau = c + \sigma \tan \phi \tag{5.1}$$

where τ is the shear stress and σ is the effective normal stress on the surface. If we recall that the associated flow rule tells us that the plastic strain rates must be normal to the yield surface, then the simple picture shown in Figure 5.4 emerges.

The plastic strain rate vector will have two components: the plastic shear strain rate $\dot{\varepsilon}_t^p$ acting tangential to the slip surface, and the plastic extensional strain rate $\dot{\varepsilon}_n^p$ normal to the slip surface. Note that while $\dot{\varepsilon}_t^p$ points in the same direction as the shear stress τ, the component $\dot{\varepsilon}_n^p$ points in the opposite direction to the normal stress σ. This is a manifestation of the dilatancy always associated with the Coulomb yield condition. The rate of energy dissipation per unit volume within the slip band is then

$$\textit{dissipation rate} = \tau \dot{\varepsilon}_t^p + \sigma \dot{\varepsilon}_n^p \tag{5.2}$$

Now suppose that we identify the velocity components of the sliding block by v_t and v_n. To be more precise, let these be the velocity components of the upper block of soil *relative* to the lower block. In Figure 5.3 the lower block

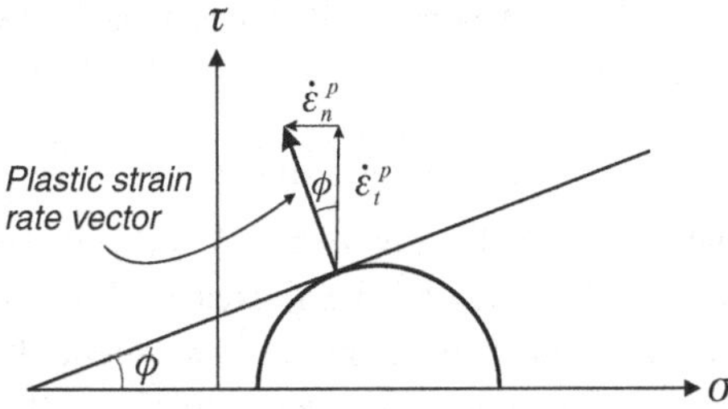

Figure 5.4. Coulomb yield criterion with a normal plastic strain rate vector.

is stationary but in other problems both blocks might be in motion and use of the relative velocity will be essential. Since the shear band is thin we can approximate the plastic strain rates by

$$\dot{\varepsilon}_t^p = v_t/h, \quad \dot{\varepsilon}_n^p = v_n/h \tag{5.3}$$

Next, let D represent the rate of dissipation per unit area of the shear surface. Since we have a plane strain problem we can consider a unit thickness of the embankment and therefore D will, in fact, be the rate of dissipation *per unit length* of the shear surface. D is obtained by multiplying the dissipation per unit volume in (5.2) by the area $h \times 1$. If we then use (5.3) we find this simple result,

$$D = \tau v_t + \sigma v_n \tag{5.4}$$

We can further simplify this by noting from the geometry of Figure 5.4 that $\dot{\varepsilon}_n^p = -\dot{\varepsilon}_t^p \tan\phi$ and hence $v_n = -v_t \tan\varphi$. Using this result in (5.4) gives

$$D = v_t(\tau - \sigma\tan\phi) \tag{5.5}$$

Finally, we use (5.1) to find

$$D = v_t c = cv\cos\phi \tag{5.6}$$

This surprisingly simple result encapsulates the dissipation on any planar shear surface. Note that the shear band thickness h cancels from the equations and, assuming that shearing happens within a relatively thin region, the rate of dissipation will be equal to D multiplied by the length of the shear surface. We will use the notation $\mathbb{D}$ to indicate the product of D with the length of the slip surface. Provided we use relative velocities, the argument will work equally well when two blocks of soil separated by the slip surface are simultaneously in motion. Things would break down if the blocks were to separate physically; not only would the condition of kinematic admissibility be violated but the result would be an unusual outcome in geotechnical engineering.

We are now in a position to investigate some particular problems of interest to the geotechnical engineer. We begin with a particularly simple example.

5.4 A vertical cut

Consider the problem illustrated in Figure 5.5. A soil with density ρ, cohesion c and angle of internal friction ϕ has been excavated to form a vertical cut. The engineer desires to know the maximum stable height of excavation. Even though loads have not been mentioned explicitly, this is a collapse load problem. The forces are supplied by gravity and we know intuitively that if the height

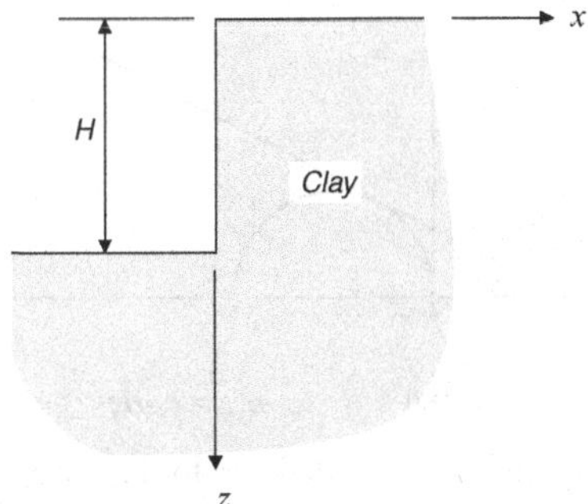

Figure 5.5. Example – vertical cut in a homogeneous clay.

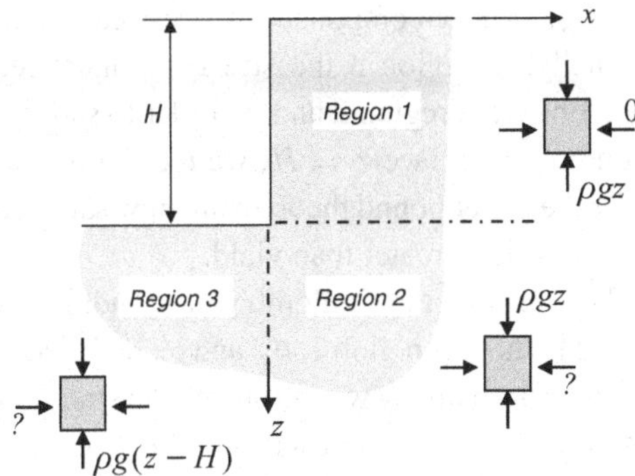

Figure 5.6. Discontinuous stress field for the vertical cut – lower bound analysis.

H is too great the vertical face will collapse. We assume the length with the cut measured perpendicular to the plane of the figure is large in comparison with the height, and hence we assume that plane strain conditions are appropriate. We can examine the problem using both lower bound and upper bound methods.

First, consider the lower bound theorem. We will need to devise a system of stresses to satisfy the boundary conditions, all of which consist of zero tractions on exposed surfaces. The stresses must also satisfy the equilibrium equations with the vertical body force b_z being equal to the gravity force ρg. The simplest system of stresses that will qualify is outlined in Figure 5.6. The stress field is divided into three regions, numbered 1, 2 and 3, separated by dashed lines. Discontinuities in some stress components occur on these lines. In region 1 the stresses are

$$\sigma_{xx} = 0, \quad \sigma_{zz} = \rho g z \tag{5.7}$$

Since the coordinate origin is located at the top of the cut, these stresses will

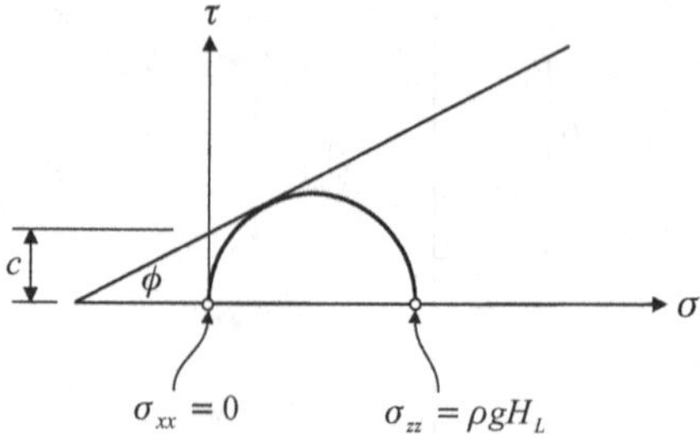

Figure 5.7. Mohr diagram for Figure 5.6.

clearly satisfy the traction-free boundaries as well as equilibrium. In region 2 the z-component of stress must be continuous across the dashed line and hence will be given by $\sigma_{zz} = \rho g z$. The x-component of stress can be left undetermined for the present. Finally, in region 3 the stress σ_{xx} must remain continuous crossing the dashed line from region 2 and, in order to satisfy the zero traction boundary condition on the surface $z = H$, we require $\sigma_{zz} = \rho g(z - H)$. All the requirements for the lower bound theorem are now satisfied, except for one, that the stresses are nowhere greater than yield.

Finally, we need to consider the Coulomb yield condition. We could use the version given in (3.35) with, for region 1, σ_m and σ_n replaced by $\sigma_{zz} = \rho g z$ and $\sigma_{xx} = 0$, respectively. Alternatively we can sketch the stress state for region 1 in Figure 5.7. The Mohr circle for region 1 will achieve its greatest diameter when $z = H$. The greatest possible height H will correspond to the Mohr circle that just touches the yield envelope as shown in the figure. Thus the critical height is given by

$$\rho g H_L = \frac{2c\cos\phi}{1 - \sin\phi} \tag{5.8}$$

where H_L represents the critical height determined from the lower bound theorem.

At this point one might ask about the undetermined horizontal stress σ_{xx} in regions 2 and 3. Can we simply disregard this stress? The answer is no. A central requirement of the lower bound theorem is that we find a stress field that satisfies equilibrium *throughout* the entire body that nowhere exceeds yield. In order for the lower bound theorem to work here we must demonstrate that a stress σ_{xx} can be found that will satisfy these conditions. This can be done by setting $\sigma_{xx} = \rho g(z - H)$ in both regions 2 and 3. This will create an isotropic stress field in region 3 and the Mohr circle for region 2 will be no bigger than that shown in Figure 5.7. For all depths greater than H the stresses in region 2 will be safe from yielding.

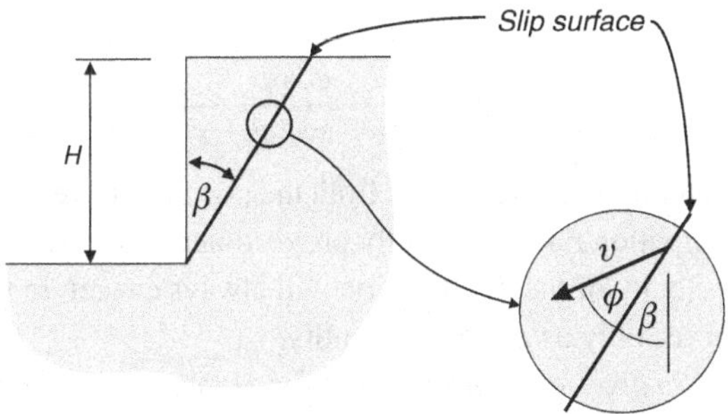

Figure 5.8. Collapse mechanism for vertical cut – upper bound analysis.

The result in (5.8) can be written in several different ways. A common way is to replace $\cos\phi$ by $\sqrt{1-\sin^2\phi} = \sqrt{(1-\sin\phi)(1+\sin\phi)}$. Then we can write the lower bound critical height as

$$H_L = \frac{2c}{\rho g}\sqrt{N} \tag{5.9}$$

where N was defined in (4.16).

Now we turn our attention to the upper bound theorem. We will require a compatible collapse mechanism. The most obvious candidate is the planar slip surface shown in Figure 5.8. Leaving the angle β undetermined for the moment, we set out an expression of energy balance between external forces and internal dissipation. For this problem the only external force is the action of gravity on the wedge of soil lying above the slip surface. The weight of soil in the failure wedge is $\frac{1}{2}\rho g H^2 \tan\beta$. Suppose we let the velocity of the wedge be v. Because the associated flow rule requires dilatancy, the direction of v will be inclined at an angle ϕ to the slip surface. Thus the vertical component of v will be $v\ \cos(\phi+\beta)$ as shown in the figure. If we now let $\mathbb{R}$ represent the rate of working of the external forces we see that

$$\mathbb{R} = \frac{v}{2}\rho g H^2 \tan\beta\cos(\phi+\beta) \tag{5.10}$$

Internal dissipation occurs on the slip surface at a rate D as shown in (5.6). The tangential component of the wedge velocity will be $v_t = v\cos\phi$. The length of the slip surface is $H/\cos\beta$ and the total dissipation rate $\mathbb{D}$ will be D multiplied by this length. If the rate of working of the external forces equals or exceeds this total dissipation rate, then we are assured of collapse. To find our upper bound we set the power and dissipation rate equal to have

$$\frac{v}{2}\rho g H^2 \tan\beta\cos(\phi+\beta) = cv\cos\phi\frac{H}{\cos\beta} \tag{5.11}$$

Solving for H gives

$$H = \frac{2c}{\rho g} \frac{\cos\phi}{\sin\beta\cos(\phi+\beta)} \tag{5.12}$$

Note that the velocity v has vanished. Both the power of the external forces and the internal dissipation rate are directly proportional to v and in comparing the two rates the velocity of the deformation will always cancel. In these problems the velocity serves only as a virtual quantity.

Equation (5.12) gives one upper bound for H. We must now find the smallest upper bound by minimising H with respect to the angle β. Setting the β-derivative of the right-hand side of (5.12) equal to zero we find this transcendental equation for the critical value of β

$$\tan(\phi+\beta_c) = \cot\beta_c \tag{5.13}$$

An aspect of the upper bound theorem is the occasional need to solve transcendental equations such as (5.13). In this case symmetry of the tangent and cotangent about $\pi/4$ immediately shows that

$$\beta_c = \frac{\pi}{4} - \frac{\phi}{2} \tag{5.14}$$

Using this value in (5.12) gives a result for the upper bound collapse height H_U,

$$H_U = \frac{4c}{\rho g}\sqrt{N} = 2H_L \tag{5.15}$$

Clearly the upper and lower bounds for H are not close. It might be tempting at this point to guess that the true answer is somewhere, perhaps halfway, between H_U and H_L, but that would be risky in any analysis. As it happens the exact collapse height for this problem, if we assume the soil cannot support tension, is precisely equal to H_L. The reason H_U is twice the exact value is due to our choice of collapse mechanism in Figure 5.8. If we assume that the soil will not support tensile stress, then a vertical tension crack will develop behind the cut face and, rather than the failure wedge in Figure 5.8, a thin slab of soil will collapse leaving a new, near vertical face. If the soil will support tension, then H_U will be closer to the true collapse load, but it is still not exact.

5.5 Shallow foundation – lower bound

Now we can turn our attention to the slightly more difficult problem of the shallow strip footing illustrated in Figure 5.9. This is in many ways a classic plane strain problem in both metal plasticity and geomechanics. The sketch in Figure 5.9 shows a long footing of width B buried in a Coulomb soil at

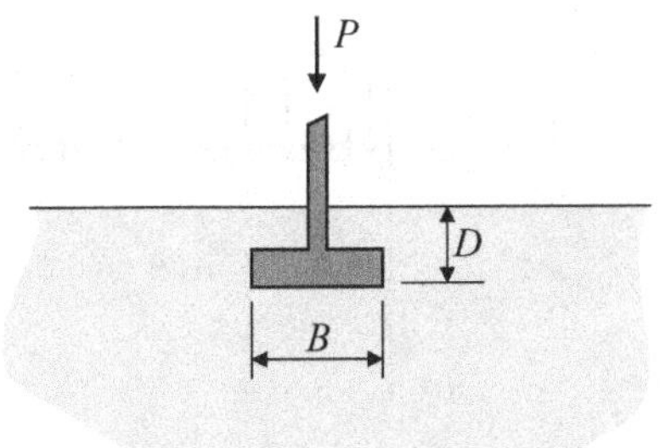

Figure 5.9. A typical shallow strip footing.

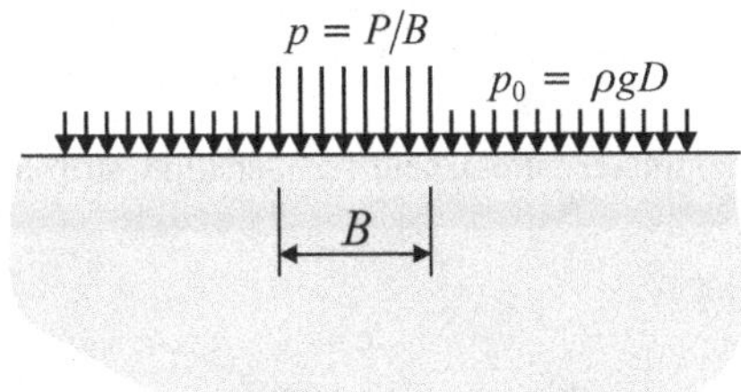

Figure 5.10. Idealised loading for a shallow strip footing.

a shallow depth D and supporting a load P per unit length. The equivalent problem in metal plasticity is a long rigid punch indenting the surface of an elastic–perfectly plastic halfspace. There have been many published analyses of the problem and, in the context of geotechnical engineering, the familiar bearing capacity equation is a well-known result.

It is possible to approach the problem in a number of different ways. In this section we will consider two relatively simple lower bound approaches based on the idealisation shown in Figure 5.10. In that figure both the footing and the soil above the horizontal plane passing through the base of the footing are represented by uniform tractions applied on the surface of a homogeneous halfspace. The uniform stress p is applied over the width of the footing and is equal to the wall load P divided by B. The stress p_0 represents the overburden surcharge and is equal to $\rho g D$. We assume that the material is a Coulomb soil.

To apply the lower bound theorem we require a statically admissible stress field for the situation shown in Figure 5.10. An obvious possibility is the biaxial stress state illustrated in Figure 5.11. The two vertical dashed lines represent stress discontinuities, similar to the discontinuity used in Figure 5.1. The halfspace is subdivided into three regions identified by the letters A and B. The stresses shown are all principal stresses. The vertical components of stress in all three regions increase with depth according to $\rho g z$. The horizontal components

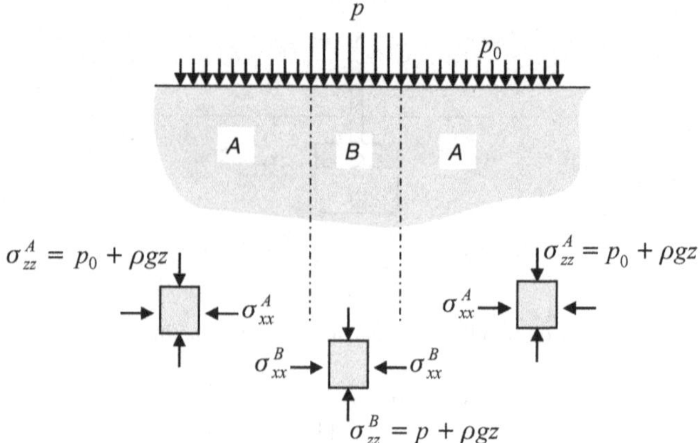

Figure 5.11. The discontinuous stress field for a shallow strip footing – lower bound analysis.

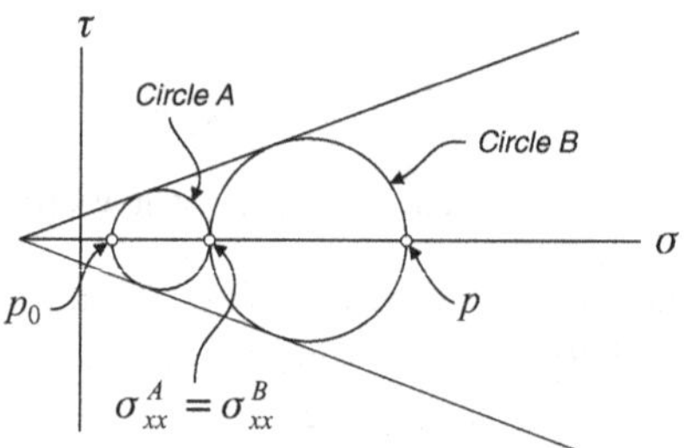

Figure 5.12. Mohr diagram for Figure 5.11.

of stress are not yet specified, but they must be equal in the different regions in order to preserve horizontal equilibrium.

Evidently there are many similarities here to the simple example we discussed in relation to Figures 5.1 and 5.2. The best lower bound will correspond to the greatest value for the stress p for which the yield condition is not violated. Therefore we will take both regions to be at their limiting states. Unlike the situation in the earlier example, the yield condition here is that of Coulomb rather than Tresca. The final stress field is summarised by the Mohr diagram in Figure 5.12, where we have taken the case for $z = 0$. The horizontal principal stress evaluated at $z = 0$ in both regions is

$$\sigma_{xx}^A = \sigma_{xx}^B = 2c\sqrt{N} + p_0 N \tag{5.16}$$

and the lower bound estimate for p is

$$p = 2c\sqrt{N} + N\left[\sigma_{xx}^A\right]_{z=0} = 2c\sqrt{N}(1 + N) + p_0 N^2 \tag{5.17}$$

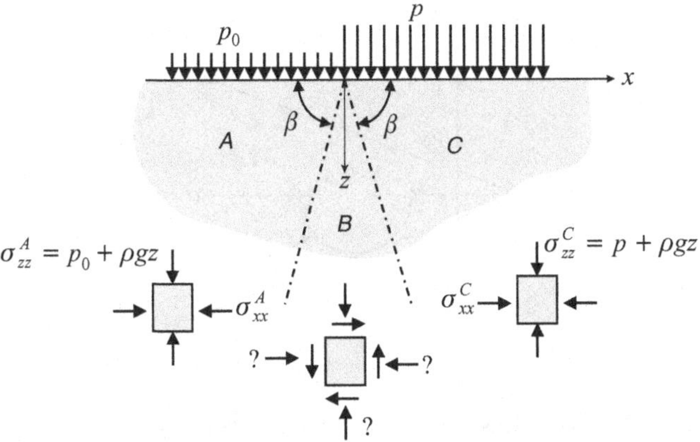

Figure 5.13. Discontinuous stress field for the step-load problem – lower bound analysis.

Suppose $\phi = 20°$. Then N is 2.04 and our lower bound estimate (5.17) can be evaluated as

$$\frac{p}{c} = 8.68 + 4.16\frac{p_0}{c} \tag{5.18}$$

For the case where $p_0 = c$, we find $p = 12.84\,c$. This is not a very good lower bound, in the sense that the true collapse load is significantly greater, but the stress field we have used is exceedingly simple and we should not expect remarkable accuracy.

Next, we can attempt to improve our estimate. One way to do this is to add additional regions to the stress field of Figure 5.11. Before we do so it is helpful to consider the somewhat simpler problem of a stepped surface traction as shown in Figure 5.13. For this problem the stress p_0 acts for all $x < 0$, while p is applied for $x \geq 0$. This problem looks like the left edge of the footing problem and we can concentrate our attention on it for the moment. The new stress field we want to investigate consists of the three regions shown in Figure 5.13.

Regions A and C are familiar from our first attempt. The principal stresses align with the coordinate directions, as they must in order to satisfy the surface boundary conditions. The horizontal components of stress in both regions are as yet not specified. The 'new' region in this situation is region B. As we will see, the principal stresses in this region will not align with the coordinate axes. The interesting point here is that the dashed lines of stress discontinuity are no longer vertical but make an angle β with the horizontal. In order to preserve equilibrium we must ensure that the normal and shear tractions on the β lines are continuous. Also, in order to maximise p, we will probably want the stress

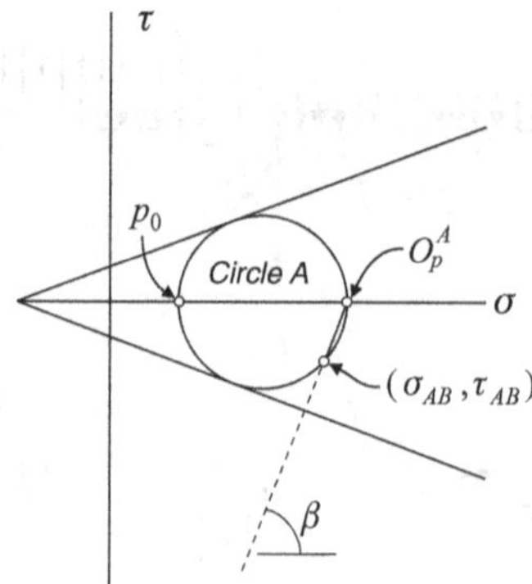

Figure 5.14. Mohr diagram for region A in Figure 5.13.

fields in each region to have as large a Mohr circle as possible, indicating the limiting yield state.

Begin with region A. As we noted above, the soil surface is a principal surface, suggesting that the stress state within the region should have horizontal and vertical principal directions. It is therefore immediately evident that the horizontal stress σ_{xx}^A should be the major principal stress, given by

$$\sigma_{xx}^A = 2c\sqrt{N} + p_0 N + \rho g z \tag{5.19}$$

This ensures that at $z = 0$ the Mohr circle for region A will look as shown in Figure 5.14. Note that since the stress $\sigma_{zz}^A(z = 0) = p_0$ acts on a horizontal surface, the pole for the Mohr circle will lie at σ_{xx}^A as shown. We identify the pole as O_p^A.

Now consider the stresses that act on the surface separating regions A and B. On circle A, those stresses will lie at the point where a line making an angle β with the horizontal and passing through the pole O_p^A intersects the circle. The stresses are identified in Figure 5.14 by (σ_{AB}, τ_{AB}). These are, of course, the normal and tangential components of stress acting on the surface separating regions A and B and, in order to satisfy equilibrium, they must act in region B itself. Thus the point (σ_{AB}, τ_{AB}) must lie on both the Mohr circle for region A *and* the circle for region B.

It is now straightforward to completely define the stresses in region B. We know that the Mohr circle must pass through (σ_{AB}, τ_{AB}) and, in order to maximise p, the circle must be at its limiting state. Of course, two limiting circles can be constructed through any single stress point, but it would make little sense to reuse circle A, therefore the Mohr circle for region B must be as shown in Figure 5.15. In this figure we can also identify the pole for region B. It lies on the intersection of the left-hand β line with the circle and is marked as O_p^B. Note

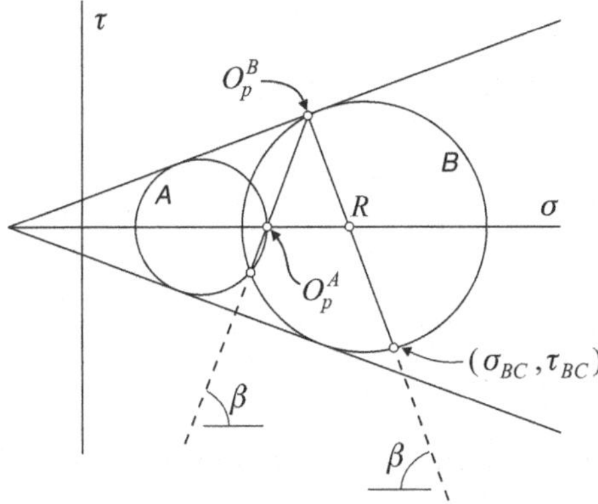

Figure 5.15. Mohr diagram for regions A and B in Figure 5.13.

that while O_p^B falls near the point where the circle is tangential to the failure envelope, this is a coincidence and has no relevance to the solution. Clearly, since the pole does not coincide with either of the principal stresses on circle B, there will be normal and shear stresses on horizontal and vertical surfaces within the region. We could easily determine those by drawing horizontal and vertical lines through the pole O_p^B.

The last step is to determine the stress state in region C. Just as with region B, we can immediately see that the Mohr circle for region C must pass through the point marked (σ_{BC}, τ_{BC}). Also, in order to maximise our result for p, the Mohr circle should be at the limiting state. These two things alone are sufficient to fully determine a Mohr circle for region C, but there is still one more condition we must satisfy. Since the ground surface is a principal surface, the normal stresses in region C must be principal stresses. Therefore the pole for circle C *must* coincide with the minor principal stress. Referring to Figure 5.15, the circle for region C must pass through not only point (σ_{BC}, τ_{BC}), but also the point marked R, and it should be tangential to the Coulomb yield surface. At first glance the problem appears intractable since we have more conditions than degrees of freedom available. In fact, there is one remaining parameter we have not yet specified, the angle β. The way to proceed is to adjust the angle β in such a way that the three conditions on circle C can be met. The result is illustrated in Figure 5.16. All three circles are now shown together with the poles for each and the common stress points. The greatest principal stress for circle C is our estimate for the lower bound collapse load.

Note that the situation shown in Figure 5.16 applies at the ground surface only. If $z > 0$ then the normal stresses in all three regions are increased by $\rho g z$. This has the effect of shifting the Mohr circles in the figure to the right, but

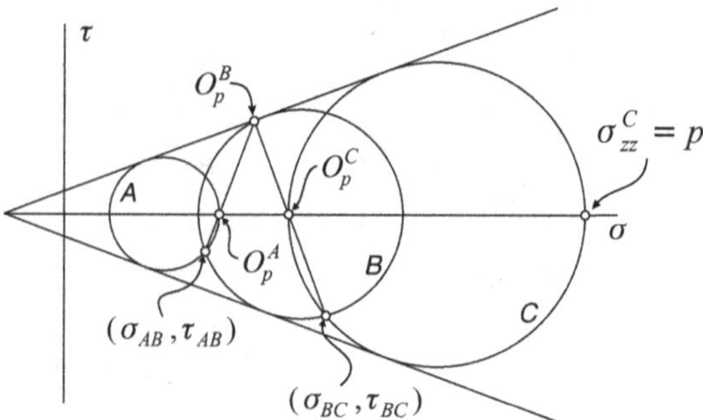

Figure 5.16. Mohr diagram for all regions in Figure 5.13.

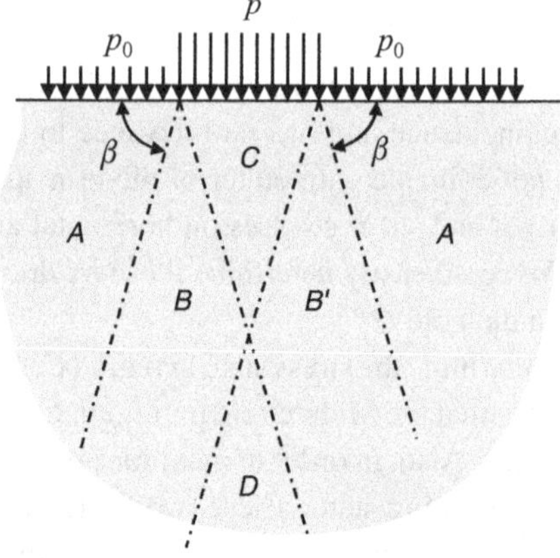

Figure 5.17. Extending the step load problem to solve the shallow strip footing – lower bound analysis.

not altering their size or their relationship to one another. In this way there is no danger of violating the yield condition at some greater depth within the halfspace. We can always add an isotropic stress to the normal stress components in problems such as this where the ground surface is horizontal. Also, it might appear from Figure 5.13 that region B will vanish when $z = 0$, but this is not a problem. The stresses in each region will vary continuously for all values of z and the limiting values as $z \to 0$ will be as shown.

Is this solution relevant to the shallow foundation problem? It is, since we can now place a new region similar to region B at each edge of the footing load. That is, instead of the simple stress field in Figure 5.11, we now wish to consider the situation shown in Figure 5.17. The two outer regions A are exactly as in the stepped traction problem, and so is the region B at the left-hand edge of the footing. On the right-hand edge of the footing we now have a new region B'. This is nearly the same as region B, but the direction of the β line separating B' from A is reversed. Referring to Figure 5.16, the Mohr circle for region B' will be exactly the same as for region B with the exception that the pole will lie near the bottom of the circle rather than the top. Region C is unchanged. We also find a new region D where B and B' intersect. We should check to see that the yield condition is not violated in this region. The Mohr circle for region D must share one point with both regions B and B'. Each point is exactly the same as the point shared between B and $B'$with region A. The two points lie directly above one another and hence there is no unique stress state for region D. One possible solution is to make region D the same as region A. Clearly this is a safe result in the sense that the yield condition will not be violated and we see that the complete stress field is statically admissible.

For the purposes of comparison, if $\varphi = 20°$, the angle β is roughly 69.2° and the lower bound estimate for p is found to be $12.36\,c + 5.50\,p_0$. In the special case where $p_0 = c$ this gives $p/c = 17.86$, an improvement of roughly 38% on our first estimate based on Figure 5.11. The new result is considerably closer to the true collapse load, but the analysis is correspondingly more complex. The reason why the lower bound stress field shown in Figure 5.17 gives a significantly better result than that shown in Figure 5.11 can be sensed intuitively. The stress field beneath the actual footing will have a natural tendency to 'distribute' or 'spread' as z increases. This is not possible in Figure 5.11, but the stresses in Figure 5.17 do permit spreading of a certain kind and hence give a better answer. It is possible to add yet more regions to the situation in Figure 5.17 and this may lead to further improvements in our lower bound estimate, but the increased accuracy entails an increase in computation complexity, and there is significantly more work required. One way to avoid this extra effort is to let a computer do the work by employing finite elements. A statically admissible, discretised stress field may be obtained through an application of finite-element theory to solve for the lower bound collapse load in a range of more complex problems.

5.6 Shallow foundation – upper bound

Now we can attempt to establish an upper bound for the shallow foundation problem sketched in Figure 5.10. The things we need are, first, a kinematically admissible deformation and second, an energy balance expression similar to (5.11). We will work through the problem in a series of stages beginning with the most simple case of a cohesive soil with no internal friction. Levels of additional complexity will be added one at a time.

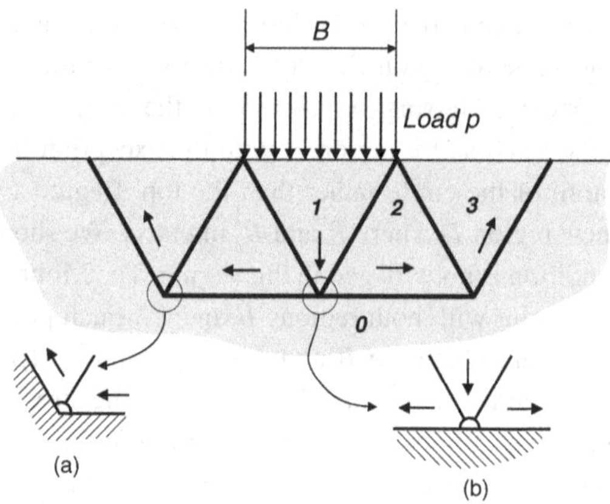

Figure 5.18. Collapse mechanism for shallow strip footing – upper bound analysis.

Case 1. $\phi = 0°$, $p_0 = 0$, *zero body forces*

This is the simplest case for a shallow foundation. The sole source of strength is the cohesion c, there is no surcharge effect due to depth of burial and the effects of gravity are omitted. One possible deformation field is shown in Figure 5.18. The heavy lines depict shear surfaces that delineate five rigid blocks shaped as equilateral triangles with side dimension B. This is a particularly simple deformation field and it will not yield especially accurate results, but it is very convenient from the standpoint of computations and will be useful for illustrating the upper bound theorem.

Recall that we can always think of the shear surfaces as thin bands of intense shearing. Inspection of the figure shows that we will expect the triangle marked 1 to move vertically downward as a rigid block due to the footing load p. The region marked 0 surrounding the five triangles remains stationary. The remaining triangles will move as rigid bodies and there is mirror symmetry about the footing centre line. The triangle marked 2 will move horizontally to the right to make way for triangle 1. The triangle marked 3 will move upward and to the right to make way for triangle 2. A point worthy of note relates to the movement of the rigid blocks at points of contact with the stationary mass 0. Clearly, the triangles cannot penetrate into the basement material without violating compatibility. One way of addressing this issue is to incorporate small 'cut-outs' such as those illustrated in insets (a) and (b) in Figure 5.18. This permits small movements of the blocks without compromising compatibility. Any change in the calculation of external work or internal dissipation caused by the cut-outs can be neglected.

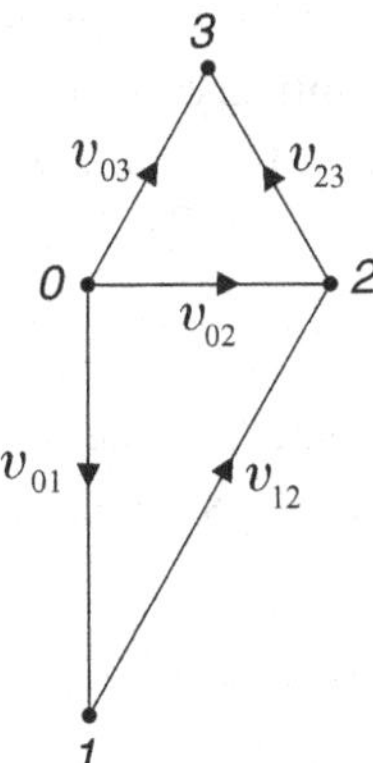

Figure 5.19. Velocity hodograph for the collapse mechanism in Figure 5.18.

For the case of zero internal friction, there will be no dilatancy. Therefore the direction of motion of the rigid blocks will be parallel to the shear surfaces. In order to calculate both the internal dissipation rate $\mathbb{D}$ and the external power $\mathbb{R}$, we will need to know the relative velocities for all of the numbered regions in Figure 5.18. An easy way to determine all the necessary velocities is to construct a velocity diagram or *hodograph*. This is a graphical representation of all the velocities shown in Figure 5.19. In the figure the velocity of each region is measured relative to each adjoining region. For example v_{01} represents the velocity of region 1 relative to the (stationary) region 0. To construct the hodograph we begin by drawing v_{01} vertically downward to an arbitrary scale. The point 0 is the origin and represents no motion. Point 1 represents the downward velocity of block 1. Next, we realise that block 2 must move horizontally relative to region 0; but must also move relative to block 1, and its relative motion will be upward and to the right on a 60° angle. Drawing two lines: one from point 0 in the horizontal direction, and one from point 1 in the 60° direction, the intersection identifies point 2, and the velocities v_{02} and v_{12} are shown. Finally, we find point 3 by drawing lines through point 0 and point 2 parallel to the slip surfaces that separate block 3 from regions 0 and 2. By constructing the diagram in this way we are assured of creating a compatible deformation in the sense that the blocks will move without creating gaps or overlaps.

The magnitudes of all the relative velocities are easily determined in terms of the velocity v_{01}. The geometry of the hodograph shows that

$$v_{02} = v_{03} = v_{23} = v_{01} \tan 30^\circ = 0.577 v_{01}, \quad v_{12} = v_{01}/\cos 30^\circ = 1.155 v_{01} \tag{5.20}$$

The internal dissipation rate will be the sum of the dissipation occurring on each slip surface. Because of symmetry we can work this out for the right-hand half of the deformation and double the result. Following (5.6) and remembering that $\phi = 0$, we find

$$\mathbb{D} = 2 \times Bc(v_{02} + v_{12} + v_{03} + v_{23}) = 5.77Bcv_{01} \tag{5.21}$$

To find our upper bound we equate $\mathbb{D}$ to the power of the external forces. The only external load is the applied stress p and this will move downward with velocity v_{01}. Therefore,

$$\mathbb{R} = pBv_{01} = 5.77Bcv_{01} = \mathbb{D} \tag{5.22}$$

So we find that our upper bound estimate is $p = 5.77c$. Once again, note that v_{01} has cancelled from the energy balance equation. The magnitude of the slip velocity is a virtual quantity and has no effect on the collapse load.

The exact solution to this problem is well known. It will be described in detail in Chapter 6. The exact collapse load is $(2 + \pi)c = 5.14c$, roughly 12% lower than our upper bound estimate. It is also straightforward to carry out a lower bound analysis for these conditions. Using the multiple stress fields from Figure 5.14 we can find that the lower bound is $4.83c$, roughly 6% less than the exact result.

Case 2. $\phi = 0°$, $p_0 \neq 0$, zero body forces

It is a simple matter to incorporate the surcharge due to the depth of burial of the footing. This creates a new external force that must be considered in the energy balance equation. The rate of work of the applied stresses now becomes

$$\mathbb{R} = pBv_{01} - 2 \times p_0Bv_{03} \sin 60° \tag{5.23}$$

Note that negative work is done by these forces since the block 3 is moving upward. The presence of the surcharge on both sides of the footing accounts for the factor of 2 in (5.23). If we now equate the power $\mathbb{R}$ to the rate of dissipation from (5.21) we find a new upper bound collapse load

$$p = 5.77c + p_0 \tag{5.24}$$

Note that the footing width B does not appear. The result in (5.24) is especially simple due to the simplified geometry assumed for the collapse mechanism in Figure 5.18.

Case 3. $\phi = 20°$, $p_0 \neq 0$, zero body forces

Next we will introduce friction. In order to compare this with our lower bound results we set ϕ equal to 20°. This will have a dramatic effect on the collapse

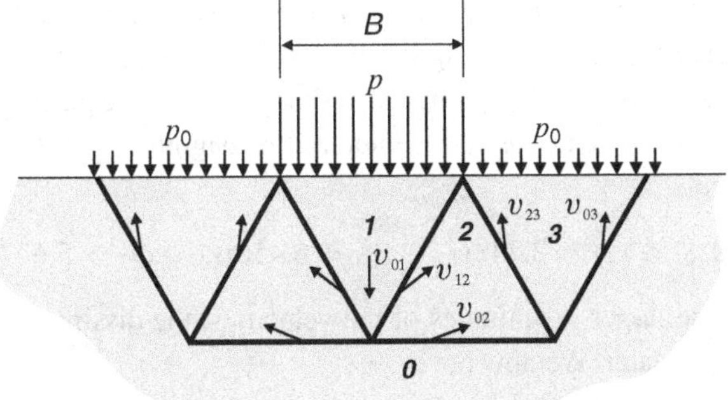

Figure 5.20. Collapse mechanism for a shallow strip footing showing the effect of friction on the relative velocity directions.

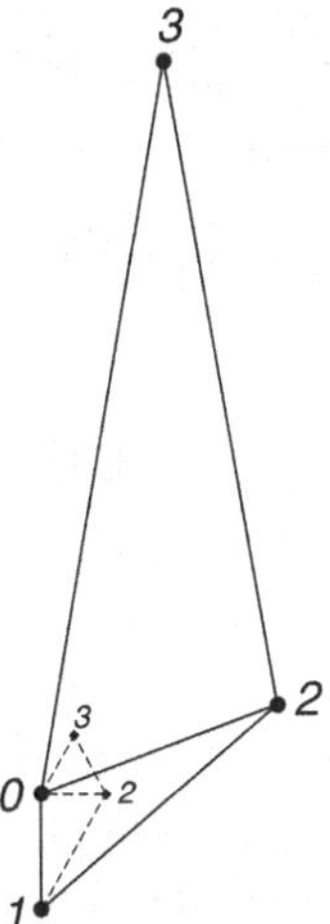

Figure 5.21. Velocity hodograph for Figure 5.20.

load. The reason is dilatancy. Figure 5.20 redraws the collapse mechanism of Figure 5.18, but now shows the direction of relative motion associated with each of the slip lines.

The velocity v_{01} is unchanged, but in every other case the direction of relative motion is altered by an angle ϕ. Dilatancy forces block 2 to move upward, away from the stationary region *0*. There is also a new component of relative motion on the line separating blocks 1 and 2, and blocks 2 and 3. In each case the velocity vector is oriented at an angle ϕ to the slip line. We construct a new hodograph as shown in Figure 5.21.

In Figure 5.21 the original hodograph of Figure 5.19 is shown as a dashed line. The new hodograph is significantly expanded due to the effect of dilatancy.

In the figure the lines representing v_{02}, v_{12}, v_{03} and v_{23} all lie at an angle ϕ to the direction of the corresponding slip line. This moves point 2 well to the right from its original position and point 3 is pushed far above the place it occupied in Figure 5.19. From the hodograph geometry the magnitudes of the new relative velocities are

$$v_{02} = 2.240v_{01}, \quad v_{12} = 2.748v_{01}, \quad v_{03} = 6.450v_{01}, \quad v_{23} = 5.672v_{01} \tag{5.25}$$

Because of the larger magnitudes of the velocities, the dissipation rate is now considerably greater. We now have

$$\mathbb{D} = 2 \times Bc(v_{02} + v_{12} + v_{03} + v_{23})\cos\phi = 32.16\, Bcv_{01} \tag{5.26}$$

Note that we must multiply all of the velocities in (5.26) by $\cos\phi$ in order to have the tangential velocity component shown in equation (5.6).

The power of the footing load p is unchanged for this case, but the surcharge term now moves with a velocity equal to the vertical component of v_{03}. The energy balance equation now becomes

$$\mathbb{R} = pBv_{01} - 2p_0Bv_{03}\cos 10^\circ = 32.16\, Bcv_{01} = \mathbb{D} \tag{5.27}$$

Solving for p gives

$$p = 32.16\,c + 12.70\, p_0 \tag{5.28}$$

Clearly internal friction has had an important effect on our upper bound, but note that as yet there is no influence of the footing width B.

Case 4. $\phi = 20^\circ$, $p_0 \neq 0$, gravity effects included

The final step is to let gravity come into play. This will provide another external force that must be accounted for in the energy balance equation. Each of the triangular blocks has the same area and hence the same weight, given by $\rho g B^2 \sin 60^\circ \cos 60^\circ = 0.433\rho g B^2$. The gravity force does positive work in block 1 since it moves downward, but negative work is done in both blocks 2 and 3. The power of all external forces now becomes

$$\mathbb{R} = pBv_{01} - 2p_0Bv_{03}\sin 80^\circ + 0.433\,\rho g B^2(v_{01} - 2v_{02}\sin 20^\circ - 2v_{03}\sin 80^\circ) \tag{5.29}$$

When this is equated to the rate of internal dissipation $\mathbb{D}$ we find an upper bound estimate of

$$p = 32.16\,c + 12.70\, p_0 + 5.73\rho g B \tag{5.30}$$

Now we finally see a contribution to p that is proportional to the footing width B.

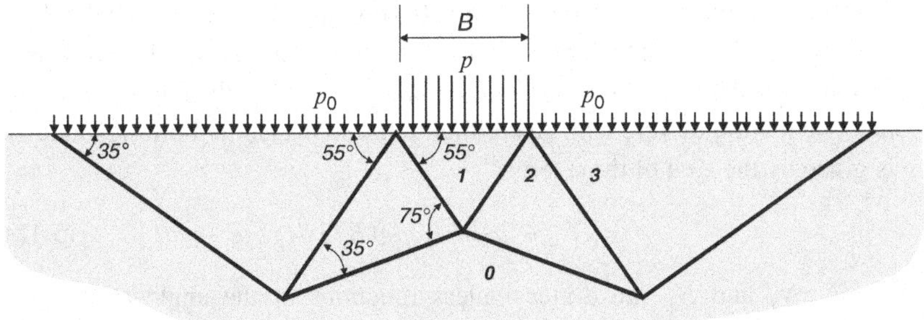

Figure 5.22. An alternative collapse mechanism for a shallow strip footing.

Comparing the result in (5.30) with the lower bound estimate, we find that they differ by a considerable amount. First, the lower bound had no dependence on the footing width. Even if we disregard the term involving B in (5.30), the two results are still significantly different. For the case where $p_0 = c$, the lower bound gave $p/c = 17.86$. Ignoring the term involving B, equation (5.30) gives $p/c = 32.16 + 12.70 = 44.86$. The reason for this large discrepancy lies primarily with our upper bound calculation. The collapse mechanism given in Figure 5.18 is too different from the true collapse mechanism to provide a good estimate. It is relatively easy to improve on the result in (5.30) by simply assuming a different collapse mechanism that is closer to the true mechanism. As an example, consider the mechanism shown in Figure 5.22. Here the 60° angles of the equilateral triangles have been replaced with either 35°, 55° or 75° angles at the places indicated. If we carry through the calculation for the case where $\phi = 20°$, $p_0 \neq 0$ and gravity effects are included, we find the following result:

$$p = 18.00\, c + 7.55\, p_0 + 4.18\, \rho g B \tag{5.31}$$

This estimate for the collapse load is significantly smaller than that in (5.30) and therefore will be closer to the actual value. If we set $p_0 = c$ and ignore the term involving B, we find $p/c = 25.55$, roughly a 75% improvement. The patterns of the collapse mechanisms of Figures 5.18 and 5.22 are quite similar, but the resulting upper bound has changed significantly.

5.7 Shallow foundation – discussion

We can pause here for a moment to consider some other aspects of the shallow footing problem. It is a problem that has been studied by many researchers over a period of 80 or more years and, because of the relatively large amount of information that has accumulated, it is beyond the scope of this textbook to summarise all that is known. We can, however, point out a few relevant facts.

The most widely used solution for the shallow footing is the familiar bearing capacity equation of Terzaghi. According to his analysis, the limiting value of p is given by the sum of three terms

$$p = cN_c + p_0 N_q + \rho g(B/2)N_\gamma \tag{5.32}$$

Here N_c, N_q and N_γ are dimensionless functions of the angle of internal friction ϕ

$$N_q = N\, e^{\pi \tan\phi}, \quad N_c = (N_q - 1)\cot\phi, \quad N_\gamma = 1.8(N_q - 1)\tan\phi \tag{5.33}$$

where $N = (1 + \sin\phi)/(1 - \sin\phi)$. These so-called *bearing capacity coefficients* N_c, N_q and N_γ will be discussed in more detail in Chapter 6. At this point we wish only to draw out enough of the detail underlying (5.32) to provide comparisons with our limit theorem results. Terzaghi's analysis was based on a collapse mechanism similar to that shown in Figure 5.23. In the figure we see a system of shear surfaces not too dissimilar to the mechanisms we have used in our upper bound analysis. The triangular regions aba' and acd can be assumed to translate as rigid bodies, while the region abc deforms in such a way that the requirement for kinematic admissibility is satisfied. Region abc must maintain contact both along planes ab and ac and on the curved surface bc. The angles marked α and β on the figure have values $(45° + \varphi/2)$ and $(45° - \varphi/2)$, respectively. We will meet the collapse mechanism shown in Figure 5.23 again in Chapter 6 where it will be studied in considerable detail.

The region abc is different from anything we have considered thus far. It is called a region of *radial shearing*. In such a region, energy is dissipated both throughout the region, as well as along the boundary bc. One might expect that since the region abc rotates, the associated shear surface would be circular, but this is not the case. Dilatancy requires that the moving block must not only

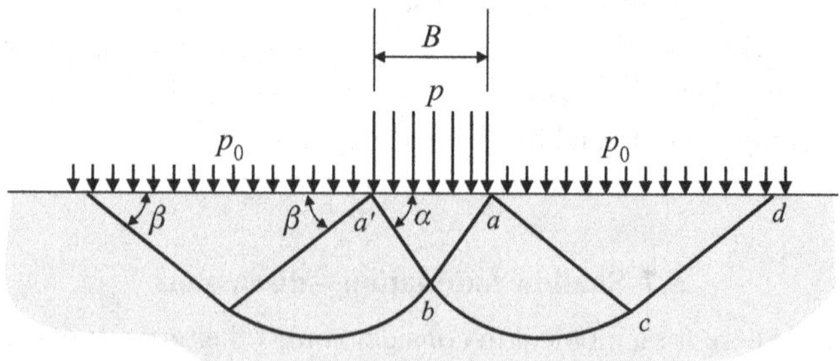

Figure 5.23. Terzaghi analysis for shallow strip footing (see the detailed discussion in Chapter 6).

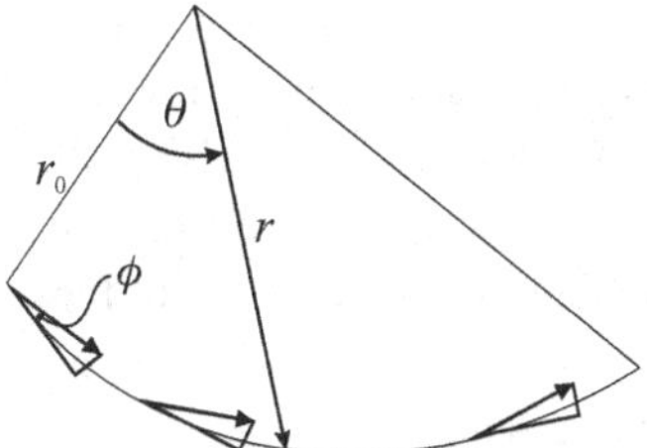

Figure 5.24. Construction of a logarithmic spiral to define the zone of radial shearing in Figure 5.23.

rotate but also shift away from the underlying soil. A soil particle initially on the shear surface cannot move tangentially to the surface, but must instead move at an angle ϕ to the tangent as shown in Figure 5.24. Clearly a circular surface will not work since compatibility will be violated. The shape of surface that will preserve compatibility is called a logarithmic spiral. It has the form

$$r = r_0 e^{\theta \tan \varphi} \tag{5.34}$$

where r and θ are, respectively, the radius and angle shown in the figure. In the upper bound analyses we have done so far, only translational motions have been considered. Rotational motions can also be considered, but, if $\phi \neq 0$, the associated slip surfaces must be logarithmic spirals.

If we compare (5.32) with (5.31) or (5.30), we see that that all three equations have a similar form, each with three terms proportional to c, p_0 and $\rho g B$. The three terms in Terzaghi's equation have roles similar to the terms in (5.31) and (5.30). The term proportional to c arises due to the strength of the soil. If the soil possessed no cohesion, there could be no internal dissipation and the c-term would vanish from (5.31) and (5.30). The term proportional to p_0 results solely from the surcharge due to the depth of burial. The term proportional to $\rho g B$ occurs because of the effect of gravity acting on the blocks of soil within the collapse mechanism. It happens that the coefficients N_c and N_q are exact results for the special case of a weightless soil, while N_γ is an approximation introduced to account for the effects of gravity. The form shown for N_γ in (5.33) was suggested by J.B. Hansen in 1961. Note that each of these contributions to the limit load depends upon ϕ.

Using (5.33) we can evaluate the bearing capacity coefficients for any value of ϕ. If we set $\phi = 20^\circ$, we find (5.32) becomes

$$p = 14.84\, c + 6.40\, p_0 + 1.77 \rho g B \tag{5.35}$$

This can be regarded as a reasonably accurate result. There are a great many other solutions for the problem that consider different aspects, such as whether the footing is rigid or flexible, or whether the base of the footing is smooth or rough, but the resulting collapse loads are all roughly similar to (5.35). We can use (5.35) as a basis for comparison with our lower and upper bound estimates.

First, consider our lower bound result. Using the multiple stress fields of Figure 5.17 we arrived at a lower bound estimate of $p = 12.36c + 5.50\,p_0$. Comparison with (5.35) suggests our coefficient for c is 17% too small, while that for p_0 is 14% too small. Of course there is no term proportional to the footing width B. It seems reasonable to conclude that the lower bound analysis gives a good result except for the inability to account for the effect of gravity on the soil, and hence the footing width B. In contrast, our upper bound analyses do include the effect of footing width, but provide somewhat less accurate estimates for the collapse load. Our best upper bound result used the collapse mechanism sketched in Figure 5.22 and gave the collapse load shown in (5.31). The three coefficients in (5.31) are too large by factors of 21, 18 and 133%, respectively. Of the three coefficients only the third one appears to be grossly in error. This third coefficient N_γ is by far the least well understood of the three. The reason for this uncertainty will become clear in the next chapter when we attempt to incorporate the effects of gravity within an exact solution of the two-dimensional plasticity equations.

The main point to be made here does not really concern the lower and upper bound analyses we have carried out thus far. Any geotechnical engineer confronted with the footing problem in Figure 5.10 would immediately turn to Terzaghi's, or another similar, solution. There would be no reason to carry out an analysis based on the collapse load theorems since other accurate solutions already exist. The collapse load theorems are of limited use in regard to the analysis of such a straightforward problem. Lower and upper bound analyses come into their own, however, when problems of a more specialised nature arise. Some examples are illustrated in Figure 5.25. Problems such as these

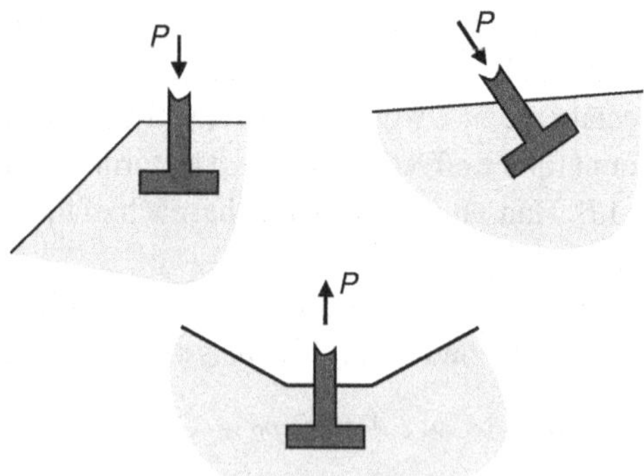

Figure 5.25. Foundation problems where handbook solutions will not be useful.

will rarely have ready-made solutions and the collapse load theorems provide an immediately useful approach for the geotechnical specialist to follow. The procedures can readily yield results that will be enhanced by the ability of the practitioner to identify plausible load carrying paths within the soil mass and plausible collapse mechanisms.

5.8 Retaining walls

When in 1773 Coulomb published his first memoir, he was concerned with the strength of the common building materials of his day: masonry, timber and soil. It was natural that he should focus attention on retaining walls as, at that time, they constituted a major unsolved problem. In his first attempt he provided the solution we continue to use today. Many things have changed including the way we set out the problem and the notation we use, but Coulomb's original analysis remains basically unchanged. Although he would have been unaware of all the modern plasticity theory discussed above, his solution was based on an intuitive understanding of the principles associated with the upper bound theorem.

Coulomb realised that the retaining wall problem had two natural collapse loads depending on how the wall itself might move. If the wall is free to move slightly away from the backfill, the resulting force on the wall will be minimised. Conversely, if the wall moves toward the backfill the thrust will be maximised. Today we refer to these two conditions as *active* and *passive states* respectively. All students of geomechanics will be familiar with the connotation of active and passive in the sense that the active loads represent the weight of the backfill and associated surface loads actively forcing the wall to move whereas the passive forces are mobilised in resisting the movement of the wall towards the soil.

The retaining wall problem lends itself naturally to upper bound methods. We will use the upper bound theorem to examine the problem sketched in Figure 5.26. A retaining wall with vertical back of height H supports a horizontal backfill with cohesion c and friction angle ϕ. Like Coulomb, we will assume a collapse mechanism based on a failure plane making an undetermined angle β measured from the back of the wall. This isolates a failure wedge behind

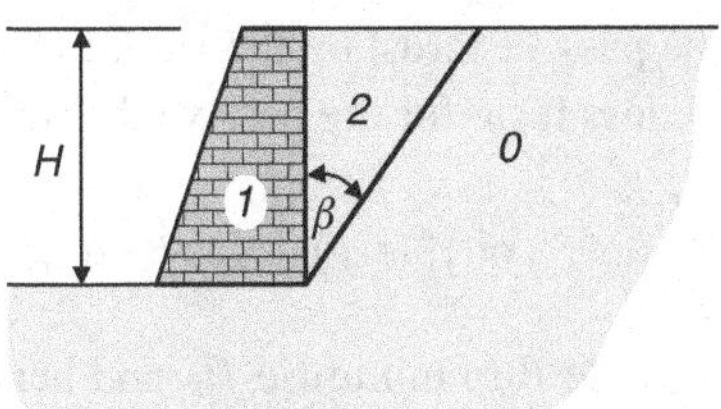

Figure 5.26. Collapse mechanism for a retaining wall – upper bound solution.

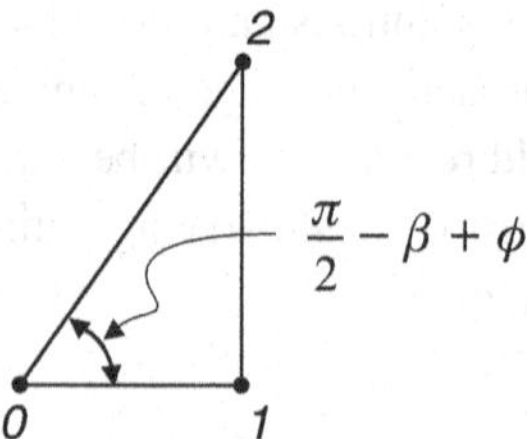

Figure 5.27. Velocity hodograph for Figure 5.26 – passive case.

the wall. We will identify the stationary soil mass outside the failure wedge as region 0, the wall itself will be region 1 and the failure wedge region 2.

To begin, consider the passive case where the wall moves to the right in Figure 5.26 and we initially assume the back of the wall is smooth so that, although there is relative motion between the wall and the failure wedge, no dissipation occurs there. The hodograph is as shown in Figure 5.27. The wall velocity relative to the stationary mass is v_{01}, the other velocities have magnitudes

$$v_{02} = v_{01} \csc(\beta - \phi), \quad v_{12} = v_{01} \cot(\beta - \phi) \tag{5.36}$$

From Figure 5.26 we see that the length of the slip surface is

$$L = H \sec \beta \tag{5.37}$$

The weight of the failure wedge is then found to be

$$W = \frac{\rho g H^2}{2} \tan(\beta) \tag{5.38}$$

The rate of dissipation is

$$\mathbb{D} = cLv_{02} \cos \phi = cHv_{01} \sec \beta \csc(\beta - \phi) \cos \phi \tag{5.39}$$

The power of the external forces is

$$\mathbb{R} = P_P v_{01} - W v_{01} \cot(\beta - \phi) \tag{5.40}$$

where P_P represents the passive thrust on the wall. Equating $\mathbb{R}$ and $\mathbb{D}$ we find the following dimensionless form for the passive thrust:

$$\frac{P_P}{cH} = \frac{\rho g H}{2c} \tan \beta \cot(\beta - \phi) + \sec \beta \csc(\beta - \phi) \cos \phi$$

Now we can adjust the angle β to minimise P_P and hence find the best upper bound. When we do this we find that the critical value for β is independent of H and c. Its value is $(\pi/4 + \phi/2)$. This is, of course, the same as Coulomb's

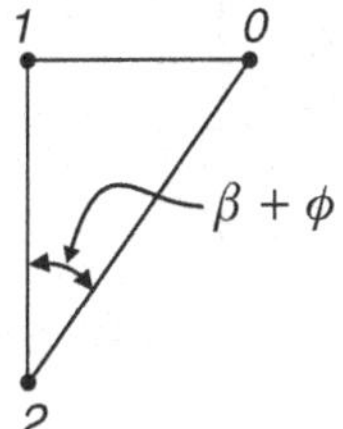

Figure 5.28. Velocity hodograph for Figure 5.26 – active case.

result. If we set $\varphi = 20°$, the upper bound estimate becomes

$$\frac{P_P}{cH} = 2.04\frac{\rho g H}{2c} + 2.856 \tag{5.41}$$

For the active case the orientations of the relative velocities change. The wall moves to the left in Figure 5.26 and the failure wedge moves down and to the left. The hodograph is shown in Figure 5.28. The relative velocity magnitudes are now given by

$$v_{02} = v_{01} \csc(\beta + \phi), \quad v_{12} = v_{01} \cot(\beta + \phi) \tag{5.42}$$

Note that for the active condition, the motion of the wall is away from the backfill, opposite to the direction of the wall force P_A. Therefore the wall force does negative work. The rate of work of the external forces now becomes

$$\mathbb{R} = -P_A v_{01} + W v_{01} \cot(\beta + \phi) \tag{5.43}$$

while the dissipation rate is

$$\mathbb{D} = cHv_{01} \sec\beta \csc(\beta + \phi) \cos\phi \tag{5.44}$$

Solving for P_A we find

$$\frac{P_A}{cH} = \frac{\rho g H}{2c} \tan\beta \cot(\beta + \phi) - \sec\beta \csc(\beta + \phi) \cos\phi \tag{5.45}$$

Unlike the passive case we now must maximise P_A. The upper bound is less than the true collapse load in the active condition. When we do maximise P_A we find the critical value of β is $(\pi/4 - \phi/2)$, just as Coulomb did. The upper bound for the active case with $\varphi = 20°$ is

$$\frac{P_A}{cH} = 0.49\frac{\rho g H}{2c} - 1.40$$

When Coulomb carried out his calculations he did not use an energy balance equation as we have done here. Instead he examined the forces that acted on the

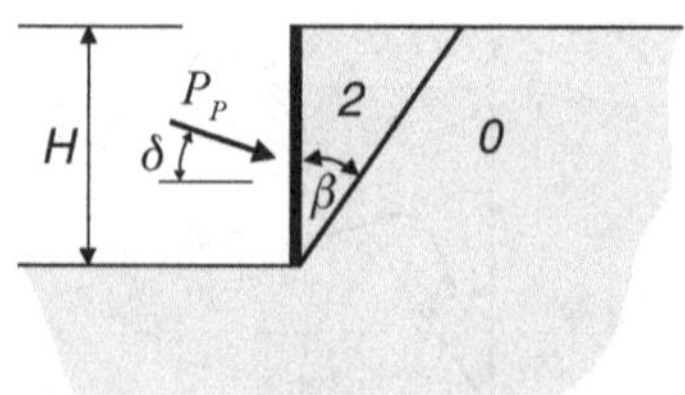

Figure 5.29. Effect of wall friction on retaining wall problem.

failure wedge and used equilibrium to establish a relationship between the thrust (either active or passive), the reaction force on the slip surface and the gravity force. He then minimised or maximised the thrust by varying the geometry of the slip surface. This method is equivalent to the upper bound energy balance approach as is shown in Appendix I.

There are many variations to this problem. A useful example is the situation where the back surface of the wall is rough. The roughness is quantified by an angle called the angle of wall friction, usually denoted by δ. For this case the thrust will no longer be horizontal. Figure 5.29 illustrates the passive problem. In the figure the wall is represented by the heavy vertical line. We now have the condition where horizontal translation of the wall produces the hodograph in Figure 5.27, but there is an extra source of internal dissipation due to slip on the back face of the wall. This introduces a term proportional to the vertical component of the thrust, $P_P \sin\delta$, into the dissipation rate.

$$\mathbb{D} = cLv_{02}\cos\phi + P_P v_{12}\sin\delta \tag{5.46}$$

The external work is the same as (5.40) except that we must use the horizontal component of the wall thrust, $P_P \cos\delta$.

$$\mathbb{R} = P_P v_{01}\cos\delta - W\,v_{01}\cot(\beta - \phi) \tag{5.47}$$

Energy balance then gives

$$\frac{P_P}{cH} = \frac{\dfrac{\rho g H}{2c}\tan\beta\cot(\beta-\phi) + \sec\beta\csc(\beta-\phi)\cos\phi}{\cos\delta - \sin\delta\cot(\beta-\phi)} \tag{5.48}$$

Now, when we minimise P_P, we find that the result depends on δ. For example, if $\phi = 20°$ and $\delta = 15°$, the critical value of β is roughly 65°, which is considerably greater than the case of the smooth wall where $\beta = \pi/4 + \phi/2 = 55°$. The revised upper bound, with $\phi = 20°$, is

$$\frac{P_P}{cH} = 3.03\frac{\rho g H}{2c} + 4.45$$

Comparison with our result for the smooth wall (5.41) indicates that wall friction may have a significant effect on the calculated thrust.

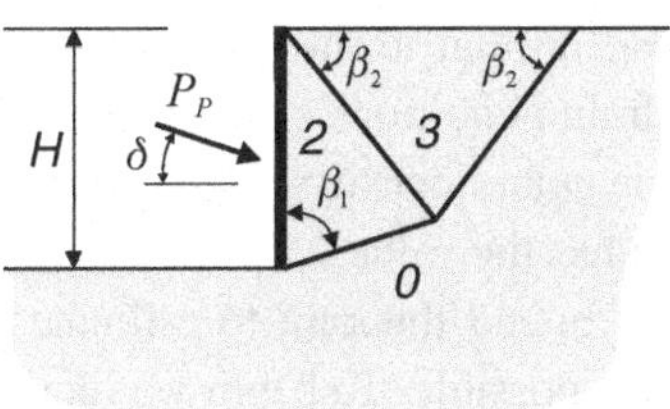

Figure 5.30. A more complex collapse mechanism for the retaining wall problem.

Of course, it is also possible to use more complex collapse mechanisms. For example, instead of the simple failure wedge in Figure 5.29, we could investigate the mechanism shown in Figure 5.30. We now have two triangular regions that depend upon the two angles β_1 and β_2. This problem becomes more difficult since both β_1 and β_2 must be varied simultaneously to minimise P_P. Fortunately, there are a number of multivariate optimisation packages in many of the symbolic manipulation computer codes and they can offer help with a problem such as this.

Still greater complexity arises in cases where the backfill surface is no longer horizontal or where the back face of the wall is not vertical. Solutions have been tabulated for many of these situations in several textbooks. For example, see the book by W.-F. Chen cited at the conclusion of this chapter.

5.9 Arching

Next, we will briefly deal with the flow of a Coulomb material through a chute or hopper of restricted size. If the opening through which the flow passes is too small, the flow may clog, forming a natural arch such as that shown in Figure 5.31. In the figure the parallel sides of the channel are vertical and the

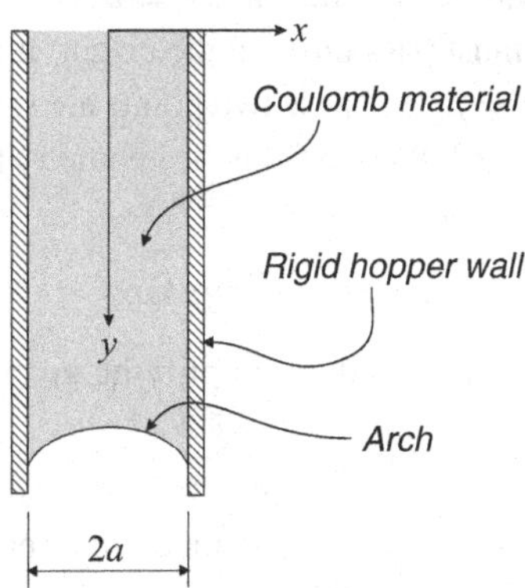

Figure 5.31. Arching of granular media in a narrow hopper.

dimension of the channel normal to the plane of the drawing is assumed to be large so that plane strain conditions exist. The material in the channel has cohesion c, friction angle ϕ and density ρ. The opening dimension is $2a$ and the angle of wall friction has the value δ.

We will use the lower bound theorem to estimate the greatest value of a for which a natural arch is possible. To begin we require a suitable stress field for the material in the channel. For the geometry shown in Figure 5.31, the equilibrium equations are

$$\frac{\partial \sigma_{xx}}{\partial x} + \frac{\partial \sigma_{xy}}{\partial y} = 0$$

$$\frac{\partial \sigma_{xy}}{\partial x} + \frac{\partial \sigma_{yy}}{\partial y} = \rho g$$

Many measurements made in channels and hoppers suggest that the horizontal component of stress σ_{xx} is approximately constant everywhere in the channel. Thus we set $\sigma_{xx} = p_0 =$ constant. Equilibrium for the x-direction then shows that σ_{xy} must be independent of y. Equilibrium for the y-direction is a bit more problematical. We cannot simply let σ_{yy} be proportional to $\rho g y$ as is the case in a natural soil deposit. If we did so, the traction-free surface at the arch would not be possible. Instead we assume σ_{yy} to be independent of y. Then equilibrium shows that $\sigma_{xy} = \rho g x$. In this way the weight of material in the channel is supported by shear tractions at the channel walls.

The greatest value of $|\sigma_{xy}|$ occurs at $x = \pm a$. That value cannot exceed the available frictional resistance $p_0 \tan\delta$. Therefore in the limiting condition we must have

$$p_0 = \frac{\rho g a}{\tan\delta} \tag{5.49}$$

Now consider the Mohr circle for the stress state at the point on the arch where it comes into contact with the wall, $x = a$. Since the arch is a traction-free surface, the circle must pass through the origin. It must also pass through the point $\sigma_{xx} = p_0$, $\sigma_{xy} = \rho g a$. These two points are sufficient to construct the circle, shown in Figure 5.32. We see from the geometry that the major principal stress is

$$\sigma_1 = p_0 + \rho g a \tan\delta \tag{5.50}$$

Using (5.49) to eliminate p_0 and then simplifying gives

$$\sigma_1 = \frac{\rho g a}{\sin\delta\cos\delta} \tag{5.51}$$

Finally, if we apply the lower bound theorem, we must ensure that the stress field is 'below yield'. The limiting condition corresponds to $\sigma_1 = 2c\sqrt{N}$. Therefore

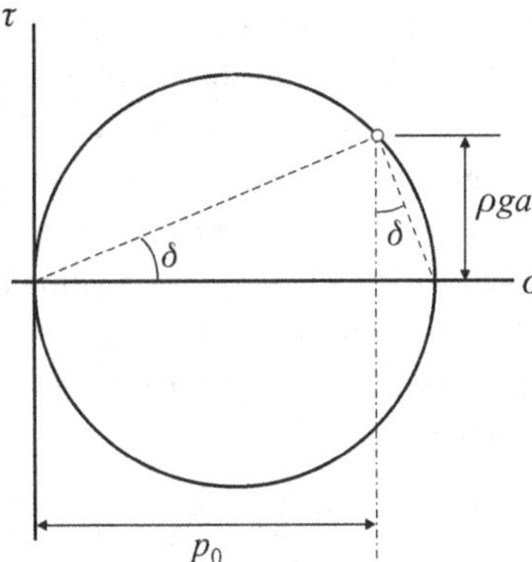

Figure 5.32. Mohr diagram for the arching problem – lower bound solution.

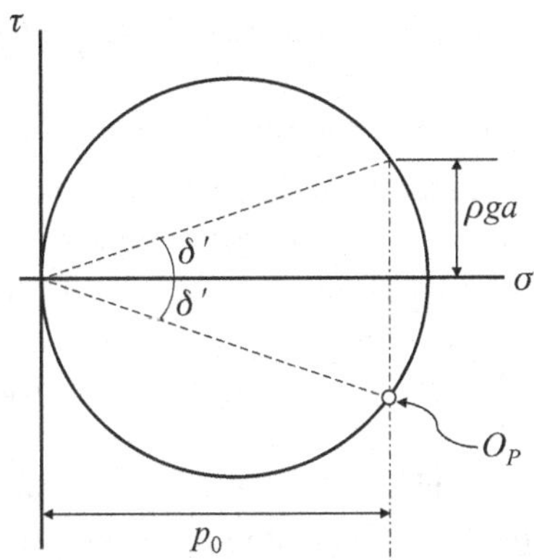

Figure 5.33. Mohr diagram for the arching problem – analysis of arch geometry.

the limiting dimension for the channel is found from (5.51) as

$$a_L = \frac{2\,c}{\rho g}\sqrt{N}\sin\delta\cos\delta \tag{5.52}$$

Note that if $\delta = 0$, then $a_L = 0$ and no arch can form when the walls are perfectly smooth. At the other extreme, the roughest wall would have $\delta = \phi$ and we would then find that a_L is equal to $(2\,c/\rho\,g)(1 + \sin\phi)\sin\phi$.

Even though our stress field is only an approximation to the true stress field, we can nevertheless use it to estimate the shape of the arch. Suppose the channel dimension a is less than a_L and an arch has formed. For a point on the hopper wall at $x = a$, let δ' represent the *mobilised* wall friction. That is, $\tan\delta'$ represents the ratio $\sigma_{xy}/\sigma_{xx} = \rho g a/p_0$. Naturally, $\delta' \leq \delta$. The Mohr circle for the point $x = a$ is now shown in Figure 5.33. Note the location of the pole on the vertical line through the stress point $(p_0, \rho g a)$. The surface supporting zero stress (the arch) will lie parallel to the line from O_P to the origin and we see that the slope of this line is $\tan \delta'$. Thus the arch intersects the wall at an angle equal to the mobilised friction angle. Now consider a point on the arch with $x < a$. The Mohr circle for this point will have a smaller diameter than that shown in Figure 5.33 since the shear stress σ_{xy} is reduced while $\sigma_{xx} = p_0$ remains unchanged. As x ranges between a and zero, the stress σ_{xy} changes from $\rho g a$ to zero and the slope of the arch changes from $\tan \delta'$ to zero. From the Mohr circle we see that the arch slope is given by

$$\frac{dy}{dx} = \frac{\rho g x}{p_0} = \frac{x}{a} \tan \delta' \tag{5.53}$$

Integrating this shows that the arch takes the shape of a parabola. Observations of actual arches in narrow channels suggest that a parabolic shape is realistic.

5.10 Non-associated flow and the upper bound theorem

Throughout this chapter there has been a peculiar phenomenon we have not yet discussed. Equation (5.6) suggests that whatever the amount of dissipation that may occur within the shearing surface, it is always directly proportional to the cohesion c. An alert student might then ask: what if c is zero? Will there be no internal dissipation? These are good questions. Obviously the soil we are concerned with might be a sand or gravel, in which case there would be no cohesion. Equation (5.6) evidently implies that there can be no dissipation, but that conclusion runs counter to our intuition. The problem lies in the associated flow rule.

Recall how the dissipation rate per unit length of slip surface D was derived. In Figure 5.4 we noted how the normality condition related the components of the plastic strain rate vector. This led to a relationship between the slip velocity components: $v_n = -v_t \tan \varphi$. Effectively, the plastic strain rate components in Figure 5.4 could be replaced by the corresponding slip velocity components and the slip velocity vector would obey a normality condition. Then calculating the dissipation rate D in (5.4) and (5.5) we find that there is positive dissipation caused by the tangential slip, but negative dissipation caused by the normal or dilatant displacement. The rate of work done by the normal stress in causing dilatancy turns out to be exactly the negative of the rate of energy dissipated by the shear stress causing tangential slip,

$$\sigma v_n = -\sigma v_t \tan \varphi = -\tau v_t \tag{5.54}$$

So, in terms of energy balance, the frictional strength of the soil can produce dissipation, but the amount is exactly counterbalanced by the work done by the normal stress against dilatancy. This leaves only cohesion as a source of strength that may result in positive dissipation, as (5.6) makes clear. If we wished to have positive frictional dissipation, then it appears there may be too much dilatancy taking place.

This does not mean that the upper bound theorem is of no value for cohesionless soils. We can and do use the theorem with c set equal to zero and, despite the fact that there is no internal dissipation, useful results are still obtained. However, we are well aware that the associated flow rule produces too much dilatancy when compared with test results for real soils, and it is clear from the above discussion that the lack of dissipation in cohesionless soils also results from too much dilatancy. One is led to the conclusion that the associated flow rule may not be the best model for plastic deformations.

Unfortunately, non-associated flow rules do not permit us to prove certain important theorems, particularly the uniqueness theorem outlined in Appendix G. Without the assurance of a unique solution, the true collapse load itself is no longer unique. Apparently the only useful result that can be obtained is that a non-associative material can be no stronger than an associative one.* Nevertheless, in one particular case we can employ a non-associated response. That is the case of plane translational collapse mechanisms where blocks of rigid soil are separated by planar slip surfaces, exactly as we have employed throughout this chapter with regard to the upper bound theorem.

To begin, we might hypothesise that the slip velocity components on some particular slip surface are related by

$$v_n = -v_t \tan \psi \tag{5.55}$$

where ψ is the angle of dilatancy discussed in Chapter 4. We anticipate that ψ will be smaller than φ so that smaller amounts of dilatancy will occur. We can also assume that the Coulomb failure condition is satisfied on the surface so that (5.1) applies. There are implications of this assumption that will be discussed further below. Using (5.55) the rate of dissipation D given in (5.4) now leads to

$$D = v_t[c + \sigma(\tan \varphi - \tan \psi)] \tag{5.56}$$

If ψ and φ are equal we re-obtain (5.6), but if they are different we see that dissipation can occur even when c is zero.

* This follows from the observation that, at collapse, the actual stress field in a non-associative soil will be statically admissible. Therefore, by the lower bound theorem, the collapse load for a non-associative material cannot exceed that for a corresponding material with the associated flow rule.

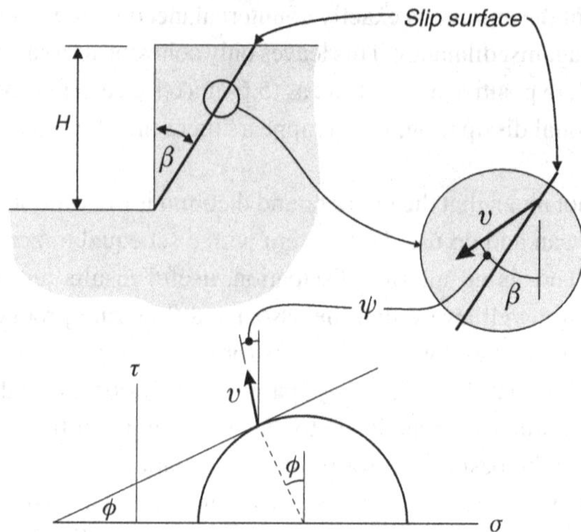

Figure 5.34. Upper bound analysis for the vertical cut problem – effect of non-associativity.

Apparently non-associativity and its result in (5.56) combine to overcome the problem of dissipation in cohesionless soils, but a new problem has now arisen. We cannot apply (5.56) unless we know the normal stress σ. Fortunately, in the case of translational collapse mechanisms, this is not a major difficulty. For any statically determinate failure mechanism we can find the normal stresses from the conditions of equilibrium. This is best illustrated by an example.

Consider once again the problem of the vertical cut illustrated in Figure 5.5. Suppose the soil has properties ρ, c and φ, and obeys a non-associated flow rule with angle of dilatancy ψ. We assume a planar failure mechanism, but we note that the slip velocity will be oriented at an angle ψ to the slip surface as shown in Figure 5.34. The Mohr diagram is also shown in this figure to emphasise that the slip velocity vector is no longer normal to the failure surface. The rate of working of the external forces is given by (cf. (5.10))

$$\mathbb{R} = \frac{v}{2}\rho g H^2 \tan\beta \cos(\psi + \beta) \tag{5.57}$$

The internal dissipation rate per unit length of slip surface is given now by (5.56) with $v_t = v\cos\psi$. Integrating over the slip surface length $H/\cos\beta$ gives the total dissipation rate $\mathbb{D}$

$$\mathbb{D} = \frac{cHv\cos\psi}{\cos\beta} + v\cos\psi\left[\int_0^{H/\cos\beta} \sigma(\tan\varphi - \tan\psi)\,d\ell\right] \tag{5.58}$$

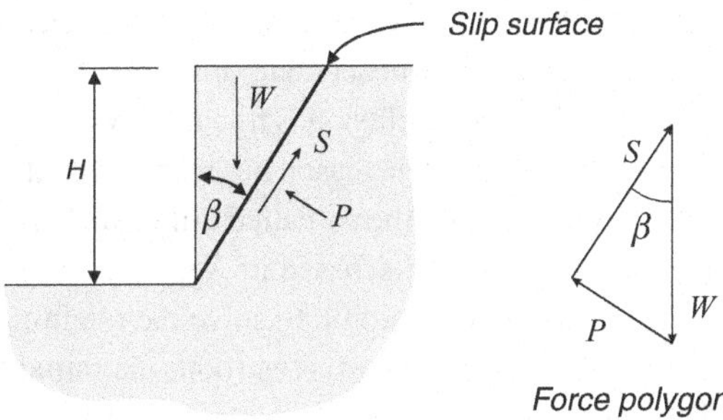

Figure 5.35. Using force equilibrium to determine normal stress on a slip surface.

Now the question arises, what is the value of σ? Equilibrium of forces on the triangular soil wedge is easily determined from the force polygon sketched in Figure 5.35. We see that the total normal force on the wedge is given by

$$P = W \sin\beta = \frac{1}{2}\rho g H^2 \tan\beta \sin\beta \tag{5.59}$$

This force is exactly equal to the stress σ integrated over the length of the slip surface, hence

$$\int_0^{H/\cos\beta} \sigma \, d\ell = \frac{1}{2}\rho g H^2 \tan\beta \sin\beta \tag{5.60}$$

and (5.58) becomes

$$\mathbb{D} = v \cos\psi \left[c \frac{H}{\cos\beta} + \frac{1}{2}\rho g H^2 \tan\beta \sin\beta(\tan\varphi - \tan\psi) \right] \tag{5.61}$$

Finally, we set $\mathbb{R} = \mathbb{D}$ and solve for H giving (cf. (5.12))

$$H = \frac{2c}{\rho g} \frac{\cos\psi}{\sin\beta\cos(\beta+\psi) - \cos\psi\sin^2\beta(\tan\varphi - \tan\psi)} \tag{5.62}$$

The usual course would now be to minimise H with respect to β in order to determine the best estimate for the upper bound. If, however, we first simplify (5.62) by expanding the $\cos(\beta+\psi)$ term, a surprising result is found. All terms involving ψ vanish! We are left with

$$H = \frac{2c}{\rho g} \frac{1}{\sin\beta(\cos\beta - \sin\beta\tan\varphi)} \tag{5.63}$$

Evidently the value of ψ does not affect the upper bound solution. Moreover, it is easy to show that (5.63) is identical to (5.12). Thus the upper bound for the non-associative case is exactly the same as that for the associative case.

The conclusion that the associative and non-associative cases have the same upper bound solution is not simply an accident or solely confined to the simple vertical cut problem. In fact, the conclusion is true for any statically determinate translational collapse mechanism. The reason lies in the equivalence of the upper bound, energy balance method with the so-called limit equilibrium method. The equivalence of the two methods is discussed in Appendix I. Limit equilibrium is the method originally used by Coulomb to solve the retaining wall problem. One uses equilibrium to determine the forces (both magnitude and direction) acting on the collapse mechanism within the soil mass. Usually a force polygon is constructed. The failure mechanism geometry is then varied to maximise (passive case) or minimise (active case) the forces tending to cause collapse. The results from limit equilibrium will always be the same as those from the upper bound theorem for any translational collapse mechanism. Hence, if the upper bound theorem is applied to a particular translational collapse mechanism, there can be no dependence on the choice of flow rule since the result can be determined solely from equilibrium considerations.

Finally, recall the assumption made following equation (5.55) that the stresses acting on the failure surface obey the Coulomb criterion. This is not a trivial assumption. In fact, if (5.1) and (5.55) both apply, then Saint-Venant's hypothesis cannot be true. That is, the principal directions of stress and plastic strain rate cannot be the same. This may not be a serious problem since, as we noted in Chapter 4, Saint-Venant's hypothesis is a convenience but is not required by any physical law. It is possible to construct a theory for which equation (5.55) and Saint-Venant's hypothesis both apply. It is then found that the stresses on the failure surface no longer obey (5.1). Instead the stress point moves nearer the top of the Mohr circle, to the point where the direction of the slip velocity vector is normal to the circle. In this modified theory the upper bound solution is no longer identical to the limit equilibrium solution. A detailed derivation may be found in a paper by A. Drescher and E. Detournay cited below.

Further reading

Two of the seminal papers by Drucker *et al.*, referred to at the beginning of the chapter, are

D.C. Drucker, W. Prager and H.J. Greenberg, Extended limit design theorems for continuous media, *Quart. Appl. Math.*, **9**, 381–389 (1952).

D.C. Drucker and W. Prager, Soil mechanics and plastic analysis for limit design, *Quart. Appl. Math.*, **10**, 157–165 (1952).

An entire book devoted to the solution of geotechnical problems using limit analysis and containing numerous useful charts, tables and an extensive list of references is

W.-F. Chen, *Limit Analysis and Soil Plasticity*, Elsevier, Amsterdam, 1975.

The use of finite elements in lower bound estimation has been studied by a number of researchers. Two interesting references are

J. Lysmer, Limit analysis of plane problems in soil mechanics, *J. Soil Mech. Fnds. Div. ASCE*, **96** (SM4), 1311–1334 (1970).
S.W. Sloan, Lower bound limit analysis using finite elements and linear programming, *Int. J. Numer. Anal. Methods Geomech.*, **12**, 61–77 (1988).

The bearing capacity coefficients given in (5.33) are derived in a variety of texts. See

K. Terzaghi, *Theoretical Soil Mechanics*, Wiley, New York, 1943.
K. Terzaghi and R.B. Peck, *Soil Mechanics in Engineering Practice*, Wiley, New York, 1948.
J.B. Hansen, A general formula for bearing capacity, *Ingeniøren*, **5**, 38–46 (1961).

Coulomb's memoir is translated and elaborated in some detail in this book:

J. Heyman, *Coulomb's Memoir on Statics – an Essay in the History of Civil Engineering*, Cambridge University Press, Cambridge, 1972.

A discussion of non-associated flow rules in the upper bound theorem may be found in

A. Drescher and E. Detournay, Limit load in translational failure mechanisms for associative and non-associative materials, *Geotechnique*, **43**, 443–456 (1993).

Exercises

5.1 Use the lower bound theorem to graphically estimate the collapse load p acting on a horizontal surface adjacent to the 20° slope shown in Figure 5.36. The soil has strength properties of $c = 10.5$ kPa and $\phi = 35°$. Disregard body forces and use two discontinuous stress fields separated by the dashed line OO'. What inclination of the line OO' is required to produce a statically admissible stress field?

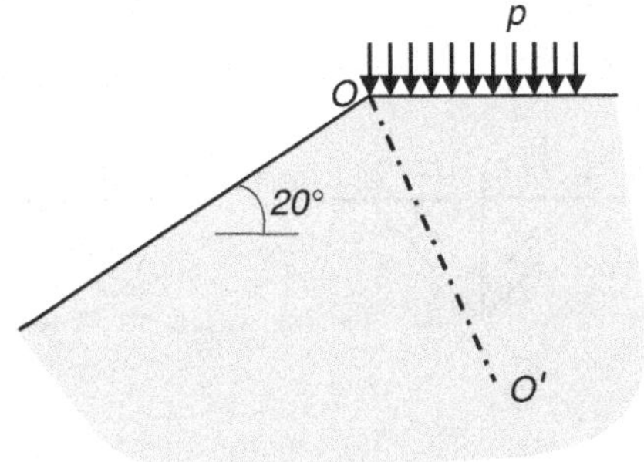

Figure 5.36.

5.2 Recalculate the lower bound from Exercise 5.1 for the case of a 45° slope.

5.3 Use the collapse mechanism sketched in Figure 5.37 to estimate the upper bound for the uniform stress p. The slope is a homogeneous soil with $\rho = 2.0\,\text{t/m}^3$, $c = 10\,\text{kPa}$ and $\phi = 15°$.

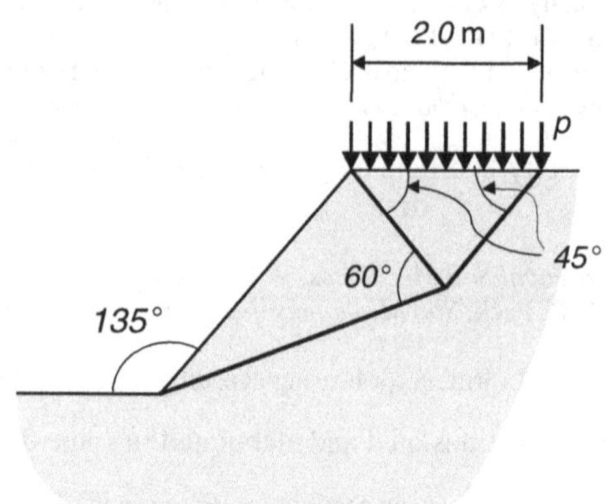

Figure 5.37.

5.4 Recalculate the upper bound in Exercise 5.3 for undrained conditions with $\phi = 0°$.

5.5 Work through the details for the collapse mechanism shown in Figure 5.22 for the strip footing problem and verify the result:

$$p = 18.00\,c + 7.546\,p_0 + 4.181\,\rho g B$$

5.6 Use the upper bound method to compare values of collapse load for the strip footing problem using the two collapse mechanisms shown in Figure 5.38. In both cases the soil has cohesion c and zero internal friction. The footing width is B and it supports a load P. The left-hand mechanism consists of two 45° wedges with planar slip surfaces. The right-hand mechanism is a semi-circular barrel-shaped slip surface with radius $R = B$.

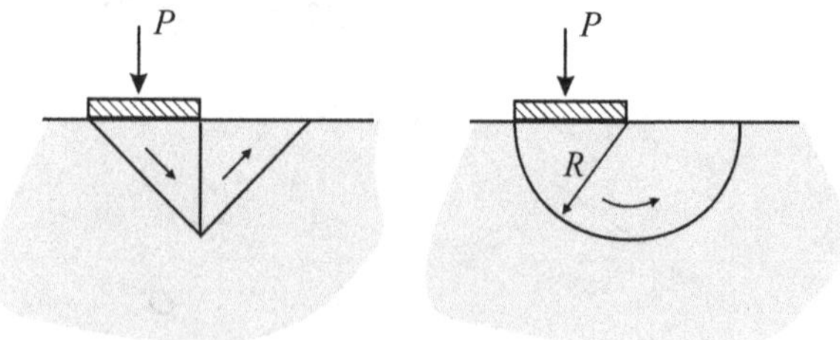

Figure 5.38.

5.7 Consider the sloping retaining wall shown in Figure 5.39. The backfill supports a uniform applied stress p_0. Given that $\phi = 15°$ and $\delta = 10°$, find the upper bound for the passive thrust P_P for the collapse mechanism shown.

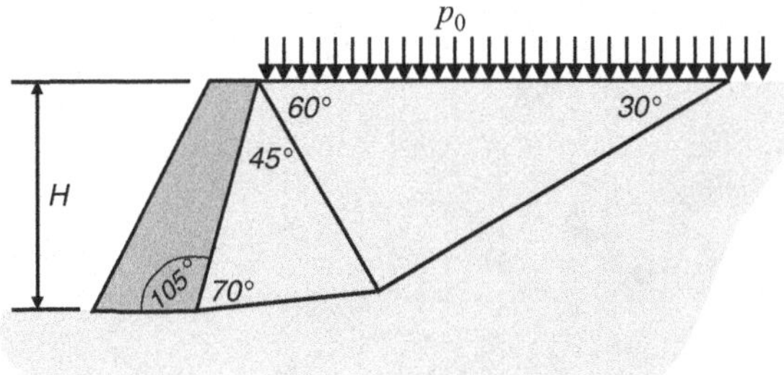

Figure 5.39.

5.8 Consider the step load problem illustrated in Figure 5.13. For the undrained case where $\phi = 0°$, prove that the value of the angle β that maximises the lower bound estimate for p is exactly 67.5°. Then show that p will equal $p_0 + 4.828\,c$ where c is the cohesion.

5.9 Consider a hopper with sloping sides as shown in Figure 5.40. The angle of wall friction is given as δ and the material filling the hopper has cohesion c and angle of friction ϕ. Use the stress field specified by

$$\sigma_{xx} = p_0 = constant, \quad \sigma_{xz} = \rho g x, \quad \sigma_{yy} = \sigma_{yy}(x)$$

to obtain a lower bound estimate for the critical opening dimension a_c in terms of the angle β. What happens if β approaches zero?

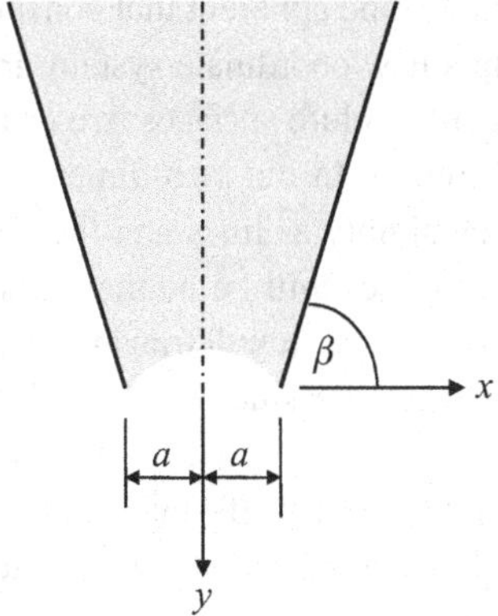

Figure 5.40.

6
Slip line analysis

6.1 Introduction

The simplicity of the collapse load theorems masks some of the more complex aspects of engineering applications involving plasticity. Solutions for fully three-dimensional elastic–plastic response will generally be difficult if not impossible to obtain in closed form. There is, however, one more class of problem for which relatively simple solutions are possible. This is the class of two-dimensional problems concerning plane plastic flow for which regions of the material are in a failure condition. The failure regions need not cover the entire body, but within the failing zone we must be assured that the yield condition is satisfied everywhere.

For these plane problems there will be three unknown components of stress: for example, $\sigma_{xx}, \sigma_{yy}, \sigma_{xy}$, where the (x, y)-plane is taken to be the plane of the problem. Within the failing region the three stresses are related by three equations: two equations of equilibrium plus the equation of the yield surface. Only first-order derivatives are involved. While this in itself does not appear overly complex, it will become apparent that considerably more simplification is possible by invoking a new coordinate system. Introducing coordinates that coincide with the potential failure surfaces, we cause the system of equations to become extremely simple. In our two-dimensional problem, the potential failure surfaces are seen simply as lines and they have come to be called *slip lines*. In general, the slip lines will be neither straight nor orthogonal, so the problem is not overly simple, but it will transpire that the geometry is often not too complex. For those who have studied partial differential equations, the slip lines are actually characteristic lines and we will be constructing characteristic solutions for this class of problem. If you are not familiar with the theory of characteristics and wish to learn more, the works cited at the end of the chapter may be useful.

Slip line analysis is applicable to a range of material models, but by far the easiest model to use to introduce the method is a Coulomb material with zero friction angle. We begin the chapter with this simple material model and investigate the modifications required for non-zero frictional strength later. Emphasis will be placed on the determination of the two-dimensional stress field that results when the yield criterion is satisfied throughout some region of the material. Calculation of the associated strain rates is considered at the end of the chapter. Slip line analysis is by nature a geometric subject and the reader is encouraged to make sketches to help understand the derivations and problems outlined below.

6.2 Two-dimensional stress states

Throughout this chapter attention is focused on the class of two-dimensional plane strain problems. Let the (x, y)-plane represent the plane of the unknown stress components. The stress matrix in two-dimensional form is written as

$$\boldsymbol{\sigma} = \begin{bmatrix} \sigma_{xx} & \sigma_{xy} \\ \sigma_{xy} & \sigma_{yy} \end{bmatrix} \tag{6.1}$$

and the three components of stress must obey the two-dimensional form of the equations of equilibrium

$$\begin{aligned} \frac{\partial \sigma_{xx}}{\partial x} + \frac{\partial \sigma_{xy}}{\partial y} &= 0 \\ \frac{\partial \sigma_{xy}}{\partial x} + \frac{\partial \sigma_{yy}}{\partial y} &= 0 \end{aligned} \tag{6.2}$$

where we have assumed no body forces are acting. We will consider the effects of gravity later in the chapter. If the material is at a state of failure, the stresses must also obey the two-dimensional form of the failure condition, i.e. an equation of the form $f(\sigma_{xx}, \sigma_{yy}, \sigma_{xy}) = 0$. To begin, consider the simple case of a *purely cohesive* material. That is, the yield function f corresponds to a Coulomb material with zero angle of internal friction (which we realise is equivalent to a Tresca material). In terms of principal stresses, we have

$$\sigma_1 - \sigma_3 = 2c \tag{6.3}$$

where σ_1 and σ_3 represent the principal stresses lying in the (x, y)-plane. The intermediate principal stress σ_2 is normal to the plane. We assume σ_2 will always remain the intermediate principal stress whatever happens to the other stresses. The stress state may be visualised by sketching the Mohr circle as shown in

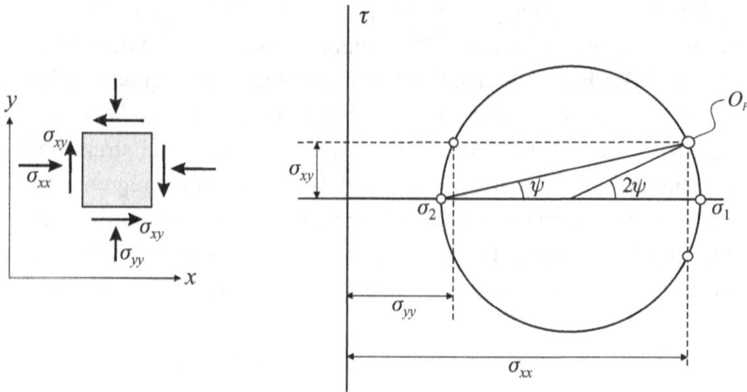

Figure 6.1. Mohr diagram for a two-dimensional stress state.

Figure 6.1. The stress points corresponding to $(\sigma_{xx}, \sigma_{xy})$ and $(\sigma_{yy}, \sigma_{xy})$ are plotted at arbitrary locations on the circle circumference (except for the fact that they lie at each end of a diameter), and the pole of the Mohr circle is also shown. It lies at the intersection of the circle with the horizontal line passing through the point $(\sigma_{yy}, \sigma_{xy})$. Since (6.3) is satisfied, the radius of the circle is the cohesion c. We have sketched the line joining the pole to the smallest principal stress σ_3. This line will lie parallel to the minor principal surface at the point in the material where the stress state applies. Let ψ be the angle between the minor principal surface and the horizontal as shown in the figure.

An interesting point can be made here. Three bits of information are required to construct the Mohr circle in Figure 6.1. Those bits could be σ_{xx}, σ_{yy} and σ_{xy}, which together contain all the relevant information; or we might consider other possible bits. Note that the radius of the circle is a constant c. It might be useful to use c as one piece of information since it does not change from place to place in the failing region. If we take c as one bit, then the centre of the circle would also be a useful bit. Let p represent the two-dimensional mean stress defined as

$$p = \frac{1}{2}(\sigma_{xx} + \sigma_{yy}) = \frac{1}{2}(\sigma_1 + \sigma_3) \tag{6.4}$$

Obviously this is not the same definition as our mean stress p for three-dimensional stress states, but it is consistent in the sense that it represents the average normal stress for our plane two-dimensional stress state, and it identifies the location of the centre of the Mohr circle. Knowing c and p we can draw the circle. The only missing information is how surfaces are oriented with regard to our (x, y)-coordinate frame. We will introduce the angle ψ as the third piece of information to satisfy this final requirement. The three numbers c, p and ψ together contain all the relevant information needed to construct the stress state at the point in question and, importantly, only two of them are variables. If we roam over the failure region in our mind's eye we see that the stress state depends only on two variables: p and ψ. If we stop at any particular point we can immediately determine the stress components from

$$\sigma_{xx} = p + c(\cos 2\psi), \quad \sigma_{yy} = p - c(\cos 2\psi), \quad \sigma_{xy} = c(\sin 2\psi) \tag{6.5}$$

Using these equations in the equilibrium equations (6.2) we have

$$\begin{aligned} \frac{\partial p}{\partial x} - 2c(\sin 2\psi)\frac{\partial \psi}{\partial x} + 2c(\cos 2\psi)\frac{\partial \psi}{\partial y} &= 0 \\ \frac{\partial p}{\partial y} + 2c(\sin 2\psi)\frac{\partial \psi}{\partial y} + 2c(\cos 2\psi)\frac{\partial \psi}{\partial x} &= 0 \end{aligned} \tag{6.6}$$

giving two equations for our two unknown variables p and ψ.

6.3 Slip lines

As they stand, equations (6.6) may still prove difficult to solve for all but simple boundary conditions. Fortunately we do not need to solve them as they stand. Instead we can introduce a new coordinate frame. Let the new coordinates be denoted by α and β and orient their directions so that they align with the potential failure surfaces. Figure 6.2 shows how the α- and β-directions are oriented. The α-direction lies parallel to the line joining the pole to the smallest shear stress. Thus the α-axis or α-line is parallel to one potential failure surface. Similarly the β-line will lie parallel to the other potential failure surface, and

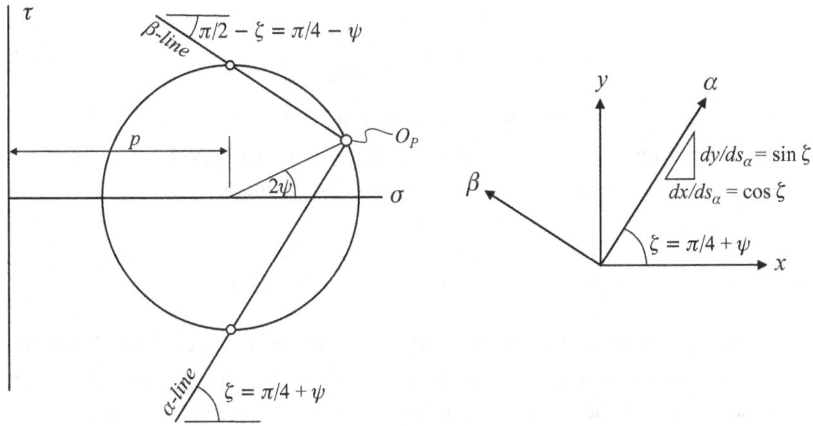

Figure 6.2. Definition of α- and β-lines on the Mohr diagram and their physical directions in the plane.

is therefore parallel to the line joining O_P to the largest shear stress. Note that the α- and β-directions are orthogonal. We will use ζ to represent the angle $(\pi/4+\psi)$ between the α- and x-directions. Finally, note that the new coordinate directions apply *only* at the point of interest in the failure region. If we move to a neighbouring point the α- and β-directions may change. Thus the α- and β-lines are not necessarily straight.

Now we will focus attention on the α direction for the moment. Let s_α denote the distance measured along the α-line and consider the directional derivative of p in this direction,

$$\frac{dp}{ds_\alpha} = \frac{\partial p}{\partial x}\frac{dx}{ds_\alpha} + \frac{\partial p}{\partial y}\frac{dy}{ds_\alpha} = \frac{\partial p}{\partial x}\cos\zeta + \frac{\partial p}{\partial y}\sin\zeta \tag{6.7}$$

We have expressions for $\partial p/\partial x$ and $\partial p/\partial y$ from the equilibrium equations (6.6). If we use those in (6.7) together with the geometric relations $\sin 2\psi = -\cos 2\zeta$ and $\cos 2\psi = \sin 2\zeta$ we find

$$\begin{aligned}\frac{dp}{ds_\alpha} = 2c\bigg[&(-\cos 2\zeta\cos\zeta - \sin 2\zeta\sin\zeta)\frac{\partial\psi}{\partial x}\\ &+(-\sin 2\zeta\cos\zeta + \cos 2\zeta\sin\zeta)\frac{\partial\psi}{\partial y}\bigg]\end{aligned} \tag{6.8}$$

Now using the double angle formulae this expression reduces to

$$\frac{dp}{ds_\alpha} = 2c\left(-\cos\zeta\frac{\partial\psi}{\partial x} - \sin\zeta\frac{\partial\psi}{\partial y}\right) = -2c\left(\frac{\partial\psi}{\partial x}\frac{dx}{ds_\alpha} + \frac{\partial\psi}{\partial y}\frac{dy}{ds_\alpha}\right) \tag{6.9}$$

So, finally, we conclude that

$$\frac{dp}{ds_\alpha} = -2c\frac{d\psi}{ds_\alpha} \tag{6.10}$$

And this will be true everywhere on the α-line.

The surprising thing about equation (6.10) is that ordinary derivatives rather than partial derivatives appear. We can immediately integrate (6.10) to see that on the α-line

$$p = -2c\psi + \text{constant} = -2c\psi + K_1 \tag{6.11}$$

Therefore, everywhere on this line p and ψ are linearly related. If we know the values of p and ψ at any single point we can determine the constant K_1. Then if we know the shape of the line in the neighbourhood of that point, we can easily find both ψ and p. In many problems of interest it will turn out that the shape of the line is easily determined. Using a similar analysis for the β-line we find a result similar to (6.11),

$$p = +2c\psi + K_2 \tag{6.12}$$

Note the difference in sign between this result and (6.11).

6.4 Slip line geometries

We can think of the α- and β-lines forming a network that covers the failing region. Every point in the region is at its limiting state in the sense that, on two particular surfaces, the shear stress magnitude is equal to the cohesion c. Physically the α- and β-lines represent the orientations of those surfaces. Embedded in equations (6.11) and (6.12) are limitations on how the surfaces may be oriented. To see this, consider two α-lines and two β-lines as shown in Figure 6.3. Aside from the fact that the lines are orthogonal we make no restrictions on either pair. The intersections at the four 'corners' of the enclosed region are denoted A, B, C and D. The angle between the α-lines and the horizontal is ζ, which we know to be $(\pi/4 + \psi)$. Let ψ_A be the value of ψ appropriate for the intersection A as shown in Figure 6.4. If p_A denotes the corresponding mean stress, then equation (6.11) asserts that $(p_A + 2c\psi_A)$ is a constant for all points on the α-line. Therefore the values of p at points A and B are related to the

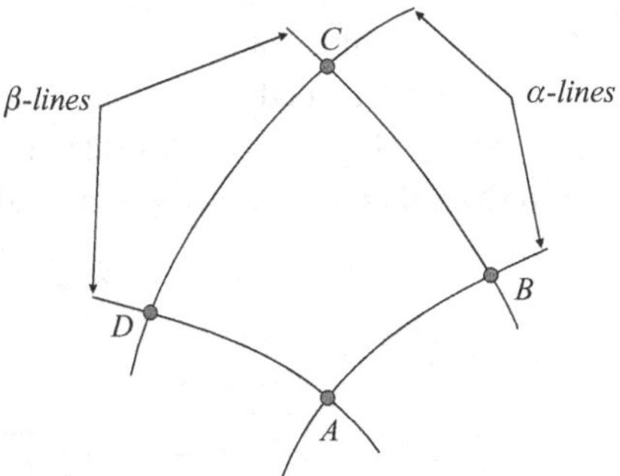

Figure 6.3. Pairs of intersecting α- and β-lines.

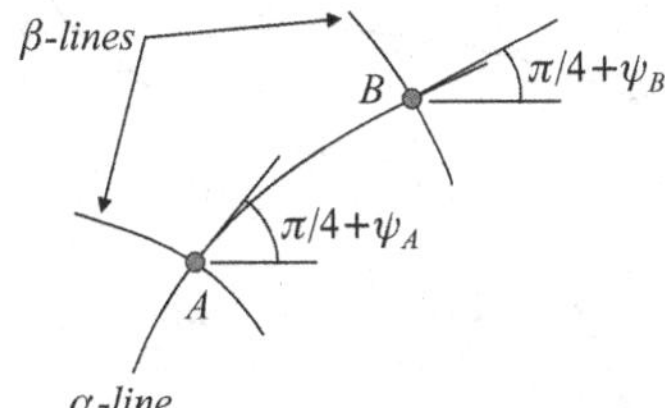

Figure 6.4. Expanded view to the lower α-line from Figure 6.3.

values of ψ at those points according to

$$p_A - p_B = -2c(\psi_A - \psi_B) \tag{6.13}$$

In the same fashion we can follow the β-line from B to C and use (6.12) to show that

$$p_B - p_C = 2c(\psi_B - \psi_C) \tag{6.14}$$

Note the difference in signs between the last two equations. Moving on around the figure we also have

$$p_C - p_D = -2c(\psi_C - \psi_D) \tag{6.15}$$

and

$$p_D - p_A = 2c(\psi_D - \psi_A) \tag{6.16}$$

If we now add the last four equations we find that all the mean stress values cancel and we are left with

$$\psi_D - \psi_A = \psi_C - \psi_B \tag{6.17}$$

Equation (6.17) is called Hencky's first theorem, named after the German mathematician H. Hencky who obtained (6.17) in 1923. The equation asserts that the angle subtended by two α-lines at their intersection with a particular β-line is the same as the angle subtended at their intersection with any other β-line within the failure region. Figure 6.5 illustrates the result. The argument is easily reversed to show that the same result must also apply for β-line

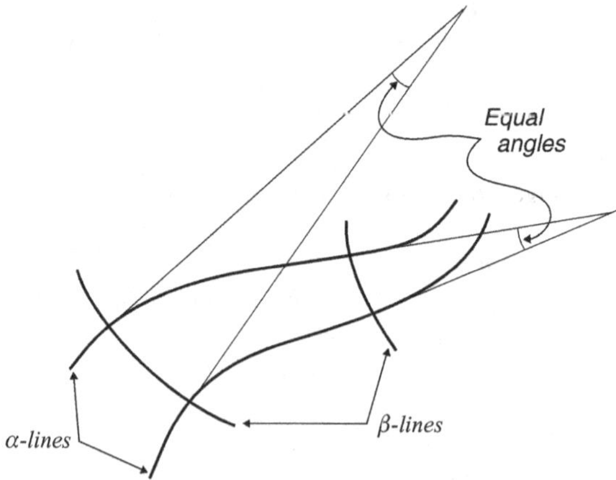

Figure 6.5. Interpretation of Hencky's theorem.

intersections with α-lines. We can also rearrange (6.17) to give

$$\psi_B - \psi_A = \psi_C - \psi_D \tag{6.18}$$

This shows that the change in the orientation of any α-line at its intersection with any two particular β-lines is the same for all α-lines throughout the failure region. In particular, if an α-line is straight between two β-lines, then all α-lines must also be straight between those two β-lines.

We can assist the use of equations (6.11) and (6.12) by considering a new diagram where we plot the mean stress p against the product $2c\psi$ as shown in Figure 6.6. If we consider the two-dimensional stress state $(p_A, 2c\psi_A)$ at some point A in the failure region, then equations (6.11) and (6.12) tell us that *all* the stress states on the α-line that passes through our point must lie on the diagonal $-45°$ line shown in the figure. Similarly, all stress states on the β-line passing through our point all lie on the $+45°$ line in the $(p, 2c\psi)$-plane.

Now consider the situation sketched in Figure 6.7. Suppose we know the values of p and ψ at points 1 and 2 inside the failure region. Let point 3 lie at the intersection of the α-line through point 2 and the β-line through point 1. We can easily determine the values of p and ψ at the new point 3. In the $(p, 2c\psi)$-plane, the stress state at point 3 must lie on the $+45°$ line through $(p_1, 2c\psi_1)$ and on the $-45°$ line through $(p_2, 2c\psi_2)$ as shown in Figure 6.7. Values of p_3 and ψ_3 are uniquely determined by the intersection of the two 45° lines. We

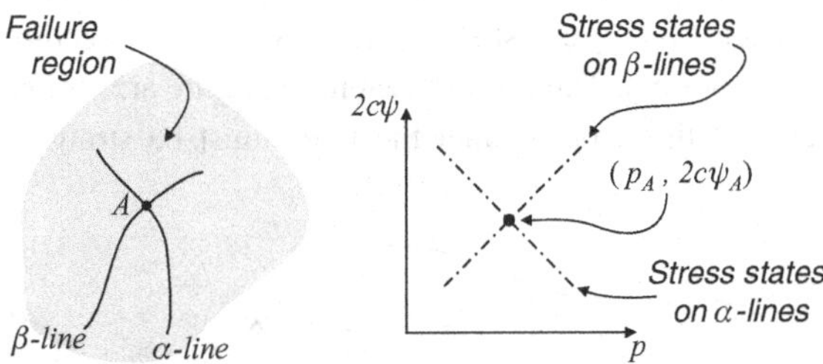

Figure 6.6. Two-dimensional stress state corresponding to a point in the failure region.

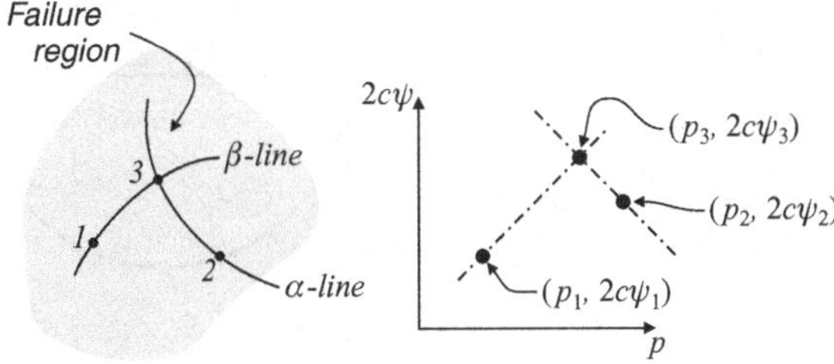

Figure 6.7. 'Marching' the slip line solution forward into the failure region.

can use this analysis to extend the solution throughout a network of α- and β-lines. Often this method is referred to as 'marching' the solution forward. In the theory of partial differential equations the α- and β-lines are referred to as characteristic lines or simply *characteristics*. So-called marching solutions are typically found for equations of the form we are dealing with here. Generally, we begin with a known stress condition on a boundary of the failure region and we march the solution into the region using our knowledge of how the characteristic lines must behave. The entire problem is reduced to correctly constructing the network of α- and β-lines throughout the failure region.

6.5 Some simple problems

We can now begin to consider some simple situations where slip lines might be used profitably. The simplest situation we could imagine would probably be a traction-free surface such as that shown in Figure 6.8. If the surface is the boundary of a failure region for a Coulomb material with $\varphi = 0$, the Mohr stress circle must be as shown in the figure. The circle must pass through the origin because of the zero-traction condition. The pole is located as shown in the figure.

How do we go about extending the given boundary information into the interior of the failure region? The answer is shown in Figure 6.8. We draw α- and β-lines through points 2 and 1, respectively so that they are parallel to the surfaces that support the least and greatest shear stresses. The intersection of the lines identifies our new point 3. The lines shown in the figure are straight. We might have drawn curved lines, but in fact the lines must be straight because the

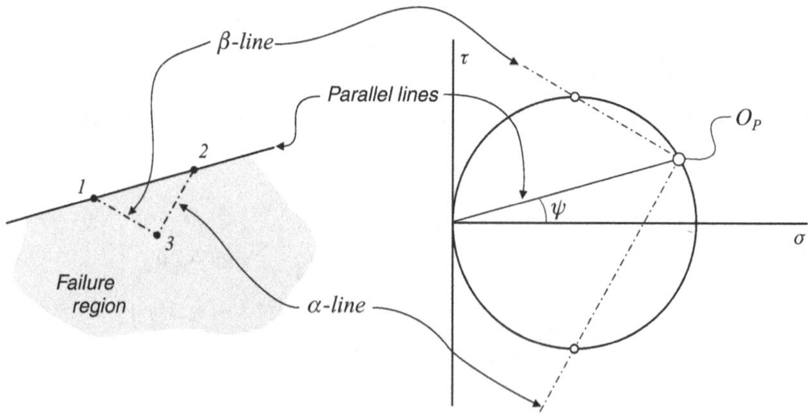

Figure 6.8. Slip line solution for a traction-free planar surface.

two boundary points both support the same stress-free condition. Thus, in the $(p, 2c\psi)$-diagram, both points 1 and 2 lie at the same point, namely at $p = c$ and $2c\psi$ where ψ is the angle of the traction free surface. In this case we see that point 3 must also lie at this same point, so that the entire triangle with vertices 1, 2 and 3 will have the same values of p and $2c\psi$.

It appears that we have uniquely solved for the stresses inside the 1, 2, 3 triangle, but this is not precisely true. There is some ambiguity concerning the Mohr circle in Figure 6.8. We have drawn the circle to represent a *compressive* stress state inside the failing region, but we could equally well have drawn a circle with radius c on the tension side of the diagram. This would produce a solution that satisfies all the conditions of equilibrium and failure as well as satisfying the traction-free boundary, but would be completely different from the solution represented in the figure. There will often be situations like this where two solutions are possible for a given problem. Usually other conditions will be present to constrain which solution is physically possible and which is not.

What if the boundary in Figure 6.8 were not traction-free but instead supported a uniform traction with components σ and τ acting normal and tangential to the surface? We could easily represent this new situation by drawing the Mohr circle with radius c that passes through the point (σ, τ). We then find the pole and determine the angle ψ as well as the orientations of the characteristic lines. Since the boundary supports a uniform traction, the α- and β-lines will be straight. We draw the lines through any two points on the boundary to establish a triangular region of uniform stress. Basically there is no additional difficulty compared with the traction-free boundary.

Once we have established a new point such as point 3 in the interior of the failure region we can use this point to extend the solution further. For the case of a planar boundary supporting a uniform traction we can easily construct a net of α- and β-lines such as that shown in Figure 6.9. All the characteristic lines represent potential failure surfaces. The lines are straight and their intersections are orthogonal.

The key to the simplicity of the solution in Figure 6.9 lies in two elements. First, the boundaries we have considered are planar. If we had a uniform traction acting on a curved boundary we would find different values of ψ at different boundary points. Second, we have only considered uniform traction conditions. Even with a planar boundary, if the specified boundary tractions were varying, then we would need to deal with different Mohr circles for different boundary points. In both cases the simplicity seen in Figure 6.9 is lost. The characteristic lines may now be curved and the stress state inside the failing region will change from place to place.

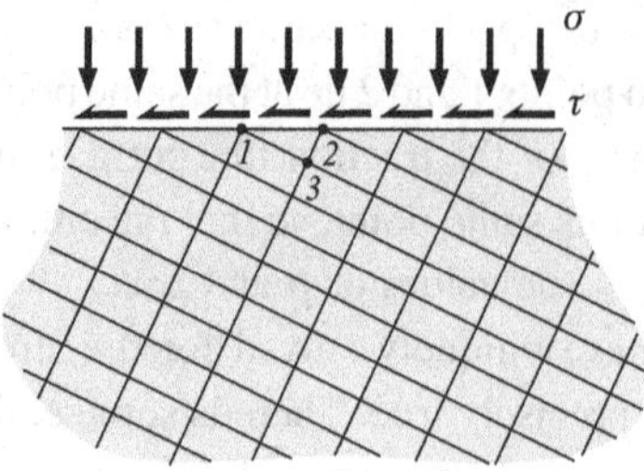

Figure 6.9. Slip line solution for planar boundary with uniform applied tractions.

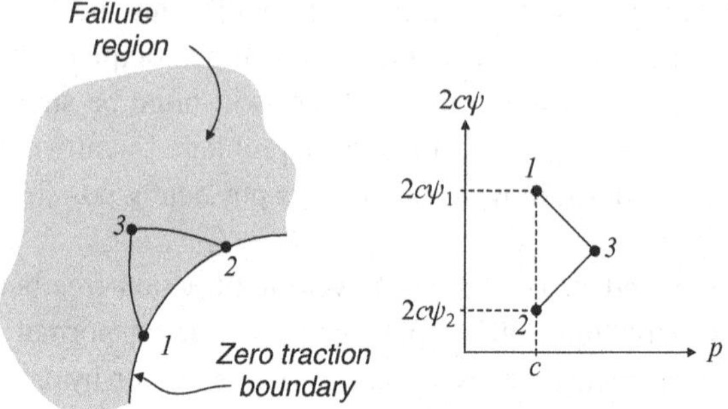

Figure 6.10. Slip line solution for non-planar boundary.

To illustrate the first point consider the curved traction-free surface sketched in Figure 6.10. If we wish to march the solution into the interior of the failing region starting from points 1 and 2, the corresponding points on the $(p, 2c\psi)$-diagram have the same ordinate but different values of $2c\psi$. The α- and β-lines are also curved since ψ changes on both lines. The values of p and $2c\psi$ at point 3 are easily determined from the intersection of the 45° lines in the $(p, 2c\psi)$-plane. Note that for both points 1 and 2 the value of the mean stress is equal to c. This results from the traction-free boundary demanding that the Mohr circles pass through the zero-stress origin. The values of ψ_1 and ψ_2 are simply the orientations of the boundary at points 1 and 2. The reader should sketch the Mohr circles for points 1 and 2 to verify that the $(p, 2c\psi)$-diagram is as shown, provided we assume a compressive stress field inside the failure region. The exact details of how the α- and β-lines change slope may still seem somewhat obscure. In practice, if points 1 and 2 are relatively close together and the boundary shape is not severely contorted then we may safely assume that both characteristic lines have constant curvature. For some problems we may be able to determine the exact shape of the lines from other conditions.

The second boundary condition that will result in curved characteristics is a non-uniform traction condition. We will illustrate this condition using the

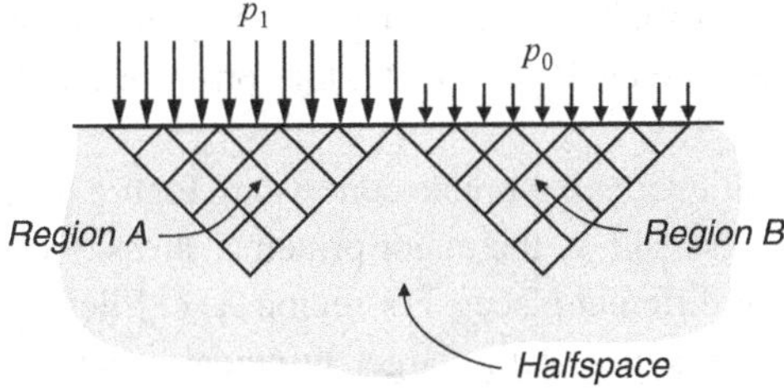

Figure 6.11. Plane strain step load problem.

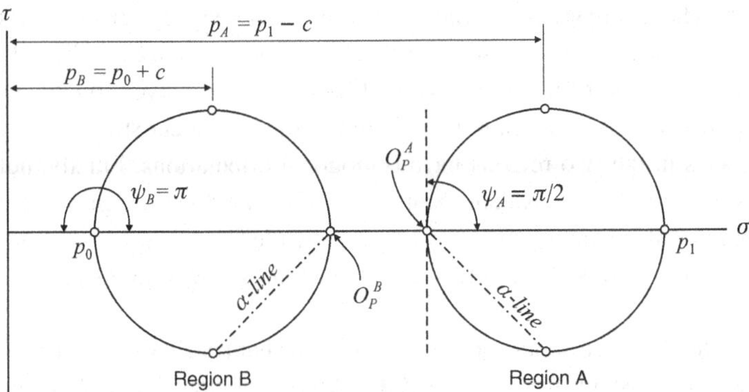

Figure 6.12. Mohr circles for failure regions A and B in Figure 6.11.

simple stepped surface traction applied to a halfspace shown in Figure 6.11. A uniform applied pressure p_0 acts on part of the halfspace surface, while a larger pressure p_1 acts on the remainder. We are familiar with the uniform traction condition and realise that the characteristics will be straight lines and the stress state inside the failure region will be constant; however, it is clear that there are two distinct regions corresponding to the two different boundary pressures. Based on our knowledge of constant boundary tractions we can sketch two sets of characteristic lines as shown in Figure 6.11. The question now is this. What happens in the region between the two sets of straight characteristics?

To begin our analysis let the area covered by the characteristics beneath the applied pressure p_1 be called region A and the area beneath p_0 be region B. It may be that p_1 extends on to the left in the figure, in which case region A would also extend away to the left. Similarly region B may extend to the right if the surface load p_0 is applied further to the right. These are details that do not concern us at this point. Our focus is on that part of the failing region between regions A and B. In order to understand what happens in this region consider the Mohr circles sketched in Figure 6.12. Both circles represent failure states and therefore have radius c. The circles have been drawn so that p_1 is the major principal stress in region A and p_0 the minor principal stress in region B. There are other possibilities such as both p_1 and p_0 being the major principal stress, but the situation sketched in Figure 6.12 is the correct one, as will become clear below.

There are several interesting points concerning Figure 6.12. Since p_0 is the minor principal stress and p_1 the major principal stress, the poles for the two Mohr circles fall at different places. For region A, O_P^A lies at the smaller principal stress, while O_P^B lies at the larger principal stress on circle B. If we construct the lines joining the two origins of planes to their respective minor principal stresses we find that for region A, $\psi_A = \pi/2$, while for region B, $\psi_B = \pi$. The fact that ψ_B is π rather than zero may seem surprising at first, but this will be the case. The reasons for this are investigated in Exercise 6.1 at the end of the chapter. Moving on, note that following our convention that α-lines are parallel to the line joining the pole to the smallest shear stress, we see that the α-lines in the two regions have orthogonal orientations. Finally, note the values of the two-dimensional mean stress associated with each Mohr circle. For region A the mean stress is $p_A = p_1 - c$, while for region B, $p_B = p_0 + c$. Each of these points is important in understanding how regions A and B are joined.

How should we go about constructing a net of characteristic lines in the region between the constant stress regions A and B? One obvious possibility would be simply to extend the straight characteristics into the intervening region, but this clearly will not work for several reasons. First, the stress states in regions A and B are different and we cannot simply merge them without altering something. Also, note that the directions of the α-characteristics in regions A and B are different. In region A the α-line has a negative slope (see Figure 6.12), while in region B the opposite is true. This fact suggests that the α-lines in region A must somehow end up as α-lines in region B. With these thoughts in mind, if we look at Figure 6.11, it might appear that bending the α-lines smoothly from region A into region B could be a good idea. The result is shown in Figure 6.13.

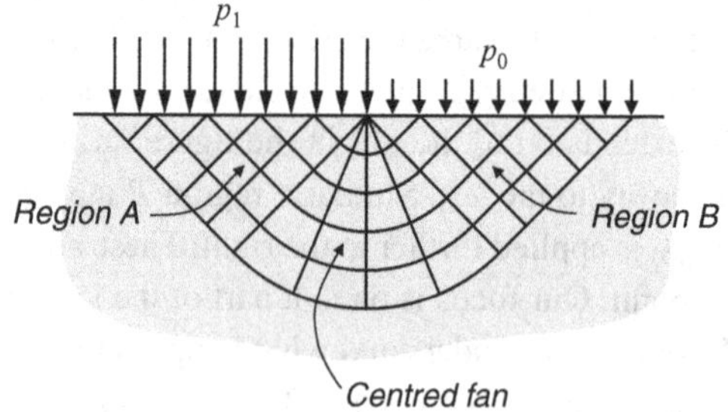

Figure 6.13. Slip line solution for the step load problem.

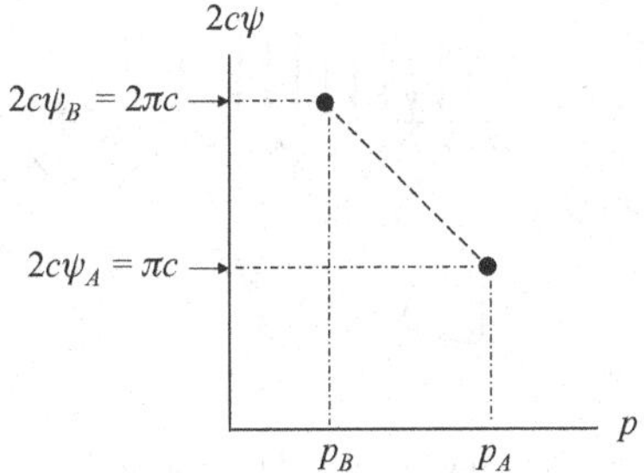

Figure 6.14. Relationship of applied tractions in the step load problem.

In Figure 6.13 α-lines in the area between regions A and B have been drawn as circular arcs. Their associated β-lines are straight lines. This solution is, in fact, the only possibility. Any attempt to construct an alternative solution will fail to meet the criteria already laid down for the characteristic line nets. The new network of lines between regions A and B is often referred to as a 'centred fan' as all the β-lines emerge from a central point.

We can now construct the $(p, 2c\psi)$-diagram for the situation in Figure 6.13. This is done in Figure 6.14. In the figure the point for region B corresponds to $\psi_B = \pi$ and $p_B = p_0 + c$. Similarly, for region A the point lies at $2c\psi_A = \pi c$ since ψ_A is equal to $\pi/2$. The mean stress p_A is also shown, but note that its value cannot be assigned independently. The points corresponding to regions A and B in the $(p, 2c\psi)$-diagram must lie on the diagonal line making an angle of $-45°$ with the horizontal axis, i.e. the condition for all α-lines. We see immediately that p_A and p_B must be related by

$$p_A = p_B + \pi c \tag{6.19}$$

If we replace p_A and p_B by their equivalent values $p_1 - c$ and $p_0 + c$, we find

$$p_1 = p_0 + (2 + \pi)c = p_0 + 5.14c \tag{6.20}$$

This equation expresses the fact that the magnitude of the pressure step on the halfspace surface must be exactly $(2 + \pi)$ times the cohesion in order that failure results. Equation (6.20) has already been noted in Chapter 5. It is well known from studies of the problem of a smooth rigid punch indenting a Tresca halfspace. We can now combine two step load problems to solve the strip footing problem once again. The resulting slip line field is shown in Figure 6.15.

To conclude this section we will consider the plane strain problem of a tunnel being driven deep in a rock mass. If the *in situ* stress in the rock is sufficiently great there will be a failure region surrounding the tunnel similar to the plastic region we found in Chapter 4 when we considered the cavity expansion problem.

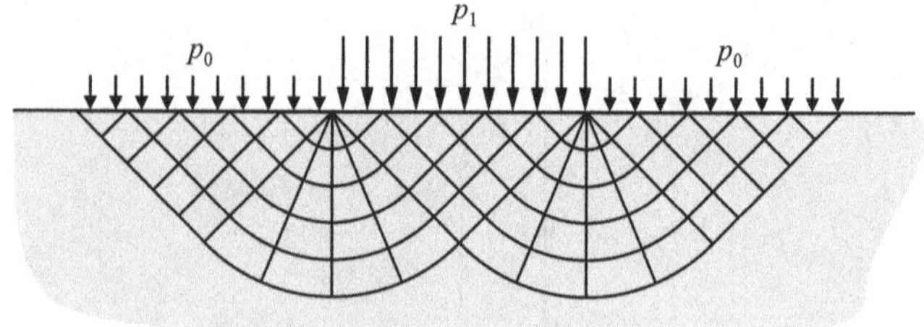

Figure 6.15. Step load solution generalised to represent the shallow strip footing.

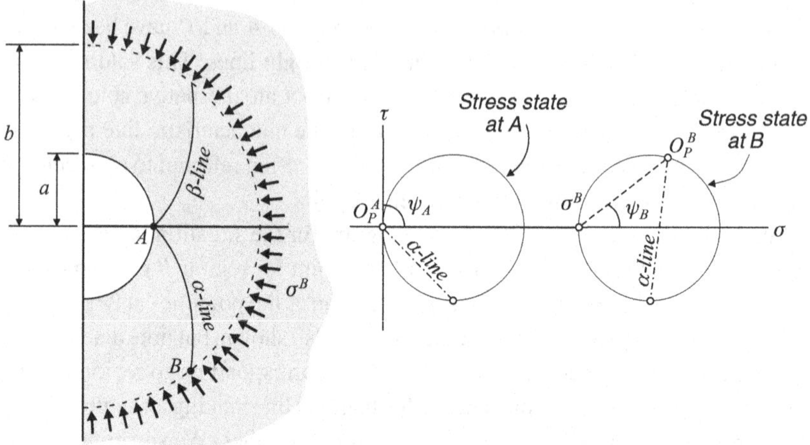

Figure 6.16. Circular tunnel in a deep rock mass.

We will approximate this unloading problem as shown in Figure 6.16. Only half the tunnel is shown due to the obvious symmetry. The tunnel radius is a and the radius of the failure region is b. We assume the boundary of the failure region supports a uniform stress σ^B as shown in the figure.

Note that the boundary of the tunnel and the boundary of the failure region are both principal surfaces. In fact, all circular surfaces in the failure region that are concentric with the tunnel boundary are principal surfaces. Mohr stress circles for point A on the tunnel boundary and point B on the outer boundary of the plastic region are also shown in Figure 6.16. For both circles the least principal stress is taken to be the stress acting on the circular boundary of the tunnel or the failure region boundary. We would intuitively expect this to be so since, assuming the rock mass initially supported an approximately uniform isotropic stress, driving the tunnel creates a reduction in the radial stress that will be greater than the reduction in the associated hoop stress.

We know that the characteristic lines will be curved in this situation since neither boundary is planar. Trial α- and β-lines are sketched on the figure. The two points A and B are chosen as the ends of our trial α-line. The particular α-line considered happens to touch the tunnel boundary at the point where the boundary tangent is vertical. As a result the pole for circle A lies at the origin of the Mohr diagram. For point B the pole lies at the point shown. Directions of the α-line are sketched on the two Mohr circles.

From our knowledge of Mohr circles we are aware that the line joining the pole to the least principal stress will always be parallel to the minor principal surface. The minor principal surface is, for any point in the failure region, the circle concentric with the tunnel boundary. Therefore the angle ψ associated with the characteristic line is the complement of the polar angle θ measured clockwise from the horizontal. That is, $\theta = \pi/2 - \psi$ for any radial direction measured from the tunnel centre. We can use the radial distance r and the angle θ to define a convenient polar coordinate system for the problem. This coordinate system is sketched in Figure 6.17.

Now consider the trial α-line in Figure 6.16 in more detail. We realise the line must be oriented at 45° to the principal surfaces, which are all radial and all circular surfaces in the failure region. This fact permits us to determine the shape of the line easily. Figure 6.17 shows a typical α-line at its intersection with two radii separated by an angular increment $d\theta$. Because the intersections must make an angle of 45°, we see that the distances dr and $r\,d\theta = -r\,d\psi$ are equal. Separating variables and integrating shows that

$$\psi - \psi_A = -\ln\left(\frac{r}{a}\right) \tag{6.21}$$

where ψ_A is the value of ψ at $r = a$. We can use (6.21) to graph all α-lines.

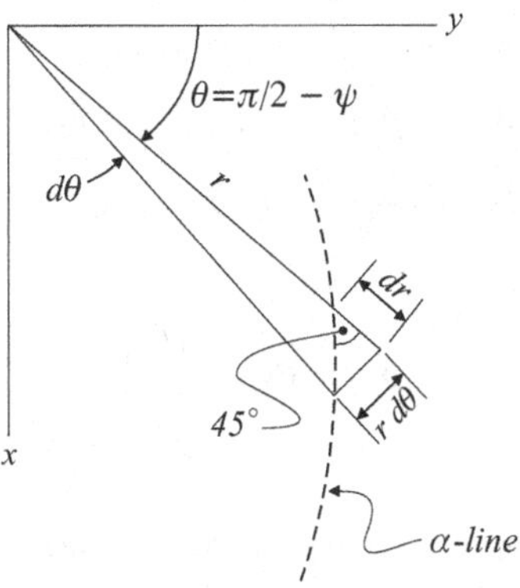

Figure 6.17. Definition of the polar angle θ associated with the α-line in Figure 6.16.

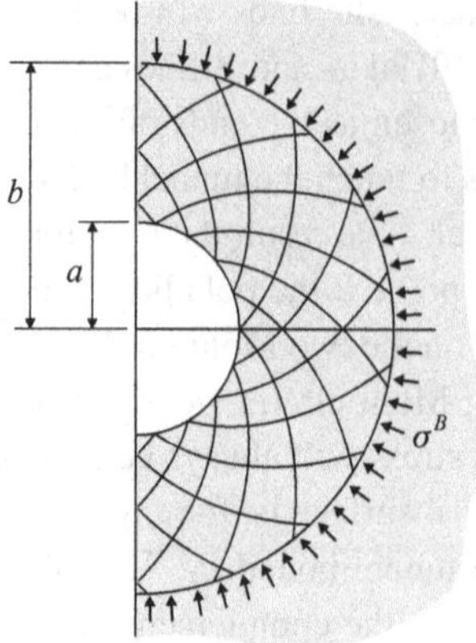

Figure 6.18. Slip line geometry for the tunnel problem.

β-lines are determined by a similar equation where the sign of $\ln(r/a)$ is reversed. Figure 6.18 shows the resulting network of characteristic lines.

Finally, note how the radial and hoop stresses σ_{rr} and $\sigma_{\theta\theta}$ vary throughout the failure region. Equation (6.11) applies for all points on an α-line. Thus for the α-line that originates on the tunnel boundary at an angle ψ_A we have

$$K_1 = p_A + 2c\psi_A = c + 2c\psi_A \tag{6.22}$$

Therefore at any other point on this line the value of the two-dimensional mean stress must be

$$p = K_1 - 2c\psi = c - 2c(\psi - \psi_A) = c + 2c\ln\left(\frac{r}{a}\right) \tag{6.23}$$

and the associated radial and hoop stresses are

$$\begin{aligned} \sigma_{rr} &= p - c = 2c\ln(r/a) \\ \sigma_{\theta\theta} &= p + c = 2c[1 + \ln(r/a)] \end{aligned} \tag{6.24}$$

If the stress σ^B is known* we can use (6.24) to determine the outer dimension b of the failure region.

$$b = a\exp\left(\frac{\sigma^B}{2c}\right) \tag{6.25}$$

* We can assume σ_B to be approximately equal to the *in situ* stress in the rock at the depth of the tunnel.

6.6 Frictional materials

Up to this point we have been content to deal with the case in which our material has no frictional strength. This results in simple equations describing orthogonal characteristics and relatively simple solutions for the problems considered. There are obvious benefits to this approach in terms of learning the basic concepts of slip line theory without becoming embroiled in too much detail. Introducing friction will clearly bring added complexity as seen at a glance by considering Figure 6.19.

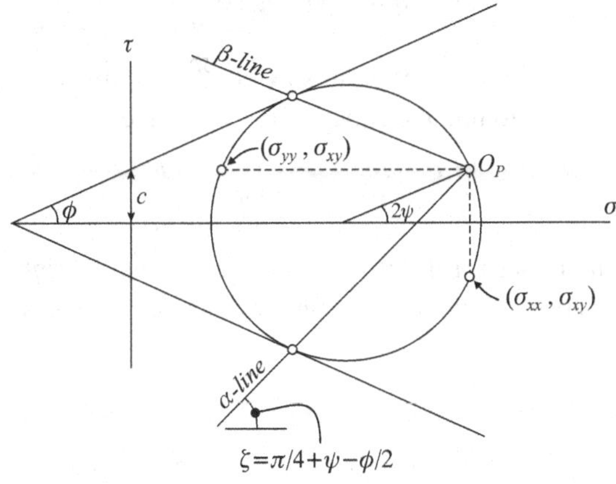

Figure 6.19. Effect of friction of slip line orientation.

With non-zero friction we no longer have a constant radius for the Mohr circle. The circle radius becomes a function of the two-dimensional mean stress p according to

$$\frac{1}{2}(\sigma_1 - \sigma_3) = \frac{1}{2}\left[(\sigma_{xx} - \sigma_{yy})^2 + 4\sigma_{xy}^2\right]^{1/2} = c\cos\varphi + p\sin\varphi \quad (6.26)$$

We also see that the α- and β-lines are no longer orthogonal but instead must intersect at an angle of $\pi/2 - \varphi$. Clearly things are not as simple as before, nevertheless we can still use p, together with the angle ψ, to characterise the two-dimensional stress state fully:

$$\begin{aligned}
\sigma_{xx} &= p + (c\cos\varphi + p\sin\varphi)\cos 2\psi \\
\sigma_{yy} &= p - (c\cos\varphi + p\sin\varphi)\cos 2\psi \\
\sigma_{xy} &= (c\cos\varphi + p\sin\varphi)\sin 2\psi
\end{aligned} \quad (6.27)$$

Using these equations to replace the stress components in the equilibrium equations (6.2) we find

$$\begin{aligned}
&(1+\sin\varphi\cos 2\psi)\frac{\partial p}{\partial x}+\sin\varphi\sin 2\psi\frac{\partial p}{\partial y}\\
&\quad=2(c\cos\varphi+p\sin\varphi)\left(\sin 2\psi\frac{\partial\psi}{\partial x}-\cos 2\psi\frac{\partial\psi}{\partial y}\right)\\
&\sin\varphi\sin 2\psi\frac{\partial p}{\partial x}+(1-\sin\varphi\cos 2\psi)\frac{\partial p}{\partial y}\\
&\quad=-2(c\cos\varphi+p\sin\varphi)\left(\cos 2\psi\frac{\partial\psi}{\partial x}+\sin 2\psi\frac{\partial\psi}{\partial y}\right)
\end{aligned}\tag{6.28}$$

which are equivalent to (6.6). We can solve (6.28) for $\partial p/\partial x$ and $\partial p/\partial y$ and use the result in (6.7) to find the equation for the α-line

$$\frac{dp}{ds_\alpha}=-2(c+p\tan\varphi)\frac{d\psi}{ds_\alpha}\tag{6.29}$$

A similar result applies for the β-line with the sign on the right-hand side of the equation reversed. Integrating we find the equivalent expressions to (6.11) and (6.12)

$$\begin{aligned}
&\alpha\text{ lines}\rightarrow\ln(c+p\tan\varphi)+2(\tan\varphi)\psi=K_1\\
&\beta\text{ lines}\rightarrow\ln(c+p\tan\varphi)-2(\tan\varphi)\psi=K_2
\end{aligned}\tag{6.30}$$

We can use equations (6.30) to relate stress states along any particular characteristic line. Even though these equations appear rather different from (6.11) and (6.12), it is easy to show that the Hencky equations (6.17) and (6.18) still apply.

Finally, note that we can construct a graph similar to the $(p, 2c\psi)$-diagram shown in Figure 6.6, but we must plot $\ln(c+p\tan\varphi)$ versus $2(\tan\varphi)\psi$. Points on any α-line will correspond to negative sloping 45° lines in the $(\ln(c+p\tan\varphi), 2(\tan\varphi)\psi)$-plane, while points on any β-line must lie on positive 45° lines in the diagram. There are obviously strong similarities between the cases of zero frictional strength and non-zero strength, and we can exploit our knowledge of the zero-friction case to revisit some of the simple problems considered earlier.

First, consider the case of a planar boundary supporting a constant normal traction shown in Figure 6.20. Just as in the purely cohesive case, there are two possible solutions but now, with frictional strength, we find different (α, β)-nets. If the boundary stress happens to be the smaller principal stress, the α- and β-lines make angles of $(\pi/4-\varphi/2)$ with the boundary. On the other hand, if the applied stress happens to be the larger principal stress, the α- and β-lines make angles of $(\pi/4+\varphi/2)$ with the boundary. Clearly, the resulting stress

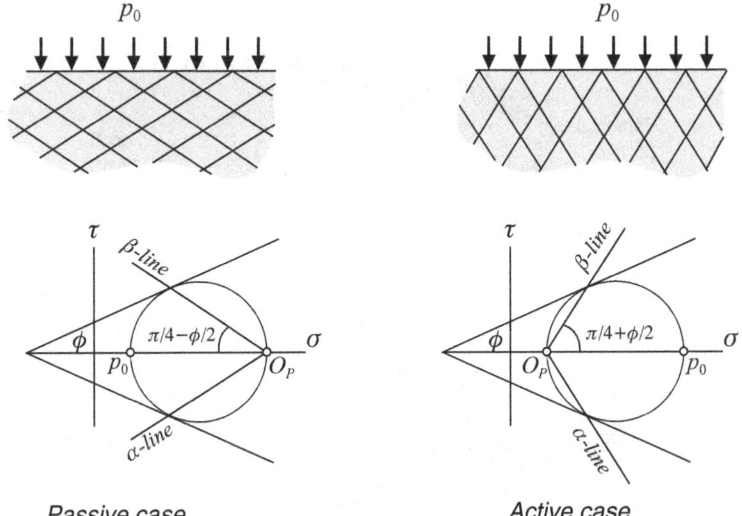

Figure 6.20. Uniform traction acting on halfspace surface–passive and active cases.

states are very different. The familiar terms *passive* and *active* are applied to the two conditions. The relation to Rankine's retaining wall analysis is obvious.

Next, consider the step load problem from Figures 6.11 and 6.13. For a frictional material we would expect to see regions of constant stress beneath each constant part of the surface load, but unlike Figure 6.11 the alignment of the characteristic lines will differ as shown in Figure 6.21. Region A corresponds to an active stress state while region B corresponds to a passive condition. Mohr circles for the two regions are also shown. The material between regions A and B is filled by a centred fan, but its geometry is also different from that in Figure 6.13. We can investigate the shape of the characteristic lines within the fan by noting that the intersections between α- and β-lines must inscribe an angle of $(\pi/2 - \varphi)$ as noted in Figure 6.22. From the geometry shown it is evident that $\tan\varphi = dr/r\,d\theta$. Integrating leads to

$$\ln\left(\frac{r}{r_0}\right) = (\theta - \theta_0)\tan\varphi \tag{6.31}$$

where r_0 and θ_0 are defined in the figure. Equation (6.31) is exactly equivalent to equation (5.34), the equation for the logarithmic spiral. All the α-lines within the fan are logarithmic spirals. The β-lines of course are all straight.

Considering once again the Mohr circles in Figure 6.21 we see that the values of the angle ψ in regions A and B are $\pi/2$ and π, respectively, exactly the same as found in Figure 6.12. We also note that $2(\tan\varphi)\psi$ and $\ln(c + p\tan\varphi)$ for

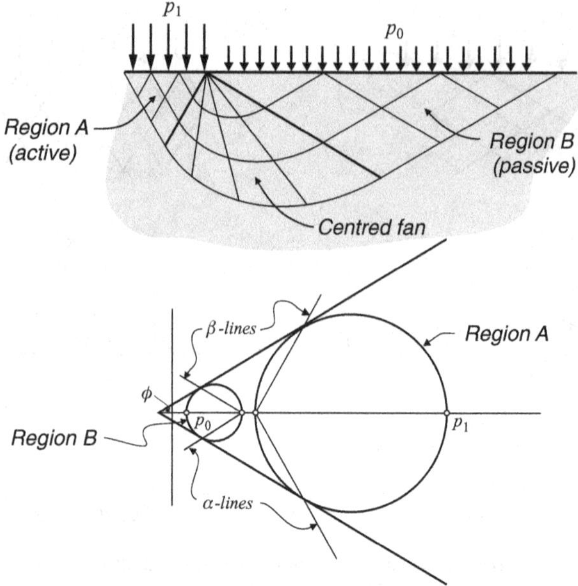

Figure 6.21. Step load problem for a frictional Coulomb soil.

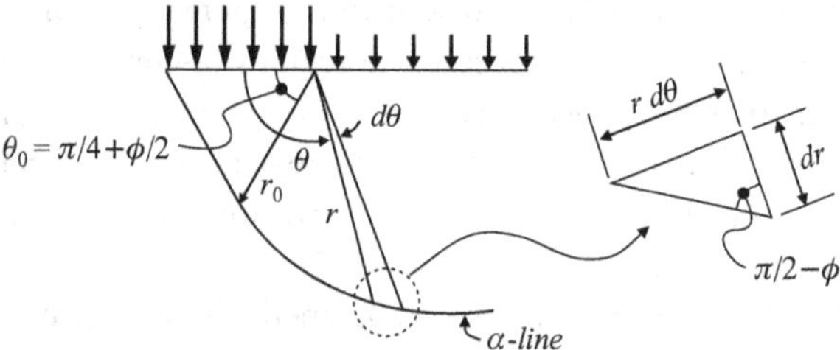

Figure 6.22. Geometry of slip lines within a centred fan for frictional material.

each region are related, lying on the $-45°$ line in $(2(\tan\varphi)\psi, \ln(c + p\tan\varphi))$-space. Thus the values of two-dimensional mean stresses p_A and p_B must be related by

$$2(\tan\varphi)(\psi_A - \psi_B) = -(\tan\varphi)\pi = -\ln(c + p_B\tan\varphi) + \ln(c + p_A\tan\varphi) \tag{6.32}$$

Rearranging this equation gives

$$p_A = c(e^{\pi\tan\varphi} - 1)\cot\varphi + p_B e^{\pi\tan\varphi} \tag{6.33}$$

We note that p_A and p_B are related to p_1 and p_0 by

$$\begin{aligned} p_1 &= c\cos\varphi + p_A(1+\sin\varphi) \\ p_0 &= -c\cos\varphi + p_B(1-\sin\varphi) \end{aligned} \tag{6.34}$$

Using these in (6.33) gives

$$p_1 = c\cot\varphi(Ne^{\pi\tan\varphi} - 1) + p_0 N e^{\pi\tan\varphi} = cN_c + p_0 N_q \tag{6.35}$$

where N carries its usual definition from (4.16) and N_c and N_q are the first and second bearing capacity coefficients defined in (5.33).

It is a simple matter to join two characteristic nets to give the complete solution for the strip footing problem. Figure 6.23 shows the familiar diagram. The figure has been drawn for the case where $\varphi = 30°$. Use of a different value for φ would clearly alter the details, but not the substance, of the (α, β) net.

The solution for the strip footing shown in Figure 6.23 as well as the relation (6.35) was first discovered by the great German engineer Ludwig Prandtl in 1920. Prandtl is best known for his work in fluid dynamics but he also made a number of important discoveries in the field of solid mechanics. Note that the solution we have obtained still represents the case of zero body forces. We will see shortly that introducing gravity will lead to significant complications.

A second kinematic failure mechanism that is equally likely to that illustrated in Figure 6.23 was suggested by Hencky and is illustrated in Figure 6.24. In this mechanism two active zones develop beneath the strip load. The value of the limit pressure p_1 for this solution is exactly the same as that given in (6.35) for the Prandtl mechanism. In an intuitive sense the Hencky mechanism may appear to be less physically realisable than that of Prandtl, but mathematically there is no reason why it should not occur.

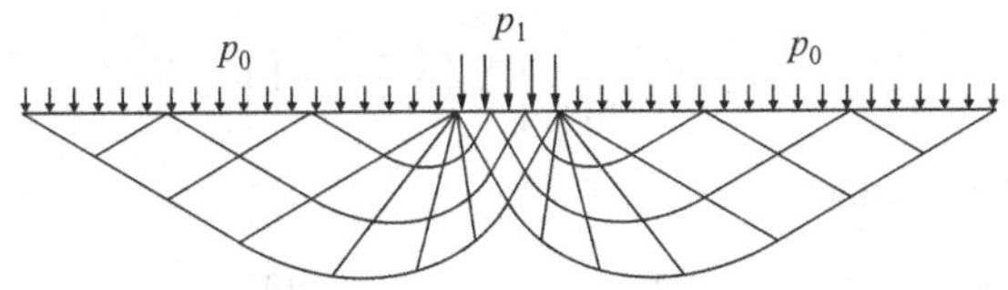

Figure 6.23. Prandtl solution for the shallow strip footing problem.

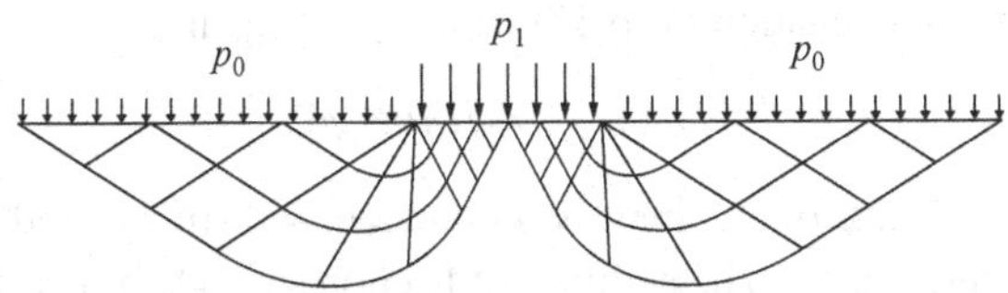

Figure 6.24. Hencky solution for the shallow strip footing problem.

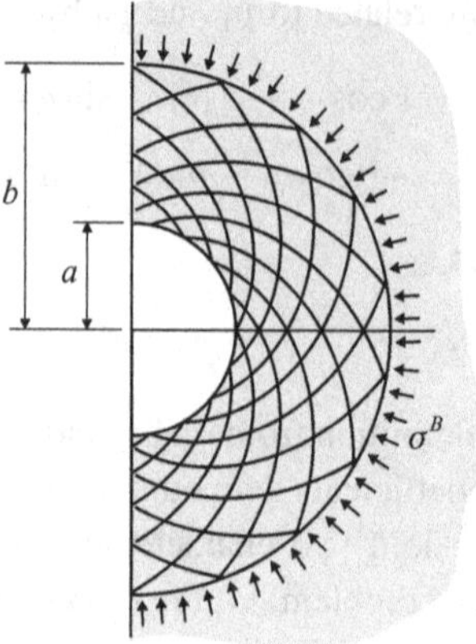

Figure 6.25. Slip line geometry for a tunnel within frictional material.

To complete this section we will briefly revisit the tunnel problem illustrated earlier in Figure 6.16. If we introduce frictional strength the results are modified as follows. We note that all circles concentric with the tunnel boundary will still be principal surfaces and will support the smallest principal stress, exactly as before. Equation (6.21), describing the geometry of α-lines, is now replaced by

$$\psi - \psi_A = -\frac{\ln(r/a)}{\tan(\pi/4 - \varphi/2)} \tag{6.36}$$

This leads to the network of characteristic lines shown in Figure 6.25. A value of 30° has been used for φ in constructing the figure. Finally, the radial and hoop stresses in the failure region are given by

$$\begin{aligned} \sigma_{rr} &= \frac{c}{\tan\varphi}\left[\left(\frac{r}{a}\right)^{\Gamma} - 1\right] \\ \sigma_{\theta\theta} &= \frac{c}{\tan\varphi}\left[N\left(\frac{r}{a}\right)^{\Gamma} - 1\right] \end{aligned} \tag{6.37}$$

where Γ is the constant

$$\Gamma = 2\tan\varphi\cot(\pi/4 - \varphi/2) \tag{6.38}$$

If $\varphi = 30°$, the value of Γ will be exactly 2. The radius b of the failure region is given by the first equation of (6.37) with σ_{rr} set equal to σ_B,

$$b = a\,[1 + (\sigma_B/c)\tan\varphi]^{1/\Gamma} \tag{6.39}$$

Equations (6.37) and (6.39) may be compared with (6.24) and (6.25) from the purely cohesive case. The addition of friction strongly affects the solution, particularly as σ_B becomes greater.

6.7 Effects of gravity

In this section we will briefly outline the effects of introducing gravity, both on the equations and on solutions. In a nutshell, gravity brings complexity, more complexity than is comfortable for an introductory text such as this. Thus we will bypass most of the complications and simply summarise the main differences from the weightless cases considered thus far. A far more complete discussion may be found in the book by V.V. Sokolovski cited at the end of this chapter.

Let us begin by considering the two-dimensional plane strain problem of a halfspace with horizontal surface supporting some applied loads. Suppose we organise our coordinate frame so that the x-axis is horizontal and the y-axis is vertical (*downwards*) and $y = 0$ identifies the halfspace surface. The equilibrium equations now become

$$\begin{aligned} \frac{\partial \sigma_{xx}}{\partial x} + \frac{\partial \sigma_{xy}}{\partial y} &= 0 \\ \frac{\partial \sigma_{xy}}{\partial x} + \frac{\partial \sigma_{yy}}{\partial y} &= \rho g \end{aligned} \tag{6.40}$$

First, we consider purely cohesive materials. The effect of gravity in this case is minimal and it provides an easy introduction to the basic ideas. For a purely cohesive material equation (6.10) is now replaced by

$$\frac{dp}{ds_\alpha} = -2c\frac{d\psi}{ds_\alpha} + \rho g \sin\zeta \tag{6.41}$$

As before ψ and ζ are related by $\zeta = \pi/4 + \psi$. If we separate variables and take note of the geometrical relationship $dy = \sin\zeta\, ds_\alpha$, we may integrate (6.41) to find the following relation must hold on α-lines:

$$p + 2c\psi = \rho g y + K_1 \tag{6.42}$$

Similarly for β-lines,

$$p - 2c\psi = \rho g y + K_2 \tag{6.43}$$

It is easy to show that Hencky's equations (6.17) and (6.18) still apply for this case. If our halfspace supports a uniform applied stress p_0, the characteristic lines are straight 45° lines, exactly as in the case with zero gravity. Adding (6.42) and (6.43) shows that, at the intersection of any pair of α- and β-lines, the two-dimensional mean stress is increased by the amount $\rho g y$, exactly as we would expect. Aside from this feature, the solution for the strip load problem shown in Figure 6.15 and equation (6.20) is unchanged. Gravity has no effect on the limiting stress for the strip footing resting on purely cohesive soil.

Intuitively we might expect an absence of dramatic effects for a purely cohesive material since there is no dependence of strength on normal stress. Clearly this is not the case for a frictional material. The strength will depend on normal stress and, if gravity is operating, normal stresses will increase with depth below the halfspace surface. So what happens when gravity acts on a frictional material? We begin again with equations (6.27). Using them in the equilibrium equations (6.40) we can eventually arrive at the following differential equations applying on the characteristic lines:

$$\begin{aligned} \cos\varphi \frac{dp}{ds_\alpha} + 2(c\cos\varphi + p\sin\varphi)\frac{d\psi}{ds_\alpha} &= \rho g \sin(\varphi + \zeta) \\ \cos\varphi \frac{dp}{ds_\beta} - 2(c\cos\varphi + p\sin\varphi)\frac{d\psi}{ds_\beta} &= \rho g \cos\zeta \end{aligned} \tag{6.44}$$

These equations are sometimes referred to as Kötter's equations, named after the German engineer F. Kötter who obtained a set of equivalent expressions in 1903. They are significantly more difficult to integrate than the corresponding equations for the zero-gravity case. In most instances numerical methods are required both to find the shape of the characteristic lines as well as the stresses at the nodes where the lines intersect. In general, the lines will not be straight. As an illustration, Figure 6.26 shows the network of α- and β-lines for the strip load problem.

There are several interesting points concerning this figure. First, note that the kinematic mechanism shown is similar to that suggested by Hencky in Figure 6.25 with two active regions beneath the strip load. The passive regions lying outside the strip load consist of straight line characteristics and are exactly the same as for the zero-gravity case. Differences only arise in the centred fans and in the active regions. All characteristics in both these regions are curved, unlike the zero-gravity case. A second important difference is found when one traces the mean stress p from the surface outside the footing (we have assumed a traction-free case in Figure 6.26) along a characteristic curve to the surface beneath the footing. It is found that the normal traction beneath the footing can no longer be uniform as we have assumed earlier. A stress distribution similar

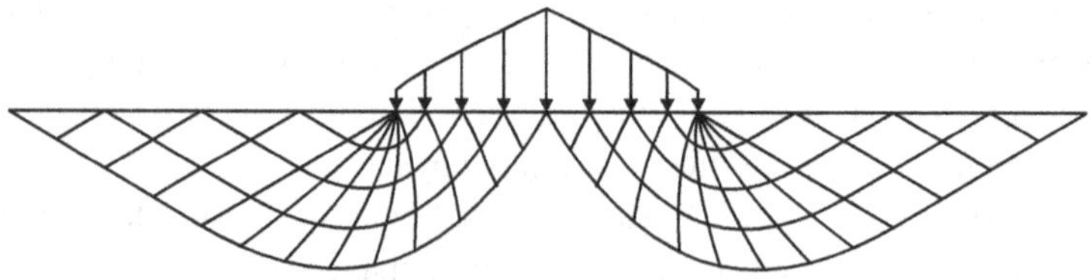

Figure 6.26. Slip line geometry for strip load with gravity.

to that shown is required to produce failure throughout the network of α- and β-lines. The exact details of both the characteristic net and the distribution of normal load depend on the values of c, φ and ρ. However, the details are not of great concern. The points we wish to emphasise here are the differences from the zero-gravity case.

We can now clearly see why the third bearing capacity coefficient N_γ in (5.33) is the most difficult to specify. All attempts to formulate an equation of the form of (5.32) are founded on the concept that the effects of gravity may be separated from the effects of soil strength and surcharge. In effect equation (5.33) represents a superposition of three different solutions; a superposition which, of course, is theoretically impossible because of the nonlinear nature of the problem. All bearing capacity formulae of the type shown in (5.32) are approximations. The first two terms represent the exact solution to the strip load problem *without gravity* while the third term approximates the effect of soil weight. This overlooks the fact that, when gravity is included, the weightless solution becomes invalid. The only correct solution is to solve Kötter's equations as they stand using numerical methods. This was done by Sokolovski and, more comprehensively, by H.Y. Ko and R.F. Scott.

6.8 The velocity field

To complete this chapter we will develop representations for the plastic strain rates and associated velocities that apply on the slip lines. We assume we are dealing with a perfectly plastic Coulomb material obeying an associated flow rule. We also assume the slip line network has been determined. To begin, rewrite equation (6.26) in the form of a yield equation for our two-dimensional stress state

$$f = \left[(\sigma_{xx} - \sigma_{yy})^2 + 4\sigma_{xy}^2\right]^{1/2} - (\sigma_{xx} + \sigma_{yy}) \sin\varphi = 2c\cos\varphi \qquad (6.45)$$

Then the associated flow rule tells us that

$$\begin{aligned}
\dot{\varepsilon}^p_{xx} &= \lambda \frac{\partial f}{\partial \sigma_{xx}} = \lambda \left\{ \frac{\sigma_{xx} - \sigma_{yy}}{\left[(\sigma_{xx} - \sigma_{yy})^2 + 4\sigma_{xy}^2\right]^{1/2}} - \sin\varphi \right\} \\
\dot{\varepsilon}^p_{yy} &= \lambda \frac{\partial f}{\partial \sigma_{yy}} = \lambda \left\{ -\frac{\sigma_{xx} - \sigma_{yy}}{\left[(\sigma_{xx} - \sigma_{yy})^2 + 4\sigma_{xy}^2\right]^{1/2}} - \sin\varphi \right\} \qquad (6.46) \\
\dot{\varepsilon}^p_{xy} &= \lambda \frac{\partial f}{\partial \sigma_{xy}} = \lambda \frac{4\sigma_{xy}}{\left[(\sigma_{xx} - \sigma_{yy})^2 + 4\sigma_{xy}^2\right]^{1/2}}
\end{aligned}$$

If we now use equations (6.27) here, we find the following simpler forms for the plastic strain rates:

$$\dot{\varepsilon}^p_{xx} = \lambda(\cos 2\psi - \sin\varphi), \quad \dot{\varepsilon}^p_{yy} = \lambda(-\cos 2\psi - \sin\varphi), \quad \dot{\varepsilon}^p_{xy} = 2\lambda \sin 2\psi \tag{6.47}$$

Two points emerge from these equations. First, note the plastic volumetric strain rate implied by (6.47).

$$\dot{e}^p = \dot{\varepsilon}^p_{xx} + \dot{\varepsilon}^p_{yy} = -2\lambda \sin\varphi \tag{6.48}$$

Thus for any value of φ greater than zero, the plastic volumetric strain will always result in dilation. This point was noted in Chapter 4 and is emphasised again here. Usually the amount of dilation predicted by (6.48) will be greater than appears reasonable when compared with measurements. Several researchers have proposed modifications for the plane strain equations presented here in an effort to obtain more realistic volumetric strains. The most widely accepted is probably the theory proposed by A.J.M. Spencer for which an incompressible response results. The associated flow rule is obviously not used in Spencer's theory.

The second point arising from (6.47) concerns the extensional strains in the directions of the characteristic lines. It is a simple matter to obtain the plastic extensional strain rate in any arbitrary direction in the plane of our problem. For example, suppose we wish to determine the extensional strain rate $\dot{\varepsilon}^p_{nn}$, where the n-direction is oriented at an arbitrary angle η measured counterclockwise from the horizontal x-axis. Then it is easy to show that

$$\dot{\varepsilon}^p_{nn} = \dot{\varepsilon}^p_{xx} \cos^2\eta + \dot{\varepsilon}^p_{yy} \sin^2\eta + 2\dot{\varepsilon}^p_{xy} \sin\eta \cos\eta \tag{6.49}$$

If we let $\eta = \zeta$, the angle of the α-line, then (6.49) will give $\dot{\varepsilon}^p_{\alpha\alpha}$, the extensional strain in the α-direction. Replacing η by ζ and using (6.47) we find the following result:

$$\dot{\varepsilon}^p_{\alpha\alpha} \equiv 0 \tag{6.50}$$

That is, the plastic extensional strain rate in the direction of the α-line is everywhere exactly zero. A similar result applies for the β-lines. This condition of *inextensibility* tells us that all of the deformation associated with the α- and β-directions is pure shear. If we consider other directions such as the x-direction, for example, there will be both shearing and extensional deformation as is clear from (6.47).*

Even though the plastic extensional strain parallel to the α- and β-directions is identically zero, the plastic strains perpendicular to the slip lines can never be zero except in the case of a purely cohesive material. The plastic strain rate

* Of course if the x-axis happens to coincide with the α-direction, then we would have $\zeta = 0$ and (6.47) would show that $\dot{\varepsilon}^p_{xx} = 0$. The same would apply to the β-line.

perpendicular to either slip line will be given by $-2\lambda \sin\varphi$. This result may be obtained by transforming the coordinate frame as we did above to find $\dot{\varepsilon}_{\alpha\alpha}$, or by simply assuming that the x-axis happens to align with the α-direction in equations (6.47). The corresponding shear strain rate is given by $-2\lambda \cos\varphi$. The ratio of the strain rates is $\tan\varphi$. This fact agrees precisely with comments made in Chapter 5 leading up to the derivation of equation (5.5). Whenever $\varphi > 0$ and the associated flow rule is in use, we will always find dilation involved with plastic flow.

Particles within the failure region will deform plastically according to the plastic strain rates given above. These strains will result in a field of velocity vectors covering the failure region. Let the velocity at a point be $\mathbf{v} = \mathbf{v}(x, y)$ with components v_x and v_y in the x- and y-directions. Of course, there may be velocities associated with elastic deformation or with rigid-body motion, but we will assume that these are zero and all velocities are associated with plastic flow. It will be convenient to let v_α and v_β be components of $\mathbf{v}$ in the α- and β-directions. A typical configuration showing v_α as well as v_x and v_y is shown in Figure 6.27. A similar construction may be made for the v_β component. Of course, the α- and β-directions are not orthogonal and the components v_α and v_β will not be either. We can visualise either component as the orthogonal projection of the vector $\mathbf{v}$ on the α- or β-direction. The different velocity components are related as follows:

$$v_\alpha = v_x \cos\zeta + v_y \sin\zeta, \quad v_\beta = v_y \cos(\zeta + \varphi) - v_x \sin(\zeta + \varphi) \tag{6.51}$$

$$\begin{aligned} v_x &= \frac{1}{\cos\varphi}[v_\alpha \cos(\zeta + \varphi) - v_\beta \sin\zeta], \\ v_y &= \frac{1}{\cos\varphi}[v_\alpha \sin(\zeta + \varphi) + v_\beta \cos\zeta] \end{aligned} \tag{6.52}$$

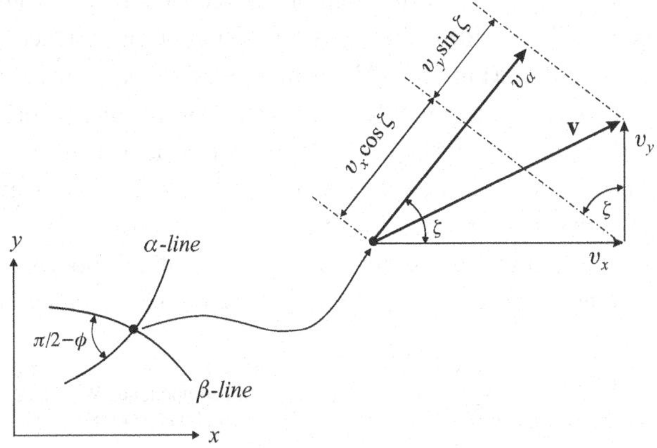

Figure 6.27. Components of the velocity vector in the slip line directions.

Next, consider how v_α varies in the α-direction

$$\begin{aligned}\frac{dv_\alpha}{ds_\alpha} &= \frac{d}{ds_\alpha}(v_x\cos\zeta + v_y\sin\zeta)\\ &= \left(\frac{\partial v_x}{\partial x}\cos\zeta + \frac{\partial v_x}{\partial y}\sin\zeta\right)\cos\zeta + \left(\frac{\partial v_y}{\partial x}\cos\zeta + \frac{\partial v_y}{\partial y}\sin\zeta\right)\sin\zeta\\ &\quad + (-v_x\sin\zeta + v_y\cos\zeta)\frac{d\zeta}{ds_\alpha} \qquad (6.53)\end{aligned}$$

The first two terms on the right-hand side of the second equals sign actually sum to zero because of the inextensibility condition.* Only the third term on the right-hand side does not vanish. If we now use (6.52) to replace v_x and v_y we find

$$\cos\varphi\frac{dv_\alpha}{ds_\alpha} = (v_\alpha\sin\varphi + v_\beta)\frac{d\zeta}{ds_\alpha} \qquad (6.54)$$

A similar development for the change of v_β in the β-direction leads to

$$\cos\varphi\frac{dv_\beta}{ds_\beta} = -(v_\alpha + v_\beta\sin\varphi)\frac{d\zeta}{ds_\beta} \qquad (6.55)$$

These two equations fully define the velocity components v_α and v_β on the characteristic lines.

We can immediately use (6.54) and (6.55) to investigate the velocity fields associated with the slip line networks we determined earlier. First, if the characteristic lines are straight such as those illustrated in Figure 6.20, the angle ζ is constant and consequently the velocities v_α and v_β must also be constant. This does not mean that v_α and v_β are constant *everywhere*, however. Each velocity component need only be constant along its particular slip line. Thus if we consider two parallel α-lines separated by some normal distance h_α, the velocity v_α may be completely different on the second slip line from the value found on the first. It is evident that v_α is a function of the distance h_α and we can write $v_\alpha = v_\alpha(h_\alpha)$. Similarly, if h_β is measured normal to the β-lines, then $v_\beta = v_\beta(h_\beta)$. Clearly it is possible to construct quite complex deformations from the combined action of the two velocity components even in this simple case of constant stress. The complete velocity field is said to be a superposition of two *shear flows*. In fact, there is not even a requirement that the functions $v_\alpha(h_\alpha)$ and $v_\beta(h_\beta)$ should be continuous. If a discontinuity occurs in $v_\alpha(h_\alpha)$, for example, then there will be discontinuous velocities on either side of one

* Referring to equation (6.49), if we replace η with ζ the result is zero as shown in (6.50). Also $\dot{\varepsilon}^p_{xx} = \partial v_x/\partial x$ with similar expressions for the other strain components. We see that the terms in (6.53) are exactly the same as those in (6.49), with $\eta = \zeta$, and therefore the sum is zero as stated above.

particular α-line. There might be zero velocity on one side of the line and non-zero slip on the other side. The relation to our upper bound theorem solutions in Chapter 5 is clear.

If lines of discontinuity do exist, either within the failure region or bounding the failure region, we must think of them as thin shear bands. The reason lies in the fact that dilatancy will always be associated with slip. If the shear band possessed no thickness it would be impossible to accommodate dilation. This point was also made in Chapter 5. Note that if we construct a local Cartesian coordinate system parallel and perpendicular to a line of discontinuity, the components of relative velocity normal and tangent to the line will have the ratio $\tan\varphi$. The relative velocity vector itself will always lie at an angle φ to the line of discontinuity as was pointed out in the derivation of equation (5.5). Of course, if a non-associated flow rule is used, these constraints no longer apply.

Finally, let us consider the velocity field associated with a centred fan of slip lines like that in Figure 6.21. In the fan shown in that figure the β-lines are straight while the α-lines are logarithmic spirals. Because the β-lines are straight we know that v_β will be constant on any particular line, but may take on different values for different lines. This fact can be expressed by noting that

$$v_\beta = g(\theta) \tag{6.56}$$

where g is an arbitrary function of the angle θ defined in Figure 6.22. Using (6.56) in (6.54) and integrating leads to the following expression for the velocity v_α associated with the α-lines:

$$v_\alpha e^{-\zeta \tan\varphi} = \sec\varphi \int g(\zeta + \varphi) e^{-\zeta \tan\varphi}\, d\zeta + C_1 \tag{6.57}$$

Here we have used the fact that $\theta = \zeta + \varphi$. C_1 is the constant of integration associated with the particular α-line of interest. In the special case where $g(\theta) = 0$, we see that v_α will behave exponentially with ζ.

The usual interpretation of the velocity field for the situation shown in Figure 6.21 is that regions A and B move as rigid bodies. The outermost α-line represents a discontinuity separating the failure region from the surrounding body which remains stationary. Region A moves vertically downward, the fan rotates and region B moves upward and to the right. This corresponds to the picture presented in Chapter 5 for the same problem. Note that the actual magnitude of the velocities is not specified. This is consistent with the assumption of perfect plasticity in that the strain rates are determined only to within the multiplier λ. It is only possible to determine the velocity field completely if the velocity of a boundary point is specified initially. For example, we might specify that the footing is to move vertically downward with a velocity v_0. Then all the velocities within the failure region could be calculated in terms of v_0. These thoughts are elaborated in Exercise 6.6 below.

Further reading

Early works referred to in this chapter may be found in

H. Hencky, Über einige statische bestimmte Fälle der Gleichgewichts in plastischen Koerpern, *Zeit. Angew. Math. Mech.*, **3**, 241–251 (1923).

F. Kötter, Die Bestimmung des Druckes an gekrümmten Gleitflächen, eine Aufgabe aus der Lehre vom Erddruck, *Berlin Akad. Wiss.*, Berlin, 229–233 (1903).

L. Prandtl, Über die Eindringungs-festigkeit (Härte) plastischer Baustoffe und die Festigkeit von Schneiden, *Zeit. Angew. Math. Mech.*, **1**, 15–20 (1921).

Two works that present complete analyses of slip line theory applied to Coulomb materials are

V.V. Sokolovski, *Statics of Soil Media*, (translated by D.H. Jones and A.N. Schofield), Butterworths, London, 237pp., 1960.

A. Nadai, *Theory of Flow and Fracture of Solids*, Vol. 2, McGraw-Hill, New York, pp. 435–475, 1963.

Comprehensive numerical solutions for Kötters' equation may be found in

H.Y. Ko and R.F. Scott, Bearing capacities by plasticity theory, *J. Soil Mechan. Found. Div., ASCE*, **99**, SM1, 25–43 (1972).

The theory for non-associated flow by Spencer is described in the reference below. Spencer's work was significantly extended in the second citation.

A.J.M. Spencer, A theory of the kinematics of ideal soils under plane strain conditions, *J. Mech. Phys. Solids*, **12**, 337–351 (1964).

M.M. Mehrabadi and S.C. Cowin, Initial planar deformation of dilatant granular materials, *J. Mech. Phys. Solids*, **26**, 269–284 (1978).

Seminal works in determination of the strain rate and velocity fields for the case of associated flow of perfectly plastic Coulomb materials are

D.C. Drucker and W. Prager, Soil mechanics and plastic analysis or limit design, *Quart. Appl. Math.*, **10**, 157–165 (1952).

R.T. Shield, Mixed boundary value problems in soil mechanics, *Quart. Appl. Math.*, **11**, 61–75 (1953).

An introduction to the method of characteristics, applied to soil mechanics as well as other fields, is given by

M.B. Abbott, *An Introduction to the Method of Characteristics*, Thames and Hudson, London, 1966.

Exercises

6.1 In Figure 6.12 the angle ψ_B was set equal to π. An alternative interpretation of the situation depicted in Figure 6.11 would have $\psi_B = 0$. Using this alternative interpretation, carry through the analysis to determine p_1 in terms of p_0. Discuss your conclusion in light of the initial assumption that p_1 is the greater applied stress.

6.2 Consider the problem illustrated in Figure 6.28. An embankment with slope angle ξ consists of a purely cohesive material. It supports an applied stress p_0 on its horizontal surface. Ignore gravity.

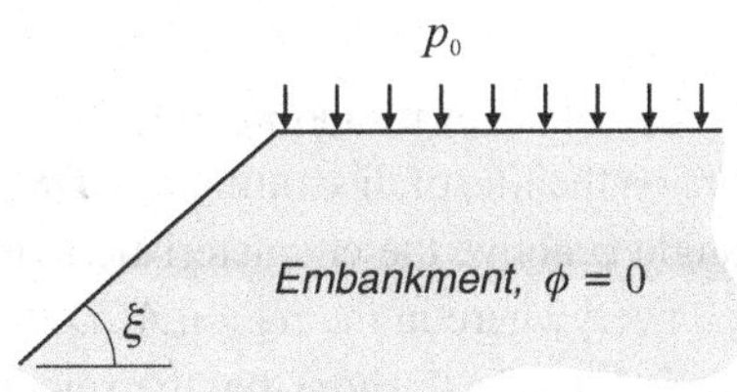

Figure 6.28.

(a) Sketch the network of α- and β-lines that would result if the embankment was in a failure state.

(b) Determine the magnitude of the load p_0 required to cause failure in terms of the cohesion c and the angle ξ.

6.3 Solve equations (6.28) for $\partial p/\partial x$ and $\partial p/\partial y$. Use your result to complete the derivation of (6.29).

6.4 A halfspace of purely cohesive material supports a system of applied surface tractions that produce the network of α- and β-lines shown in Figure 6.29. Determine all possible systems of surface tractions that might produce the network shown.

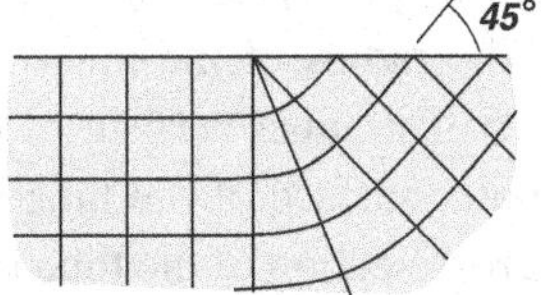

Figure 6.29.

6.5 Note from Figure 6.26 that, in the passive region outside the strip load, both α- and β-lines are straight. Thus the derivatives of ψ in Kötter's equations (6.44) will vanish. Use Kötter's equations to show that the vertical stress component everywhere within the passive zone is equal to $\rho g y$.

6.6 Figure 6.30 shows the right-hand side of the strip footing characteristic line network. Assume that the region ABC moves vertically downward as a rigid body with velocity v_0.

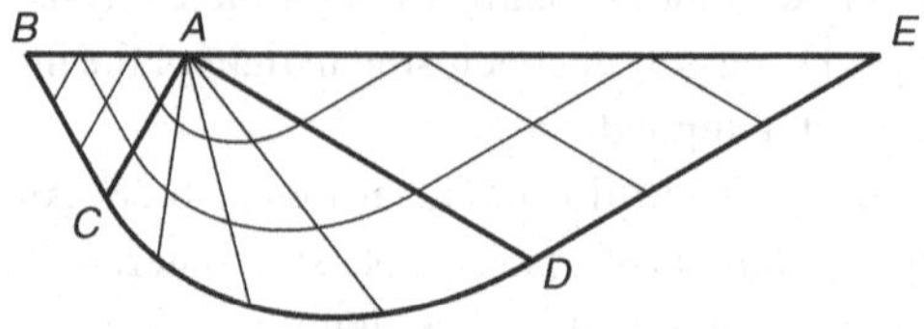

Figure 6.30.

(a) Determine the values of v_α and v_β in terms of v_0 in the region ABC.
(b) Everywhere on the line of discontinuity CDE the velocity vector must lie at an angle φ above the characteristic line. Use this fact to prove that $v_\beta = 0$ everywhere in the region $ACDEA$.
(c) Use the result from (b) to show that the velocity vector at every point in the region $ACDEA$ must be perpendicular to the β-line passing through the point.
(d) Crossing the line of discontinuity AC, the relative velocity (or change in velocity) must lie at an angle φ to the line. Use that fact together with the result from (c) to prove that on the line AC the velocity of particles in $ACDEA$ must have magnitude $v = v_0/2 \sec(\pi/4 + \varphi/2)$.
(e) Use the result from (d) to show that the value of v in the fan ACD is given by

$$v = \frac{v_0}{2} \sec\left(\frac{\pi}{4} + \frac{\varphi}{2}\right) e^{\theta \tan \varphi}$$

where θ is measured from the line AC.
(f) What is the velocity of the ground surface AE?

6.7 A thick-walled tube of purely cohesive material is deforming in plane strain conditions. The radius of the inner surface of the tube is a, and the outer radius is b. The inner surface of the tube supports an applied pressure p_0. The outer surface of the tube supports an applied pressure p_1. Assuming the cross-section of the tube is in a fully yielded condition, discuss all the possible values that p_0 and p_1 may take.

7
Work hardening and modern theories for soil behaviour

7.1 Introduction

This final chapter presents a collection of ideas related to work hardening, as well as some thoughts on modern descriptions of the mechanical response of soils. Recall from Chapter 3 that work hardening materials, in contrast to perfectly plastic materials, may change their response during yielding. These changes are accomplished by altering, in some fashion, the shape or size of the yield surface *as* plastic flow occurs. Initially the concept of work hardening was introduced to give a better representation of the stress–strain response of metals. The ideas involved are straightforward, although there is a price to pay in terms of increased levels of complexity. We have avoided the topic until now, not because it is unimportant, but because it plays such an important role in the modern theories of soil plasticity that are also considered in this chapter.

Geotechnical engineers have found the general concept of work hardening extremely useful whenever there is a need for response calculations that are more detailed than is possible with perfect plasticity. In particular, the closed yield surfaces described in Chapter 3 would be nearly useless without work hardening. With a closed yield surface there is a possibility of plastic response under increasing isotropic stress, but one cannot arbitrarily limit the amount of stress increase by establishing a fixed yield surface at some arbitrary stress level. We must be able to increase the isotropic stress, to move the stress point out along the space diagonal, without limitation. This implies that the yield point must also change and hence, work hardening must occur.

The whole concept of how a soil yields under increasing isotropic stress is one of the two central points (the other being frictional strength) that discriminates between metal plasticity and soil plasticity. There has been ample evidence for soil yielding due to isotropic stress for many decades, but only since the 1960s have there been plasticity theories that incorporate the effect coherently. We group these theories under the general heading of critical state soil mechanics. Even newer are theories that attempt to account on a microscopic scale for the grain crushing and rearrangement that lead to plasticity effects.

The 40 or so years since the introduction of critical state soil mechanics has seen a proliferation of similar material models all attempting to reproduce more closely the observed stress–strain response measured in laboratory tests. At times these theories seem to provide more detail than is either necessary or desirable. To reproduce every twist and turn in a collection of laboratory data may seem an admirable goal from the standpoint of academic accuracy but the practicalities usually require that the model be based on a large number of parameters. Models with as many as 20 or more material parameters have been proposed. The problem of determining parameter values for a particular application becomes a daunting task. Moreover, the variability of natural soil deposits suggests that even when the parameters are well defined, they may provide a model which is at best applicable in only a small part of the total soil affected. In many practical situations the geotechnical engineer has more need of robust approximations than of extreme but fragile precision.

In the preceding six chapters we have mostly taken the rough but robust approach that characterises the engineering approximation in contrast to the delicate but precise approach that is possible with modern critical state theories. That is not to suggest that we see no value in critical state soil mechanics. The exact opposite is true. Nevertheless, it is in the best interest of geotechnical engineering that practitioners have a solid background knowledge of the nuts and bolts of all of the theory of plasticity. The aspects of plasticity covered in many modern textbooks as well as in many geotechnical engineering courses begin and end with critical state theories, with the result that graduates have little contact with what might be called the classical aspects of plasticity: metal plasticity, collapse load theorems and slip line fields. We believe these aspects form an essential underpinning for the appreciation and critical analysis of what the modern theories are capable of producing.

This chapter will introduce some of the detail involved in critical state theories, but no more than is necessary to provide a basic understanding of the goals of the theory and how the earliest models worked. We will also attempt to investigate one of the rudimentary theories for a micromechanical description of soil response. Unlike the previous six chapters some of the material presented here may become dated or even obsolete because of discoveries yet to be made. We begin, however, with some basics.

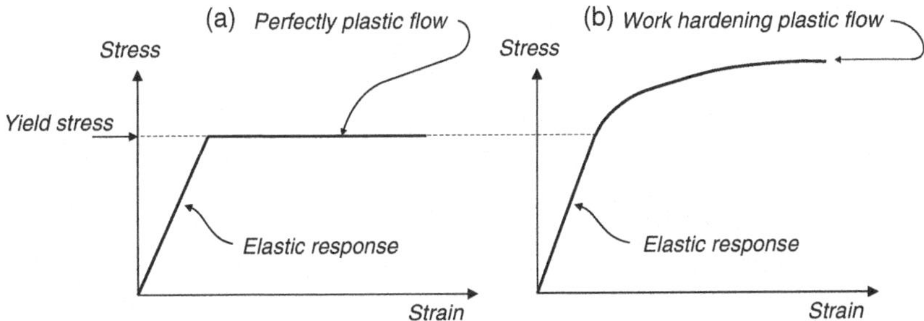

Figure 7.1. Stress–strain response for perfect and work hardening plastic materials.

7.2 Work hardening for metals

The post-yield response of a typical ductile metal is usually more like that illustrated in Figure 7.1(b) than the perfectly plastic response shown in Figure 7.1(a). Yielding and flow are accompanied by an apparent increase in strength called *hardening*. The term *strain hardening* is often used since the strength increase happens as the plastic straining increases. The term *work hardening* is also used to describe the phenomenon.

Figure 7.1 is a bit vague about what measures of stress and strain are in use, but we can think of the data as arising from a simple tension test. In that case our concerns would be the principal stress σ_1 and the corresponding axial strain ε_1. The yield stress would be the tensile strength σ_T. For the work hardening case we see an initial yield stress similar to the perfectly plastic case, but as the strain increases the yield stress also increases. A relatively painless way to account for that behaviour is to make σ_T grow as the amount of plastic strain or the quantity of plastic work increases.

The above ideas may be generalised as follows. Suppose we wish to use a Tresca yield condition which, following (3.10), we might write as

$$\sigma_1 - \sigma_3 = \sigma_T = \sigma_T(W_p) \tag{7.1}$$

Here the yield stress σ_T is written as a function of the *plastic work* W_p. In a general sense, the yield function f in (3.1) and (3.2) has now become a function of W_p as well as a function of the components of stress. The plastic work is found by integrating equation (4.10) or (4.11)

$$W_p = \int \dot{W}_p \, dt = \int tr(\boldsymbol{\sigma}\dot{\boldsymbol{\varepsilon}}^p) \, dt = \int \left(\sigma_1 \, d\varepsilon_1^p + \sigma_2 \, d\varepsilon_2^p + \sigma_3 \, d\varepsilon_3^p\right) \tag{7.2}$$

where the range of integration covers the range of plastic response. Physically the plastic work is that part of the total stored energy W not produced by elastic

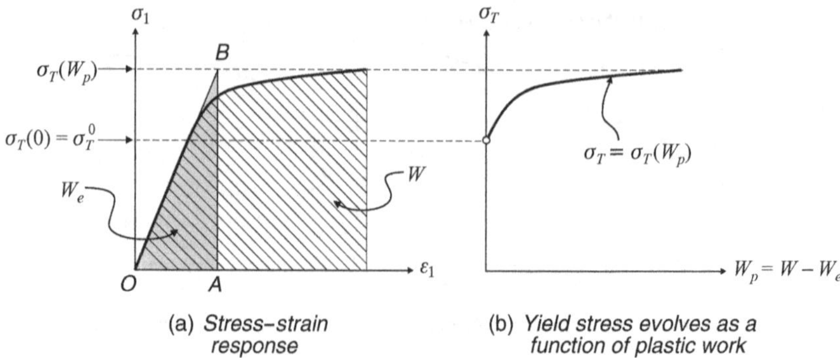

Figure 7.2. Calculation of plastic work in a work hardening material.

deformation. In the context of a simple tension test the plastic work may be written as

$$W_p = \int \sigma_1 \, d\varepsilon_1^p = \int \sigma_1 \left(d\varepsilon_1 - d\varepsilon_1^e \right) = W - W_e \qquad (7.3)$$

where σ_1 and ε_1 are the axial stress and strain. The quantities W and W_e may be visualised as shown in Figure 7.2(a). The elastic stored energy, W_e, corresponds to the grey area of the triangle *OAB*. If E is the value of Young's modulus appropriate to the material, then $\varepsilon_1^e = \sigma_1/E$ and $W_e = \sigma_1^2/2E$. The total work W is the larger cross-hatched area beneath the stress–strain curve. The difference of the two areas gives the plastic work at that stage of the process. Note that W_p is zero for an elastic response since W and W_e are identical. W_p begins to grow after initial yield occurs. The function $\sigma_T(W_p)$ is precisely the value of the stress σ_1 in simple tension mapped as a function of W_p as shown in Figure 7.2(b). The initial yield stress when $W_p = 0$ is denoted by σ_T^0.

An important issue comes up at this point. If we map the yield stress σ_T on to the simple tension stress, then does this mean that the Tresca hexagon has grown larger? The question may be put another way. If we cause the material to yield by placing it in tension, then have we increased the yield stress in compression? The question and two possible answers are sketched in Figure 7.3. In the figure the plane strain Tresca hexagon is shown in the σ_1–σ_3 plane. The dashed hexagon is the initial yield surface and the two solid hexagons are possible subsequent surfaces after hardening has occurred. A simple tension stress trajectory is the horizontal line beginning at o and ending at b. Point a is the initial yield point and the space between a and b corresponds to hardening. Part (a) of the figure shows one possible outcome. The yield surface has been inflated by the hardening process so that if we were to reverse the stress after this initial loading we would find that the yield point has been increased. Part (b)

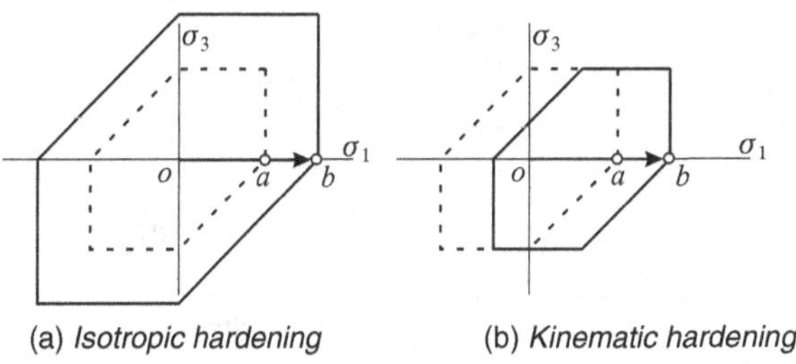

Figure 7.3. Comparison of isotropic and kinematic hardening.

shows a different possibility. The size of the yield surface has not changed and hardening has simply translated the initial surface to the right by the amount *ab*. A reversal of stress here would encounter the yield surface much earlier than in case (a). The first possibility is called *isotropic* hardening, while the second one is referred to as *kinematic* hardening. There are other possible ways to accommodate hardening. For example, we could visualise the stress point pushing out one side of the hexagon while the other side remains stationary, but the two possibilities shown in Figure 7.3 are the ones most widely used.

Which of the two hardening possibilities, isotropic or kinematic, is better? Generally the kinematic case is preferred. When most metals are tested in simple tension, post-yield hardening is observed. If the stress is then reversed, the reverse yielding usually occurs near the point predicted by the kinematically translated surface. This occurs despite the fact that the yield point in compression for an unhardened specimen would be the same as that in tension. The alteration of the yield point observed when a stress reversal occurs after hardening is called the *Bauschinger effect* named after the German engineer J. Bauschinger.

Figure 7.4 illustrates the Bauschinger effect. If we perform a simple tension test, initial yield is found at a stress denoted by σ_T^0, at the point marked A on the figure. A compression test on a similar unhardened sample would yield at the compressive stress level σ_T^0, at point B. But if we reverse the stress after some hardening has occurred in the tensile test, the new reverse yield point is found at point C. The total stress reversal required to produce reverse yielding is $2\sigma_T^0$. A similar result would be observed if the compressive test loading were reversed and a new tensile yield point determined.

Finally, we might wonder just how a kinematically hardened yield surface would move if the loading were more complex than simple tension or compression. The usual answer is that the surface should translate as if the stress point were a 'frictionless roller' pressing against the inside. The surface is guided to translate freely in any direction but is not permitted to rotate. Clearly this concept is applicable to the von Mises yield surface equally as well as to Tresca's.

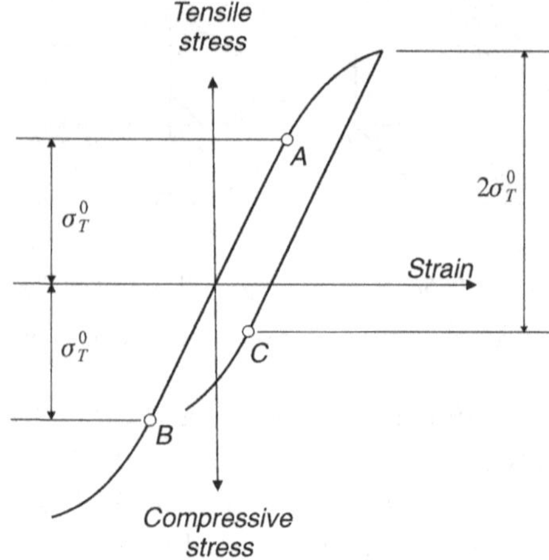

Figure 7.4. The Bauschinger effect.

Despite the fact that kinematic hardening appears to better represent actual test data, isotropic hardening is still widely used. The reason lies in the ease with which isotropic hardening may be implemented mathematically. Of course, in applications where stress reversals are not anticipated, the choice of model makes little difference.

7.3 Cam Clay

The simple work hardening theories for metals are not sufficient to describe in detail the complex behaviour exhibited by real soils. In the second half of Chapter 2 we described some of the aspects of inelastic behaviour of soils that are commonly observed. The reader is encouraged to return to Chapter 2 to review the points raised there. In summary, we noted first that irreversible effects may occur in the absence of shearing stress. Typical oedometer test results were shown in Figure 2.6 where a yielding response is shown as the applied normal stress is increased. Of course, there are shear stresses present in the oedometer sample, but a similar response may be obtained under purely isotropic stress. It is particle crushing and rearrangement due to normal stress increase that causes the irreversible deformation. Second, we noted that shear stress may also result in plastic effects, but these are complicated by interlocking. A densely packed sample of a particular soil may behave very differently from a loosely packed sample. Interlocking produces an increase in strength that may be broken down as dilation occurs. The resulting stress–strain curve might look like that shown in Figure 2.10 where a pronounced stress peak is apparent. Dilatancy leads to an increase in sample volume. In contrast, a loosely packed sample may exhibit no stress peak. The stress–strain response may look more 'conventional' such as that in Figure 7.1(b). The sample volume may decrease as loading progresses. It seems clear that a chasm exists between the simple perfect plasticity theories and the actual behaviour of real soils.

A reasonable question to ask at this point is should we worry about this chasm or not? The solutions developed for a perfectly plastic response are extremely powerful *and* have the virtue of simplicity. Perfect plasticity can only roughly approximate the stress–strain behaviour of a real soil, but is that rough approximation enough? The answer is yes and no. For many real applications, perfect plasticity will give reliable and robust results that may be sufficiently accurate, especially in light of the natural variability of real soils. But there are some applications where greater accuracy may be required. Particularly in sensitive projects where prediction errors may prove especially costly, or in applications where a reasonable margin of safety cannot be incorporated in the design process, the simple perfect plasticity answers may not be good enough.

Given that there is a need in some applications for a more accurate prediction of soil response, we must confront the complexities mentioned above. A theory is needed that incorporates yielding under isotropic stress increase as well as the complex behaviour observed in shearing. The theory of *critical state soil mechanics* will provide the things we desire. When the material is sheared the shape of the yield surface together with the normality condition will automatically *attract* the stress point to the critical state. Thus dense soils will dilate and loose soils will compress, and both will approach a state at which no further volume change occurs. At the same time the stress will approach the ultimate strength value. The very first critical state theory was called Cam Clay.

Cam Clay was originated by the Cambridge soil mechanics group in the 1960s. We have already described the Cam Clay yield surface in Chapter 3, but at that point, without the benefit of a flow rule, it was impossible to appreciate its versatility. Rather than beginning directly with the surface described in equation (3.31), we will attempt to present an overview of the Cam Clay model using the simplest possible loading geometry. By considering a very simple problem we can avoid a number of minor complications that have no effect on the fundamental model response. Later we can sketch out how to generalise the simple problem to other possible loading situations.

The problem to be considered is *simple shearing*, illustrated in Figure 7.5. A sample of soil is subjected to applied normal and shearing tractions, σ and τ, on its upper surface. If the sample happens to be saturated, we consider σ to be the effective stress. The soil is laterally constrained so that no extensional deformation occurs in the horizontal direction. This can be accomplished using

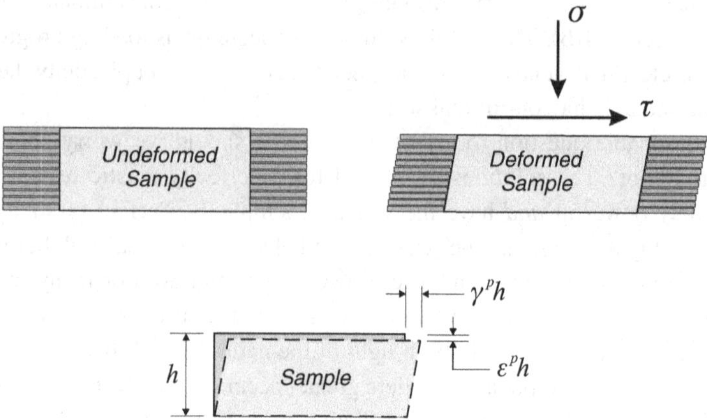

Figure 7.5. Schematic diagram showing a simple shear test.

a sort of 'pancake' container consisting of layers of rigid material that may slide over each other, permitting shearing strain but no extensional strain. It is not a simple device, but that is not our concern. The essential thing is that the only extensional strain is the vertical strain, denoted by ε^p. The only other non-zero strain is the shear strain, denoted by γ^p. We denote both strains as plastic. The theory to be derived will overlook elastic strains for the moment, but they will be considered later.

The simple shear test is an especially nice example for the purpose of developing our theory. We would expect to see all the complex effects discussed above, but there are only two components of stress and two of strain to contend with. The originators of the Cam Clay model used the triaxial test as their example problem, and it also has only two stress and two strain components, but those components are derived from invariants and are slightly more complex than the simple stresses and strains considered here.

First, we want to establish a yield surface. In the context of our example problem, this will be a function of the form $f(\sigma, \tau)$. To begin, assume the soil sample is in a yield state and write down the rate of plastic work:

$$\dot{W}_p = \sigma\dot{\varepsilon}^p + \tau\dot{\gamma}^p \tag{7.4}$$

W_p represents plastic, irrecoverable work done by the applied tractions. An approach to plasticity we have not yet mentioned takes (7.4) as its starting point and then postulates that $\dot{W}_p$ must equal a specified function called the *dissipation function*, $\dot{D}$. It can be shown that the dissipation function should be a homogeneous function of the plastic strain rates multiplied by coefficients that depend upon the stresses. If we consider the case where σ is constant then it is reasonable to assume that the plastic extensional strain ε^p is at most a function of the plastic shear strain γ^p. Then the dissipation function $\dot{D}$ can be written as a function of $\dot{\gamma}^p$ only. Also, for a frictional material, the rate of dissipation should depend on the normal stress σ. The Cambridge workers postulated a dissipation function with the form

$$\dot{D} = k\sigma\dot{\gamma}^p \tag{7.5}$$

where k is a material parameter that is constant for any particular soil. Setting the right-hand sides of (7.4) and (7.5) equal and rearranging gives

$$\frac{\dot{\varepsilon}^p}{\dot{\gamma}^p} = k - \frac{\tau}{\sigma} \tag{7.6}$$

where we assume the shear strain rate $\dot{\gamma}^p$ to be strictly positive.

Next, return to (7.4) and consider what might happen if we were to slightly alter the stresses σ and τ. Suppose we alter both stresses by small amounts $\delta\sigma$ and $\delta\tau$. Then the rate of plastic work would also be altered by some amount $\delta\dot{W}_p$. Drucker's postulate* states that, so long as the body remains in equilibrium, $\delta\dot{W}_p$ should always be equal to or greater than zero. Therefore

$$\delta\dot{W}_p = \delta\sigma\,\dot{\varepsilon}^p + \delta\tau\dot{\gamma}^p \geq 0 \tag{7.7}$$

In the limiting case where the equality holds this expression embodies the normality condition. If we take the equality, we may write

$$\frac{\delta\tau}{\delta\sigma} + \frac{\dot{\varepsilon}^p}{\dot{\gamma}^p} = \frac{d\tau}{d\sigma} + k - \frac{\tau}{\sigma} = 0 \tag{7.8}$$

where (7.6) has been used and δs have been replaced by ds. We can integrate (7.8) to give

$$\tau = \sigma\,(C_1 - k\ \ln\sigma) \tag{7.9}$$

where C_1 is a constant of integration. This expression will become our Cam Clay yield surface; however, an initial condition is still needed to find the constant C_1.

It is not immediately clear what initial condition is appropriate for (7.9), but (7.6) offers a clue. Looking at (7.6) and assuming $\dot{\gamma}^p > 0$, we see that there are three possibilities for the extensional strain rate, $\dot{\varepsilon}^p$. If $\sigma > \tau/k$, then $\dot{\varepsilon}^p > 0$ and compression is occurring. On the other hand, if $\sigma < \tau/k$, then $\dot{\varepsilon}^p < 0$ and dilation is occurring. The first condition would correspond to a loosely packed

* See Appendix E. Drucker's postulate is based on the idea that for a yielding body in equilibrium under a system of loads any small change in loading should *not* cause the body to do work on its surroundings. The postulate can be used to derive the normality condition.

soil and the second to a densely packed soil. The third possibility is $\sigma = \tau/k$. If this is the case then $\dot{\varepsilon}^p = 0$ and the sample volume is not changing. The Cambridge researchers referred to the first possibility as 'weak' yielding. The second possibility was called 'strong' yielding. Later the terms 'wet' and 'dry' were substituted for weak and strong, but the end result is the same. In the first case the soil compacts, in the second it dilates. The two possibilities are separated by the special case where no volume change occurs. This corresponds to the state where the soil is neither loosely nor densely packed. Shear strains can grow without any change in volume. We call this the critical state. Experimental evidence shows that, regardless of the initial state of packing of a sample, there is always a tendency to move toward the critical state. That is, the loose soil compacts, the dense soil dilates, and both are changing their volume in such a way as to move closer to the critical packing where no volume change occurs. If sufficiently large amounts of shearing strain occur, then the critical condition can be achieved. Recall Figure 2.10 where, at large axial strain, the rate of change of volumetric strain approaches zero. At the same time the stress deviator q has also reached its constant, ultimate value.

Now return to equation (7.9) and the constant C_1. To establish C_1 note that equation (7.6) says that when $\sigma = \tau/k$ we are at the critical state. Suppose we define a *critical state stress*, σ_c, which is equal to τ/k. In general, σ will be different from σ_c, but if they do coincide the particle packing will be at its critical state and there will be no further volume change. In that state, (7.9) would read

$$k\sigma_c = \sigma_c(C_1 - k\,\ln\sigma_c) \tag{7.10}$$

We can solve this equation for C_1 and use the result in (7.9) to find the following expression:

$$\tau + k\sigma\left[\ln\left(\frac{\sigma}{\sigma_c}\right) - 1\right] = 0 \tag{7.11}$$

This equation relates the stresses σ and τ during yielding and hence represents our Cam Clay yield surface. Comparing with (3.31), it is clear that (7.11) defines a surface similar to that shown in Figure 3.18. In fact, we can look on τ and σ as surrogates for q and p in the yield surface defined by (3.31). The parameter M in (3.31) is replaced by k in (7.11). A graph of (7.11) is shown in Figure 7.6.

We have gone through the same procedure as the original Cambridge researchers when the Cam Clay yield surface was first obtained, but our development is more specialised in that we are concerned only with the example problem in Figure 7.5. The Cambridge group cast their results in terms of stress and strain invariants and hence obtained a fully three-dimensional theory reflected

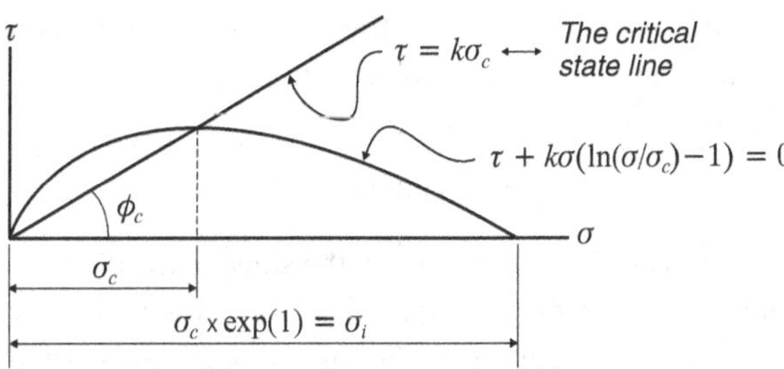

Figure 7.6. Cam Clay yield surface in a simple shear test.

in the three-dimensional nature of Figure 3.19. Our equation (7.11) mimics their result in all important details but is simpler. A number of modern textbooks reiterate the original Cam Clay derivation and there seems to be little reason to reproduce it here once again. Our aim is to bring out the essential elements of the theory in a physically intuitive way.

Suppose we now imagine the simple case where the normal stress σ is increased but no shear stress is applied. The stress point will move along the horizontal axis in Figure 7.6 until it arrives at the yield surface when $\sigma = \sigma_c \exp(1)$. What happens next? Presumably there will be plastic compression and, somehow, the surface must now expand to accommodate any further increase in σ. It is unthinkable that the surface might not expand because in that case we could never increase σ above the initial yield value, and we know that we should be able to increase σ to any value we wish. If the surface expands then evidently σ_c must increase, and we could make this happen by letting σ_c be a function of the plastic extensional strain ε^p. We will refer to this process as hardening and $\sigma_c = \sigma_c(\varepsilon^p)$ becomes our *hardening parameter*. It is the parameter that prescribes where the yield surface lies in (σ, τ) space.

We must now decide how σ_c depends on the plastic extensional strain. This will not be difficult since we can rely directly on experiment to tell us what happens when σ is increased while keeping $\tau = 0$. For a moment let $\sigma_i = \sigma_c \times \exp(1)$, i.e. the normal stress intercept of the yield surface (see Figure 7.6). Then keeping $\tau = 0$, yielding will commence when σ first reaches σ_i. As hardening occurs, σ and σ_i will move together towards the right in Figure 7.6. At the same time the yield surface will grow larger. Since there is no shear stress applied to the sample our simple experiment is exactly the same as an oedometer test. Recall the oedometer test results sketched in Figure 2.6. Loading the sample results in a decrease in the void ratio and there are usually two linear portions in the relationship between the void ratio and the logarithm of stress. The steeper portion is referred to as virgin compression or normal compression and it corresponds to a plastic response. We also realise that the sample void ratio is related linearly to the compressive strain. Therefore, to be consistent with this well-known behaviour, we need to make the plastic strain ε^p a logarithmic function of the stress σ_i,

$$\varepsilon^p = m \ln \left(\sigma_i / \sigma_i^0\right) \tag{7.12}$$

Here m is a material constant representing the slope of the ε^p–$\log \sigma_i$ line. σ_i^0 is the initial value of σ_i. It defines the initial position of the yield surface. It also corresponds to the upper point of the virgin compression curve, the so-called preconsolidation stress. Now we can see one way in which the yield surface can change. In the oedometer test the applied stress σ is first increased from zero to σ_i^0. The stress point lies inside the yield surface and an elastic response will occur. When σ equals σ_i^0, yielding commences, and a further increase in σ pushes σ_i to the right. The yield surface is enlarged and, in terms of the void ratio versus the logarithm of the stress curve, the sample state is moving down the virgin compression curve. Finally, recall that $\sigma_i = \sigma_c \times \exp(1)$. Thus (7.12) may be rewritten as

$$\varepsilon^p = m \ln \left(\sigma_c / \sigma_c^0\right) \tag{7.13}$$

where $\sigma_c^0 = \sigma_i^0 \times \exp(-1)$ is the initial value of σ_c. Equation (7.13) provides our relationship between ε^p and the hardening parameter σ_c.

Taken together, (7.11), (7.13) and the associated flow rule fully describe the plastic behaviour of our sample. Three material constants have been used, k, m and σ_c^0. The constant m is determined directly from oedometer test data. The constant σ_c^0 defines the initial shape and the position of the yield surface. It is related to σ_i^0, which is the preconsolidation stress, also determined from the oedometer test. The last constant k represents the ratio τ/σ_c. How might it be determined? Recall that for both densely and loosely packed samples, laboratory test results show that there will be an ultimate shear strength. In the case of a triaxial test, the stress deviator q approaches an ultimate strength value as the strain grows large. The applied shear force in a direct shear test does the same. The ultimate strength occurs at the same time as the sample volume change approaches zero. We say that the sample is approaching the critical state. In our simple shear test we would expect a similar response. Thus τ/σ should approach a constant value as the shear strain becomes large. But at the same time σ will approach σ_c. The value of τ/σ_c, and hence k, will be the same as τ/σ at the ultimate state. We will use φ_c to denote the ultimate or critical state friction angle for our soil, i.e. $\varphi_c = \tan^{-1} \tau/\sigma$ at failure. Thus $k = \tan \varphi_c$, and we see that the three material constants are easily determined.

Now we are in a position to solve a simple problem. This will help to illustrate the variety of responses that Cam Clay may produce. The problem envisioned

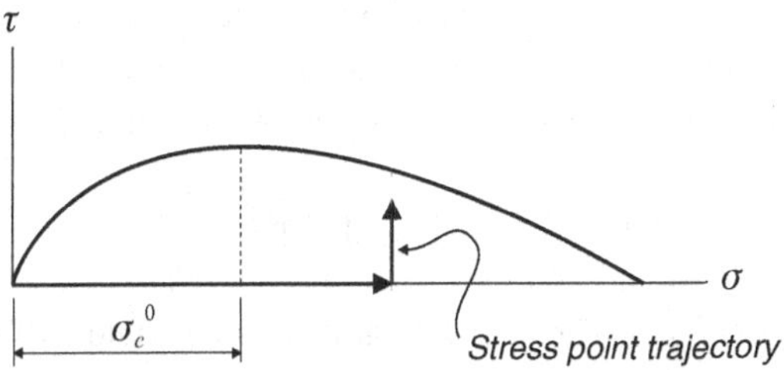

Figure 7.7. Stress point trajectory and Cam Clay yield surface for a simple shear test.

is as follows. We assume the three constants m, σ_c^0 and φ_c are known. Then the normal stress σ is increased to some specified value, and is thereafter held constant. Next, we increase the shear stress τ and take the sample to failure (to the critical state). Our aim is to determine the stress–strain curve (τ versus γ) for the soil. The initial yield surface and the trajectory of the stress point in the elastic region are shown in Figure 7.7. As yet we have not considered any elastic strains, but it is a simple matter to set

$$\gamma^e = \frac{\tau}{G}, \quad \varepsilon^e = \frac{\sigma}{E}\left[\frac{(1+\nu)(1-2\nu)}{1-\nu}\right] \tag{7.14}$$

where G, E and ν are the elastic constants. So long as the stress point lies within the yield surface the only strains will be given by (7.14). At some point however, the stress point will arrive on the yield surface. At that point plastic strains will commence and the associated flow rule tells us how they are related.

$$\begin{aligned} \dot{\gamma}^p &= \lambda\frac{\partial f}{\partial \tau} = \lambda \\ \dot{\varepsilon}^p &= \lambda\frac{\partial f}{\partial \sigma} = \lambda k \,\ln(\sigma/\sigma_c) \end{aligned} \tag{7.15}$$

where f is the yield surface defined by (7.11). Differentiating (7.13) we also have

$$\dot{\varepsilon}^p = m\frac{\dot{\sigma}_c}{\sigma_c} \tag{7.16}$$

Combining (7.15) and (7.16) to eliminate λ and $\dot{\varepsilon}^p$ gives

$$\frac{\dot{\sigma}_c}{\sigma_c} = \frac{k}{m}\ln\left(\frac{\sigma}{\sigma_c}\right)\dot{\gamma}^p \tag{7.17}$$

and, since σ is constant, we can integrate to obtain

$$\ln\left(\ln\frac{\sigma}{\sigma_c}\right) = -\frac{k}{m}\gamma^p + C_2 \tag{7.18}$$

where C_2 is the constant of integration. Our initial condition is $\sigma_c = \sigma_c^0$ when $\gamma^p = 0$; that is, plastic strains will commence when the stress point touches the initial yield surface. Using this to evaluate C_2 we find (7.18) becomes

$$\ln\left(\frac{\sigma}{\sigma_c}\right) = \ln\left(\frac{\sigma}{\sigma_c^0}\right) \exp\left(-\frac{k}{m}\gamma^p\right) \tag{7.19}$$

Finally, return to (7.11). The stress point lies on the yield surface and we can use (7.19) to replace $\ln(\sigma/\sigma_c)$. This gives

$$\tau = k\sigma\left[1 - \ln\left(\frac{\sigma}{\sigma_c^0}\right) \exp\left(-\frac{k}{m}\gamma^p\right)\right] \tag{7.20}$$

Equation (7.20) defines the shear stress–plastic shear strain response of Cam Clay in simple shear. A typical shear stress response is shown in Figure 7.8. For illustrative purposes parameter values of $k = 1.0$ and $m = 0.025$ have been used. Two cases are considered. In case (a) the normal stress σ is greater than the initial critical state stress σ_c^0. This would be the case illustrated in Figure 7.7, where the stress point first touches the yield surface to the right of σ_c^0. In

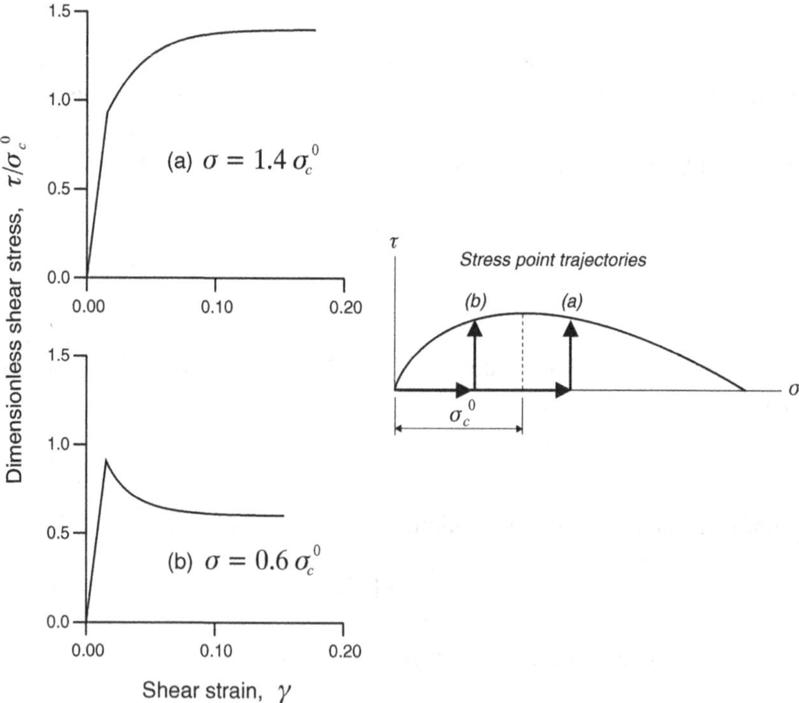

Figure 7.8. Typical Cam Clay stress–strain response in a simple shear test.

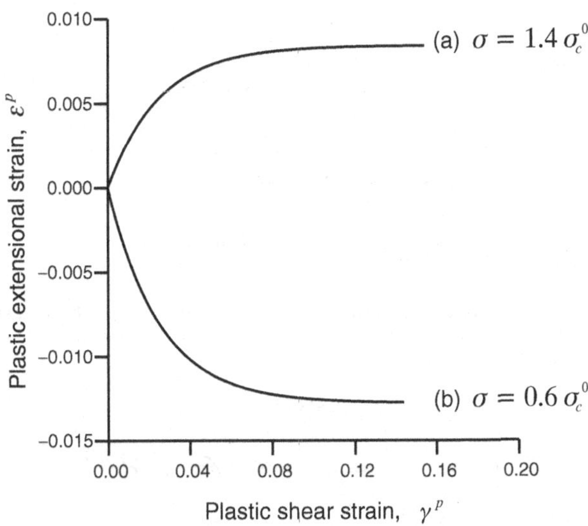

Figure 7.9. Typical extensional strain response for Cam Clay in a simple shear.

case (b) we have $\sigma < \sigma_c^0$ and the stress point contacts the yield surface to the left of σ_c^0. Elastic strains are incorporated in the figure using the first equation of (7.14). Note the crucial role played by the ratio σ/σ_c^0 in (7.20). If $\sigma > \sigma_c^0$ then $\ln(\sigma/\sigma_c^0)$ will be positive and increasing plastic strain will be accompanied by increasing shear stress. But if $\sigma < \sigma_c^0$, then $\ln(\sigma/\sigma_c^0)$ will be negative and the shear stress decreases during yielding. In both cases the shear stress approaches $k\sigma$, the critical state, for large values of strain.

The two different classes of response shown in Figure 7.8 are precisely what we expect of loosely packed and densely packed soils. The volumetric strains are also as we expect. Figure 7.9 shows plots of ε^p versus γ^p for the two situations in Figure 7.8. Note that when $\sigma > \sigma_c^0$, ε^p is positive and the sample compresses. Conversely, when $\sigma < \sigma_c^0$, ε^p is negative, indicating dilation. Each type of response reflects changes in the yield surface. When the sample compresses, the yield surface grows and the soil hardens. When the sample dilates the surface shrinks and we say the soil *softens*. Hardening and softening are characterised by the two stress–strain curves shown in Figure 7.8.

In both the hardening and softening cases the sample is being attracted towards the critical state. For case (a), where $\sigma > \sigma_c^0$, the yield surface grows and the stress path trajectory moves vertically toward the point where σ and σ_c will coincide as shown in Figure 7.10. This figure is drawn for the case where $\sigma = 2\sigma_c^0$. The reason the yield surface does grow lies in the associated flow rule. Figure 7.10 shows the stress path moving through a sequence of yield surfaces,

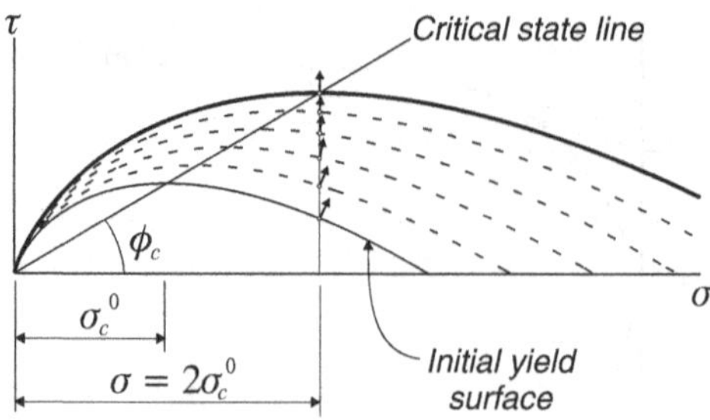

Figure 7.10. Evolution of the yield surface for hardening of Cam Clay in a simple shear.

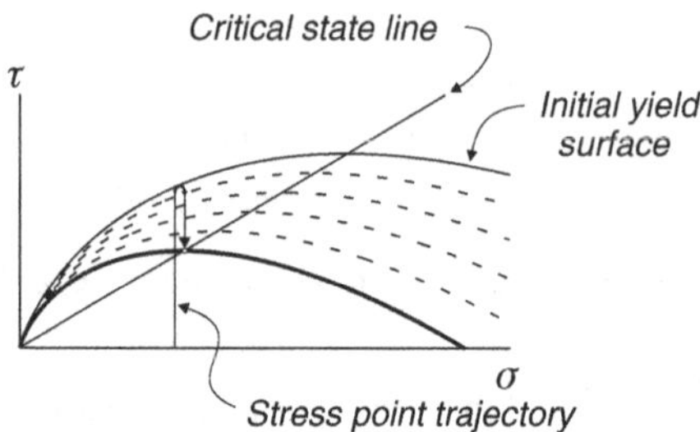

Figure 7.11. Evolution of yield surface for softening of Cam Clay in a simple shear.

and the associated plastic strain rate vectors for each. Note how the normality condition causes the vectors to gradually become more and more vertical as the stress point approaches the critical state line. When the stress point reaches the critical state line the plastic strain rate vector becomes exactly vertical. Recall that the extensional strain rate $\dot{\varepsilon}^p$ is proportional to the horizontal component of the strain rate vector. We see that the rate of plastic extensional strain is greatest at the instant yield commences and diminishes thereafter until it becomes zero when the critical state is reached. Finally, note that the size of the yield surface depends directly on ε^p through (7.13). Compressive plastic strain leads to increasing σ_c and a larger yield surface. The process continues until the plastic strain rate vector becomes vertical. A similar process occurs when σ is smaller than σ_c^0, but in that case the plastic strain rate vectors slope to the left, the plastic extensional strain is negative (indicating dilation), and the yield surface grows smaller. Yielding begins when the stress point first touches the yield surface. Then the stress is pulled down by the yield surface as illustrated in Figure 7.11. In the figure the stress path is shown doing a U-turn but this is only to make clear that the stress point moves both upward initially and downward (on the same line) once yielding commences.

It is now clear how the Cam Clay model produces the various types of response we require. The critical state has an in-built attraction for the stress point. Regardless of where the stress point first touches the yield surface, the volumetric strain will cause the surface to convey the stress point toward the critical state. Once there the volume remains constant and the shear stress is at its ultimate strength value. This is the great achievement of critical state soil mechanics.

The example of simple shearing presented above indicates all the important aspects of soil behaviour. Rather than simple shear, however, the Cambridge researchers developed Cam Clay from the perspective of a conventional triaxial test. We will roughly sketch their development. Instead of (7.4) they set down the following expression for the rate of plastic work:

$$\dot{W}_p = q\dot{\xi}^p + p\dot{e}^p \tag{7.21}$$

Here q and p are the deviatoric and mean stresses, which, in the context of the triaxial test, are given by

$$q = \sigma_1 - \sigma_3, \quad p = \frac{1}{3}(\sigma_1 + 2\sigma_3) \tag{7.22}$$

where σ_1 is the axial stress and σ_3 the radial stress. The plastic strain rates in (7.21) are

$$\dot{\xi}^p = \frac{2}{3}\left(\dot{\varepsilon}_1^p - \dot{\varepsilon}_3^p\right), \quad \dot{e}^p = \dot{\varepsilon}_1^p + 2\dot{\varepsilon}_3^p \tag{7.23}$$

The dissipation function is written as

$$\dot{D} = Mp\dot{\xi}^p \tag{7.24}$$

where M is a material constant that plays the same role as k in (7.5). The development then parallels that given above. The yield surface given in (3.31) follows. Equation (7.13) is replaced by the following:

$$e^p = \mu \ln\left(p_c / p_c^0\right) \tag{7.25}$$

where p_c is the critical state mean stress and μ is a constant similar to (but not the same as) m. The triaxial test model can be further generalised to any three-dimensional stress state as shown in Exercise 7.4 at the end of the chapter. Application of the associated flow rule is also considered in that exercise.

7.4 Beyond Cam Clay

Cam Clay revolutionised plasticity applications in soil mechanics. It produced a new paradigm for the analysis of stress and deformation in soils. A natural process in any scientific revolution is the further development and refinement of a new theory, taking it from what might be termed a state of infancy towards a state of maturity. In a sense, a new theory is shaped and honed to produce results that are closer and closer to the observed or expected response. This happened with critical state soil mechanics in the decades following the 1960s. The basic theory of a closed yield surface was exploited by many researchers in a number of novel ways. It is beyond the scope of an introductory text such as this to describe the various developments that have emerged. Only the briefest summary will be given.

One problem was immediately evident in Cam Clay. The three-dimensional yield surface illustrated in Figure 3.19 is marked by a sharp point at its intersection with the space diagonal. Corners and sharp vertices on yield surfaces are usually viewed as being inconvenient due to the ambiguity they provoke with regard to plastic strains. Two Cambridge researchers, K.H. Roscoe and J.B. Burland, provided an answer called Modified Cam Clay. The yield function was altered to produce an elliptical shape similar to that shown in Figure 3.20. Together with the associated flow rule this new surface provided a slightly more realistic response on the so-called 'wet side' of the critical state without introducing any new parameters.

Another obvious problem with Cam Clay is the abrupt loss of strength when the stress path touches the yield surface on the dry side of critical. This is illustrated in curve (b) of Figure 7.8 where the stress decreases sharply at the beginning of yield. A realistic stress–strain response usually displays a smooth transition from increasing stress to decreasing, like that shown in Figure 2.10. The volumetric strain response predicted by Cam Clay for the dry side of critical is also found to be somewhat inadequate when compared with experimental results. These as well as other deficiencies motivated many researchers to propose new critical state models that were similar to Cam Clay but differed in detail. Non-associated flow rules frequently featured in these efforts as did so-called nested yield surfaces and bounding state models. In these latter theories more than one yield surface could be employed in an effort to create a smooth transition between an increasing and a decreasing stress response for densely packed soils and to provoke a more realistic volumetric strain response. Of course, the price of better predictive ability was usually a requirement for greater numbers of material parameters.

It is interesting at this point to draw a comparison between the general features of critical state soil mechanics and the mental picture we hold of the response of a real soil. Recall from Chapter 2 how we considered plasticity effects with regard to both increasing shearing stress and increasing isotropic stress. The two situations were characterised by different microscopic effects. Under increasing isotropic stress, particles of soil fractured and crushed, permitting rearrangement into a denser configuration. In contrast, shearing a densely packed soil led to a looser particle configuration. Particle crushing probably did not occur, at least not to a great extent; but shearing did tend to break corners and asperities from particles and thus make particle rearrangement easier. These two mechanisms, fracture and rearrangement, combine to produce the irreversible effects we call plasticity. Moreover, they are ubiquitous in all soils: silts and clays as well as sands and gravels. It is relatively easy to visualise a particle of sand being fractured, but the same process may happen in clay. Clay particles are, we realise, extremely small, but they tend to form agglomerates called *peds*. A ped is like a loosely glued collection of individual particles combined to form a virtual particle. The 'glue' is provided by minute electrical forces acting on the particle surfaces. The virtual particle is generally still very small, but just like the particle of sand it may fracture into smaller peds when subjected to increasing stress. Fracture and rearrangement are fundamental processes in the deformation of any soil.

Thinking again about critical state plasticity we see that the associated flow rule combines with the closed yield surface to produce both shearing and volumetric plastic strains. It might seem reasonable to, in some way, subdivide the shear and volume straining into portions associated with fracture and with rearrangement. This has in fact been done by H.W. Chandler. All the strains may be considered. In place of the classic strain decomposition $\boldsymbol{\varepsilon} = \boldsymbol{\varepsilon}^e + \boldsymbol{\varepsilon}^p$, Chandler writes an expression equivalent to $\boldsymbol{\varepsilon} = \boldsymbol{\varepsilon}^e + \boldsymbol{\varepsilon}^d + \boldsymbol{\varepsilon}^r$. Here $\boldsymbol{\varepsilon}^e$ denotes the usual elastic strain matrix, $\boldsymbol{\varepsilon}^d$ denotes plastic strain *due to fracture or damage* and $\boldsymbol{\varepsilon}^r$ denotes plastic strain *due to rearrangement.* Chandler's theory represents a partial break away from the critical state paradigm, yet many critical state features are incorporated.

An even more complete break from conventional plasticity is possible with theories based on *micromechanics*. Micromechanics is a general term used to describe theories of material behaviour which, to some extent, avoid the assumption that the material is a continuum. Some theories completely drop the continuum assumption and attempt to consider soil explicitly as a collection of individual particles. Spherical particles have been a popular choice in these endeavours because of their ease of mathematical representation. Of course, while spherical particles may deform and slip past each other, they preclude the possibility of fracture since they would then no longer be spherical.

Theories that incorporate fracture rely on an amalgam of ideas related to mechanical properties of particles together with assumptions concerning statistics. Their ultimate aim is to provide a continuum description for the behaviour of a collection of particles, and individual particles in themselves are not considered.

This type of theory has certain advantages over the individual particle theories. Individual particle theories almost inevitably demand significant computational power since large numbers of particles must be considered in order to obtain realistic results. Statistical theories, on the other hand, may require little or no computational resources. Also, the restrictions imposed by individual particle theories on particle geometry (i.e. spherical particles) are completely avoided in statistical theories. Statistical mechanics has a long and successful history in physics, especially the physics of gases, but as yet is little used in soil mechanics. To conclude this chapter we will outline a recent statistically based theory for soil behaviour, one that has also emerged from the Cambridge soil mechanics group.

The theory is the work of two engineers, G.R. McDowell and M.D. Bolton. They set out to model soil behaviour in an oedometer test: the so-called virgin or normal compression response. Recall from Chapter 2 that we expect the volumetric strain to be a linear function of the logarithm of the applied compressive stress.* We will think of the applied stress as a characteristic stress and refer to it as the *ambient* stress in the soil. As the ambient stress increases, we anticipate that particle fracture and fragmentation will lead to volume compression.

We are aware that the collection of particles in a compressing soil will not all be loaded equally. Some particles may support far more of the ambient stress than do others. Depending on how intensely stressed the soil mass is, there may be some regions of the particle matrix that support very little stress and other regions that are heavily stressed and act effectively as an internal structure. With this in mind it may appear that any attempt at deciding rationally whether a particle fractures or not is predestined to fail, but in fact there are some general rules that can be laid down. There appear to be two important components to the question of whether a particle will fragment under a given ambient stress or not. The first question is, how strong is our particle by itself? The second question is, how well is our particle protected by other particles that surround it?

With regard to the first question, there have been several experimental studies of particle strength for different soils. Particles are crushed between rigid flat platens. The force required to cause fragmentation is measured and the strength is represented by the force divided by the square of the particle dimension δ. Since the particle is irregular the exact definition of δ is a bit vague, but we can think of it as the average diameter of the particle. A clear statistical conclusion emerges from these experiments. For a given soil type, the strength of a particle

* Here we imply the effective stress wherever appropriate.

is inversely related to its dimension δ. Roughly speaking, it is found that the particle strength is proportional to $\delta^{-1/2}$. Thus larger particles are significantly weaker. This fact is explained as follows. All particles will contain flaws in the form of micro-fissures and cracks. When our particle is crushed, the initial fracture is likely to originate in a pre-existing flaw. Larger particles will most probably contain larger flaws and these will be more vulnerable. A well-defined distribution of particle strength may be determined by carrying out sufficient numbers of experiments. The experimental data is usually well represented by a Weibull distribution.

The second question to be considered is this. How does our particle interact with other surrounding particles? In the crushing tests just described there were no other particles, just two hard plates moving together to ensure the particle will be crushed. In a real soil there will be a number of surrounding particles that make contact with our particle. The system of forces found at all the contact points holds the particle in equilibrium and, in some way, represents part of the ambient stress field in the soil. The number of neighboring particles that actually contact our particle is called the *co-ordination number*. We expect the co-ordination number will be equal to or greater then 2. Now consider two cases. Suppose the co-ordination number takes the smallest value of 2. Then the overall traction field acting on the entire surface of our particle is as *non-uniform* as possible. Our particle is effectively in a situation similar to the platen crushing test. In contrast, suppose the co-ordination number is large. Now our particle is supported by many others and there will be a much more uniform overall surface traction. Intuitively we would anticipate that our particle will be much less likely to be crushed when the co-ordination number is high. A low co-ordination number suggests greater stress concentrations and a higher probability of fragmentation. A high co-ordination number suggests a smooth stress field inside the particle and a smaller likelihood of fracture.

There are two important conclusions to be drawn from these thoughts, but first we must realise that larger co-ordination numbers will be a feature of larger particles. In a soil where a wide range of particle sizes are present, the larger particles will be expected to be surrounded by a complex matrix of other particles of lesser size. These smaller particles will cushion the large particle from the effects of the ambient stress state. On the other hand, small particles will generally have small co-ordination numbers. They will often be trapped between two larger particles and their co-ordination numbers may be as small as 2 or 3. So, there are competing influences at work inside the particle matrix. Larger particles are intrinsically weaker, but they are protected by a cushion of smaller particles. Smaller particles are intrinsically stronger, but they are more exposed to the ambient stress field. Which will win out? The answer is that larger particles will generally survive, while smaller particles will generally suffer more fragmentation. These are generalisations and we are well aware that while they are statistically accurate, any individual particle may find itself in a situation that differs from the norm and its behaviour may differ accordingly. On balance, however, it is more likely that smaller particles will fracture more often than larger ones. There is ample experimental evidence to support this fact.

Now what are the two conclusions we draw from all this? First, the fact that smaller particles suffer more breakage then larger particles implies that soils will tend to have particle size distributions that cover a broad range of particle dimensions. The process of normal compression sustained by a naturally deposited soil will lead to the creation of more smaller particles and the preservation of many larger particles, and the grain size distribution will expand towards the smaller end of the size spectrum.* If the particle distribution of a natural soil is examined in detail, it is usually found to be *fractal*. The term fractal is used to represent natural phenomena that are scale-invariant. Frequently the shape of the coastline of Britain is used as an example of a fractal quantity. If one looks at the coastline in an atlas, it has a certain characteristic roughness. If we then magnify a small part of the coast several times, it still looks about as rough as it did on the original map. If we then magnify that small portion again, the same roughness remains, right down to the scale of individual grains of sand on a beach. The roughness of the coast line is independent of the scale at which we are observing. A natural soil observed under a microscope will give a similar impression. If we use a small magnification we will see a collection of the larger particles, but the many other particles will be too small for their images to be resolved. If we then increase the magnification we see some medium-sized particles and their relative size distribution will be similar to that observed for the larger particles. Increasing the magnification again we see yet smaller particles, still having a similar relative distribution of sizes. A consequence of this fractal quality of soil is a well-defined relationship between the number of particles and the particle dimension. Observations of real soils suggest

$$N(\Delta > \delta) = A\delta^{-E} \tag{7.26}$$

Here, N represents the number of particles, Δ denotes the random particle dimension and A is a parameter that will depend on the smallest particle present.

* Other geologic processes such as transport may tend to sort the particles into more uniform gradations but, for a static soil evolving under increasing ambient stress, we can expect to find a wide range of particle sizes.

Equation (7.26) states that the number of particles with dimension greater than δ is proportional to δ raised to the power negative E. E is a constant called the *fractal dimension*. For natural soils, E is usually found to be close to 2.5.

The second conclusion to be drawn from this discussion concerns the ambient stress in the soil. Having discovered that the most likely particle to fracture is the smallest particle, it is therefore reasonable to assume that, for normal or virgin compression, the ambient stress in the soil will be close to the strength of the smallest particle. Recall that particle strength in the platen test is found to be proportional to $\delta^{-1/2}$. Therefore, if we let σ denote the ambient stress and δ_s the dimension of the smallest particle, we expect to find that the product $\sigma\sqrt{\delta_s}$ is a constant. A convenient reference stress is the value of particle strength associated with the largest particle present in the soil. Letting the largest particle dimension be δ_0 and the corresponding particle strength be σ_0, we have

$$\sigma\sqrt{\delta_s} = \sigma_0\sqrt{\delta_0} \tag{7.27}$$

In general, we expect δ_0 to remain nearly constant as normal compression occurs. Some of the larger particles will fracture but others will be preserved and the size of the largest particle will not alter greatly as the soil compresses. Equations (7.26) and (7.27) form the two important conclusions we draw from this general discussion of particle fracture.

Now consider a typical oedometer test. A sample is placed in the loading device and the load is increased. At some point the preconsolidation stress σ_i^0 is reached, particle crushing begins, and normal compression commences. We can use (7.26) to calculate the number of particles in the size range δ to $\delta + d\delta$. Let $dN = N(\Delta > \delta) - N(\Delta > \delta + d\delta)$ so that

$$dN = \frac{5}{2}A\delta^{-7/2}\,d\delta \tag{7.28}$$

where E has been set equal to a typical value of 2.5. The mass of a particle with dimension δ is given by $\rho_s\beta_v\delta^3$, where ρ_s denotes the solid density of the particle and β_v is a volume 'shape factor' that would depend on the typical particle shape for our soil. The mass of all particles with dimensions between δ and $\delta + d\delta$ is then given by

$$dM = \rho_s\beta_v\delta^3\,dN = \frac{5}{2}\rho_s\beta_v A\delta^{-1/2}\,d\delta \tag{7.29}$$

Let M_S represent the total mass of particles in our sample. Integrating (7.29) gives

$$M_S = \int_{\delta_s}^{\delta_0} dM = 5\rho_s\beta_v A(\sqrt{\delta_0} - \sqrt{\delta_s}) \tag{7.30}$$

Conservation of mass implies that M_S is constant. Therefore we see that A must depend upon the particle size range according to

$$A = \frac{M_S}{5\rho_s\beta_v}\left(\frac{1}{\sqrt{\delta_0}-\sqrt{\delta_s}}\right) \tag{7.31}$$

We will make use of this result below.

We are now drawing near the conclusion of McDowell and Bolton's analysis. The rate of plastic work in our oedometer test is given by (7.4) with $\tau = \dot{\gamma}^p = 0$. McDowell and Bolton suggested a new dissipation function $\dot{D}$ to replace (7.5).

$$\dot{D} = \Gamma\,\dot{S}_T \tag{7.32}$$

Here S_T is the *surface area* of *all* the particles in our sample and Γ is a material constant called the *surface energy*. As new surface is formed during particle breakage, energy is dissipated at the rate given by (7.32). We now want to set the rate of plastic working equal to this dissipation rate, but care must be used. In (7.32) we have the dissipation rate for the entire sample, while in (7.4) $\dot{W}_p$ represents the rate of plastic work per unit volume. We must multiply the work rate by the total volume V of the sample. Then setting the rate of plastic work equal to the rate of dissipation gives

$$V\sigma\dot{\varepsilon}^p = \Gamma\,\dot{S}_T \tag{7.33}$$

We can determine the total surface area of our sample in a similar way to the calculation of sample mass. The surface area of a particle of dimension δ will be given by $\beta_s\delta^2$, where β_s is a surface 'shape factor'. Multiplying this area by the number of particles dN from (7.28) and then integrating, we find the total surface area to be

$$S_T = 5\beta_s A\left(\frac{1}{\sqrt{\delta_s}} - \frac{1}{\sqrt{\delta_0}}\right) \tag{7.34}$$

Using (7.31) to replace A gives

$$S_T = \frac{\beta_s M_S}{\rho_s\beta_v\sqrt{\delta_s\delta_0}} \tag{7.35}$$

Now use (7.27) here. We find

$$S_T = \frac{\beta_s M_S}{\rho_s\beta_v\delta_0}\frac{\sigma}{\sigma_0} \tag{7.36}$$

Differentiating and using the result in (7.33) gives

$$\dot{\varepsilon}^p = \frac{\rho_d}{\rho_s}\frac{\beta_s}{\beta_v}\frac{\Gamma}{\sigma_0\delta_0}\frac{\dot{\sigma}}{\sigma} \tag{7.37}$$

where $\rho_d = M_S/V_T$ is the *equivalent dry density* for our sample, the total mass of particles divided by the sample volume. If the sample is dry, ρ_d is the bulk density. For a saturated sample, ρ_d represents the density the sample would have if the pore fluid were absent.

Finally, note that in the oedometer test the strain ε is the only non-zero compressional strain and hence is also the volumetric strain. If we ignore elastic strains during normal compression, then the rate $\dot{\varepsilon}^p$ is related to the void ratio $\tilde{e}$ through equation (2.31),

$$\dot{\varepsilon}^p = -\frac{\dot{\tilde{e}}}{1+\tilde{e}} \tag{7.38}$$

Now equate the right-hand sides of (7.37) and (7.38) to find

$$\dot{\tilde{e}} = -\frac{\beta_s}{\beta_\nu}\frac{\Gamma}{\sigma_0\delta_0}\frac{\dot{\sigma}}{\sigma} \tag{7.39}$$

where the identity $\rho_s = (1+\tilde{e})\rho_d$ has been used. Of course, (7.39) produces a straight line graph of $\tilde{e}$ versus $\ln(\sigma)$. McDowell and Bolton discuss realistic values for the dimensionless ratios β_s/β_ν and $(\Gamma/\sigma_0\delta_0)$ and they conclude that the slope of the $\tilde{e}-\ln\sigma$ line that emerges from their theory* has the correct order of magnitude. Linear $\tilde{e} - \ln\sigma$ response has been observed in the laboratory for many years and has been used empirically in Cam Clay and all other critical state models, but has *not* been explained rationally until McDowell and Bolton's analysis.

There are other interesting features of McDowell and Bolton's model for normal compression. We will investigate two before concluding this section. It is possible to rewrite (7.26) in the following way:

$$\begin{aligned} N &= \frac{M_S}{5\rho_s\beta_\nu}\frac{\delta^{-5/2}}{\sqrt{\delta_0}-\sqrt{\delta_s}} \\ &= \frac{N_0}{1-\sqrt{\delta_s/\delta_0}}\left(\frac{\delta_0}{\delta}\right)^{5/2} \end{aligned} \tag{7.40}$$

where (7.31) has been used and $N_0 = M_S/5\rho_s\beta_\nu\delta_0^3$ is a dimensionless constant. If we now use (7.27) to replace the ratio δ_s/δ_0, we find

$$\frac{N}{N_0} = \frac{1}{1-\sigma_0/\sigma}\left(\frac{\delta_0}{\delta}\right)^{5/2} \tag{7.41}$$

* In fact, (7.39) is not identical to the result obtained by McDowell and Bolton. Note that the work equation (7.33) assumes all dissipation occurs through particle fracture. McDowell and Bolton include another term in their work equation to account for frictional dissipation as well. This leads to a small difference in the final void ratio–stress relationship, but the essential aspects of their argument are as presented.

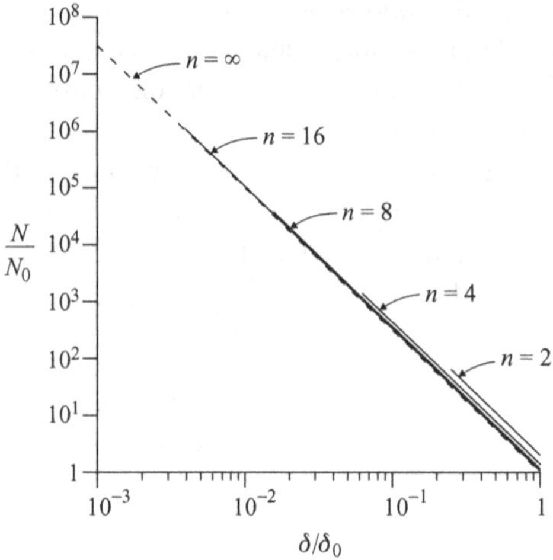

Figure 7.12. Evolution of the number distribution with increasing stress.

This gives a size distribution of number of particles that is always fractal, although the graph moves slightly as σ increases. Letting $\sigma = n\sigma_0$, results for N/N_0 plotted versus δ/δ_0 for values of n between 2 and 16 are shown in Figure 7.12. As n grows larger, δ_s grows smaller and therefore the graph covers more of the horizontal axis. For example, for $n = 2$ the smallest particle dimension is 0.25 times the largest, while for $n = 16$ the ratio is 0.0039. The individual graphs move downward as n grows, but the lines quickly become so close together that they merge into a single plot. In the limit as $\sigma \to \infty$, (7.41) becomes $N/N_0 = (\delta_0/\delta)^{5/2}$, and this is the limiting line on the graph. We conclude then that the number distribution is always fractal and nearly independent of stress.

Next, consider the mass distribution. Change the limits of integration in (7.30) to obtain the mass of particles with diameters smaller than δ,

$$M(\Delta \leq \delta) = \int_{\delta_s}^{\delta} dM = 5\rho_s \beta_v A(\sqrt{\delta} - \sqrt{\delta_s}) \tag{7.42}$$

Now use (7.31) to replace A, and let $P = M(\Delta \leq \delta)/M_S$ be the fraction of particles with diameters smaller than δ. Note that P is exactly the same quantity usually used to plot the grain size distribution curve for a soil. Then, making use of (7.27) and (7.31),

$$P = \frac{\sqrt{\delta/\delta_0} - \sqrt{\delta_s/\delta_0}}{1 - \sqrt{\delta_s/\delta_0}} = \frac{\sqrt{\delta/\delta_0} - \sigma_0/\sigma}{1 - \sigma_0/\sigma} \tag{7.43}$$

Graphs of P versus δ/δ_0 in semi-log and log–log forms are shown in Figure 7.13

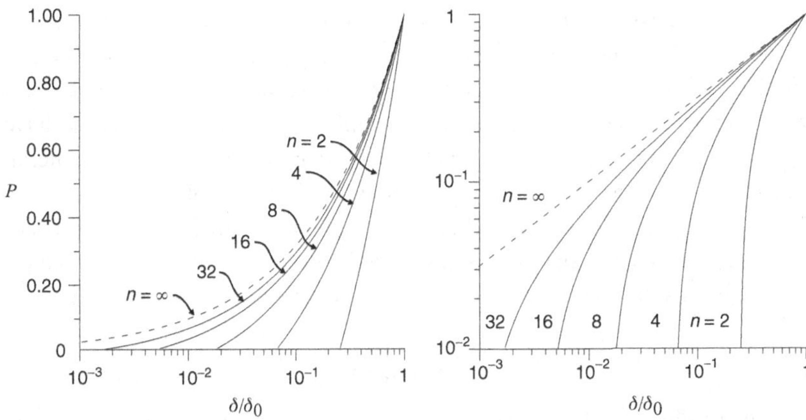

Figure 7.13. Evolution of the grain size distribution with increasing stress.

for values of $n = \sigma/\sigma_0$ between 2 and 32. For small values of stress the mass distribution is not fractal, but it evolves toward a fractal distribution as n increases. In the limit as $\sigma \to \infty$, $P \to \sqrt{\delta/\delta_0}$ and the distribution is completely fractal with a fractal dimension of 0.5. The form of equation (7.43) leads naturally to this behaviour, which appears to agree with experimental results.

7.5 Last words

In this chapter we have considered a broad range of ideas concerned with modelling plastic response. Most have emerged within the past 40 years and one, the crushing model for normal compression, is both recent and novel. The entire subject of plasticity of soil is clearly still evolving and the concepts given here will no doubt be superceded in future years. Development of powerful, inexpensive desktop computers is the major impetus driving this change. Modern theories for plastic behaviour are often far too complex to permit closed-form solutions of even very simple boundary value problems. Finite-element and finite-difference methods must be used generally, and computers are an essential tool. In a great many ways this is a positive development. More complex material models may yield more accurate solutions, especially for sensitive applications where errors may be very costly. Development of these models occasionally leads to a new insight or understanding that might not have appeared otherwise. But there is an unfortunate natural tendency to use these computational tools to solve all problems, not just sensitive ones, and excessive reliance on computer solutions has a blinkering effect on the practitioner. Analytical solutions are not simply elegant anarchisms. They are useful in a great many ways. The study of analytical solutions enhances problem solving skills and gives insight into a whole range of problems that have similar attributes to the problem being considered. In many cases analytical methods provide totally appropriate solutions for the purposes of design. Their intrinsic accuracy may be less than that of computer solutions, but the level of accuracy required may not be great and analytic methods may be cheaper to employ. Often analytic solutions have a degree of accuracy that is consistent with the degree of knowledge of material properties or of the boundary conditions in a design problem. And analytic solutions provide the only true check on numerical solution methods. It is our hope that the methods of analysis described in this book will inspire an interest in the use of simple solutions and lead to a further study of classical plasticity.

Further reading

See the reading list of Chapter 3 for references to the origins of Cam Clay and critical state soil mechanics. Readers interested in the development of new scientific theories might wish to consult this famous book:

T.S. Kuhn, *The Structure of Scientific Revolutions*, University of Chicago Press, 226pp., 1997.

The evolution of ideas behind kinematic hardening is described in

W. Prager, The theory of plasticity: a survey of recent achievements, *Proc. Inst. Mech. Eng.*, London, **169**, 41–57 (1955).

Many of the difficulties associated with modern plasticity theories for soils are discussed in this article by Spencer,

A.J.M. Spencer, Deformation of ideal granular materials, in *Mechanics of Solids: the Rodney Hill 60th Anniversary Volume* (eds. H.G. Hopkins and M.J. Sewell) pp. 607–652, Pergamon Press, Oxford, 1982.

A discussion of the mathematical aspects of dissipation functions in development of plasticity theories may be found in

J.B. Martin, *Plasticity: Fundamentals and General Results*, MIT Press, Cambridge, MA, 1975.

The work by Chandler using plastic damage and rearrangement strains is described in

H.W. Chandler, Homogeneous and localised deformation in granular materials: a mechanistic model, *Int. J. Engng. Sci.*, **28**, 719–734 (1990).

There are a number of micromechanical models based on collections of spherical particles. Two papers of particular interest are

R.D. Mindlin and H. Deresiewicz, Elastic spheres in contact under varying oblique forces. *J. Appl. Mech.* ASME, **20**, 203–208 (1953).

P.A. Cundall and O.D.L. Strack. A distinct element model for granular assemblies. *Geotechnique*, **29,** 47–65 (1979).

Development of the fractal void ratio–logarithm of stress model is summarised in

G.R. McDowell and M.D. Bolton, On the micromechanics of crushable aggregates, *Geotechnique*, **48**, 667–679 (1998).

Exercises

7.1 A *linear* work hardening elastic–plastic material responds in simple tension following the bi-linear stress–strain curve sketched in Figure 7.14. Derive expressions for the total, elastic and plastic works, W, W_e and W_p: (i) in terms of stress σ and (ii) in terms of strain ε. Show that the current yield stress may be written in terms of the plastic work as

$$\sigma_T^2 = \left(\sigma_T^0\right)^2 + 2W_p\left(\frac{EE^*}{E - E^*}\right)$$

where E and E^* are the moduli shown in Figure 7.14.

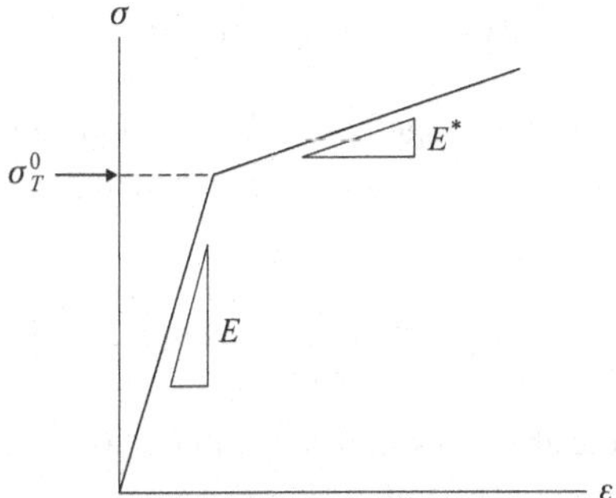

Figure 7.14.

7.2 In an *undrained* test, a fully saturated soil sample is assumed to respond by maintaining constant volume. In the context of the simple shear test shown in Figure 7.5, an undrained test would imply $\varepsilon = 0$ and hence $\varepsilon^p = -\varepsilon^e$. Use this fact together with the properties of Cam Clay to show that the *effective* normal stress σ at *critical state* in an undrained simple shear test is given by the transcendental equation

$$\sigma - \ln\left(\frac{\sigma}{\sigma_c^0}\right)^{-mD} = 0$$

where D is used to represent the uniaxial compression elastic modulus $E(1-\nu)/(1+\nu)(1-2\nu)$. What is the pore pressure in the sample at the critical state? Discuss when the pore pressure will be positive and when negative and explain why.

7.3 Modified Cam Clay uses the yield surface described in (3.32). In the context of the simple shear test described in this chapter, the equivalent yield surface for Modified Cam Clay would be written as

$$\tau - k\sqrt{\sigma(2\sigma_c - \sigma)} = 0$$

Follow through the steps in equations (7.15)–(7.20) to show that for Modified Cam Clay the simple shear stress–plastic strain response is given by

$$\tan^{-1}\left(\frac{\tau}{k\sigma}\right) = \frac{k\sigma}{4m}\gamma^p + \tan^{-1}\left(\frac{\tau_0}{k\sigma}\right)$$

where $\tau_0 = k\sqrt{\sigma(2\sigma_c^0 - \sigma)}$. Compare the Modified Cam Clay response with that for Cam Clay.

7.4 For the fully three-dimensional response of Cam Clay we must replace the mean and deviatoric stresses given in (7.22) by their three-dimensional counterparts from Chapter 3,

$$p = \frac{1}{3}(\sigma_1 + \sigma_2 + \sigma_3), \quad q = \frac{1}{\sqrt{2}}[(\sigma_1 - \sigma_2)^2 + (\sigma_2 - \sigma_3)^2 + (\sigma_3 - \sigma_1)^2]^{1/2}$$

Use these definitions together with the yield function (3.31) and the associated flow rule to show that the principal plastic strain rates are given by

$$\dot{\varepsilon}_k^p = \lambda\frac{\partial f}{\partial \sigma_k} = \lambda\left[\frac{3}{2}\frac{\sigma_k - p}{q} + \frac{M}{3}\ln\left(\frac{p}{p_c}\right)\right], \quad \text{for } k = 1, 2, \text{ or } 3$$

Finally, use the above result to show that plastic volumetric and deviatoric strain rates, defined as follows:

$$\dot{e}^p = \dot{\varepsilon}_1^p + \dot{\varepsilon}_2^p + \dot{\varepsilon}_3^p, \quad \dot{\xi}^p = \frac{\sqrt{2}}{3}\left[\left(\dot{\varepsilon}_1^p - \dot{\varepsilon}_2^p\right)^2 + \left(\dot{\varepsilon}_2^p - \dot{\varepsilon}_3^p\right)^2 + \left(\dot{\varepsilon}_3^p - \dot{\varepsilon}_1^p\right)^2\right]^{1/2}$$

are given by the associated flow rule according to (cf. (7.15))

$$\dot{e}^p = \lambda\frac{\partial f}{\partial p} = \lambda M \ln\left(\frac{p}{p_c}\right) \quad \text{and} \quad \dot{\xi}^p = \lambda\frac{\partial f}{\partial q} = \lambda$$

7.5 Investigate values for the ratio β_s/β_v for the geometric shapes that would emerge from successively splitting a sphere into two similar parts. That is, consider a sphere, hemisphere, half-hemisphere, quarter-hemisphere and so on. In each case use the sphere radius for δ. Does β_s/β_v vary with the angularity of the particle? What would be a typical value of β_s/β_v for a beach sand? For a clay? (Is the radius the most appropriate value for δ?)

Appendix A

Non-Cartesian coordinate systems

The formulation of any specific boundary value problem in geomechanics is greatly facilitated first by considering the specific attributes as they pertain to the geometry of the domain of interest. Other aspects of the formulation and solution can also include a consideration of features such as material symmetry and other geometrical features of the loading and boundaries of the domain. For example, a two-dimensional plane strain problem involving the surface loading of a halfspace region by a concentrated line load (Figure A.1) is most conveniently formulated with reference to a plane polar coordinate system, whereas the plane strain problem involving surface loading by a distributed loading (Figure A.2) is formulated most conveniently in reference to a Cartesian coordinate system.

Also, referring to Figure A.3, the axisymmetric surface loading of a halfspace region by a concentrated load is most conveniently described in relation to a system of spherical polar coordinates, whereas the axisymmetric surface loading of a halfspace region is best formulated in relation to a system of cylindrical polar coordinates (Figure A.4).

While in the examples just cited, the choice of the coordinate system is largely dictated by the mode of loading, there are other situations where the geometrical boundaries of the domain of interest have a decided influence on the choice of the coordinate system. For example, the problem of a deeply embedded tunnel (Figure A.5) can be most conveniently examined using a plane polar coordinate system, whereas that of a shallow tunnel is most conveniently formulated with reference to a system of bi-polar coordinates (Figure A.6).

The most commonly encountered coordinate systems in geomechanics are the rectangular Cartesian, plane polar and spherical polar coordinates. The objective of this Appendix is to outline briefly the basic mathematical operations, which can be utilised to express the various relationships between coordinate systems. These operations are well documented in many standard texts in continuum mechanics and applied mathematics and the reader is referred to the references cited at the end of this Appendix for further details. The primary aim here is to consider the appropriate expressions for displacements, strains, stresses and equations of equilibrium referred to the rectangular Cartesian coordinates as given and to develop a methodology whereby these results can be expressed in an alternative orthogonal curvilinear coordinate system. Many such orthogonal curvilinear coordinate systems are possible; here, however, purely for the purposes of illustration, we shall select the cylindrical polar coordinate system as the preferred choice.

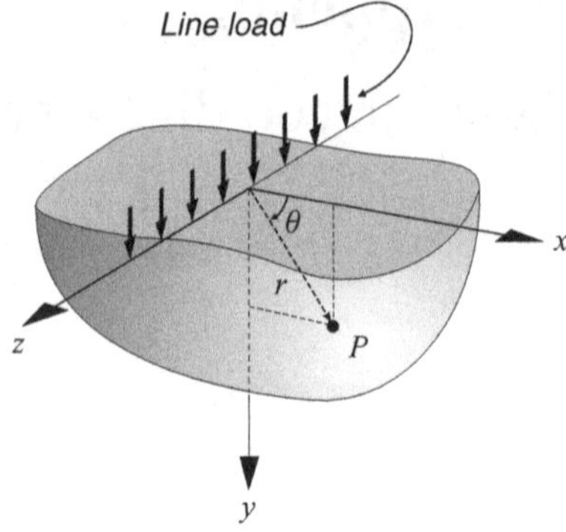

Figure A.1. Action of a line load on a halfspace – plane problem: polar coordinates.

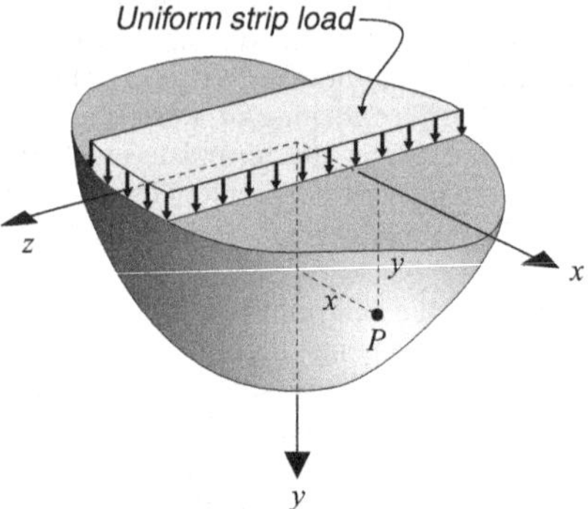

Figure A.2. Surface loading of a halfspace by a uniform strip load – plane problem: Cartesian coordinates.

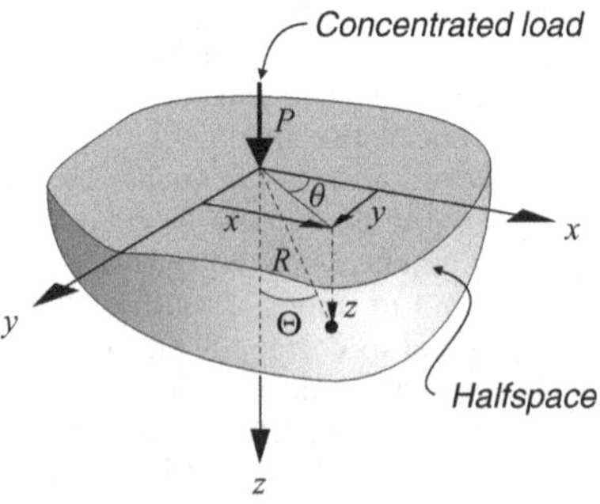

Figure A.3. Concentrated surface loading of a halfspace.

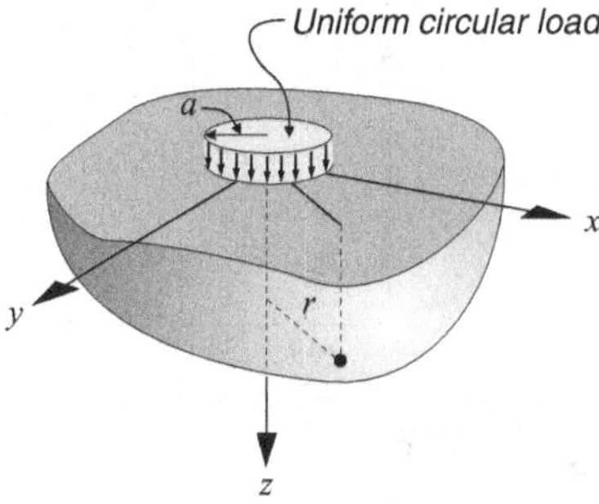

Figure A.4. Surface loading of a halfspace by a uniform circular load.

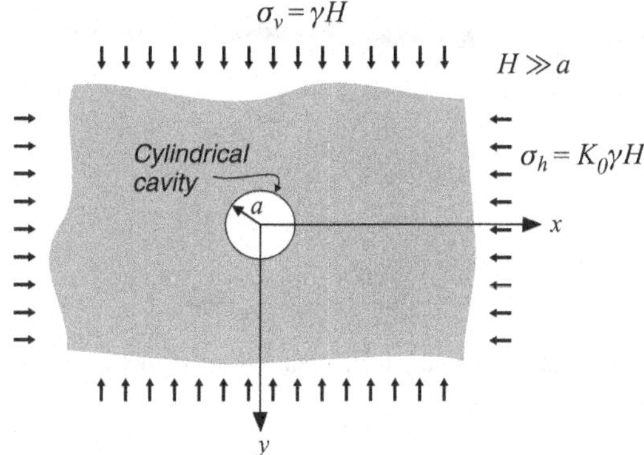

Figure A.5. Deeply embedded cylindrical cavity in a geostatic stress field (γ = unit weight of geomaterial; K_0 = coefficient of earth pressure at rest).

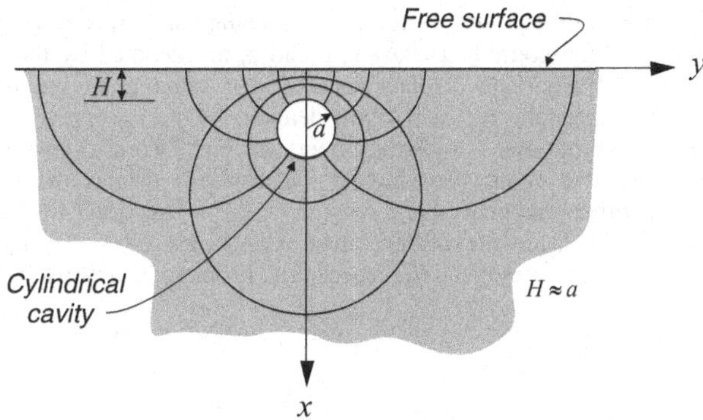

Figure A.6. Shallow tunnel in a halfspace region – bipolar coordinates.

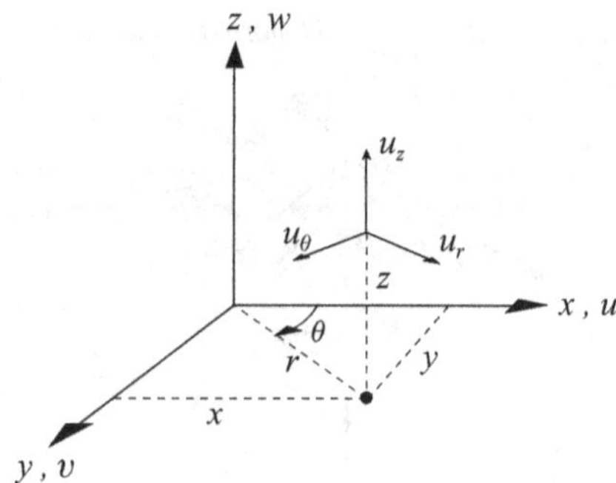

Figure A.7. Rectangular Cartesian and cylindrical polar coordinate systems.

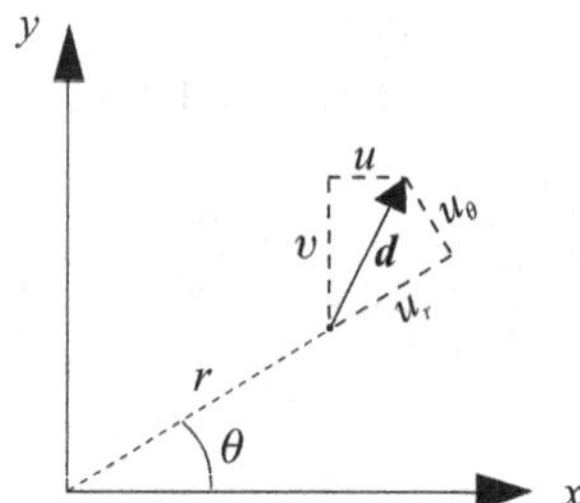

Figure A.8. Components of the displacement vector $\boldsymbol{d}$ in Cartesian and plane polar coordinate systems.

We consider a system of rectangular Cartesian coordinates defined by (x, y, z) and the associated cylindrical polar coordinate system (r, θ, z) (Figure A.7). The relationships between the coordinates in the (x, y)-plane are as follows:

$$x = r\cos\theta; \quad y = r\sin\theta; \quad r^2 = x^2 + y^2 \tag{A.1}$$

The displacements of a point referred to the rectangular Cartesian coordinate system and referred to the coordinate axes x, y and z, are denoted by the components u, v and w, respectively. The displacements of the same point but now referred to the coordinate directions r, θ and z are denoted by u_r, u_θ and u_z, respectively. Furthermore, we assume that the displacements are considered to be positive in the directions in which the coordinates increase. Consider a displacement vector $\boldsymbol{d}$ in the (x, y)-plane, with components u and v in the x- and y-directions, respectively (Figure A.8). The same displacement vector $\boldsymbol{d}$ can be expressed in terms of u_r and u_θ referred to the plane polar coordinates r and θ, respectively. From geometry, we have

$$u_r = u\cos\theta + v\sin\theta; \quad u_\theta = -u\sin\theta + v\cos\theta \tag{A.2}$$

with

$$u_z = w \tag{A.3}$$

We can combine (A.2) and (A.3) and represent these equations in the form of a matrix equation

$$\begin{Bmatrix} u_r \\ u_\theta \\ u_z \end{Bmatrix} = \begin{bmatrix} \cos\theta & \sin\theta & 0 \\ -\sin\theta & \cos\theta & 0 \\ 0 & 0 & 1 \end{bmatrix} \begin{Bmatrix} u \\ v \\ w \end{Bmatrix} = [\mathsf{H}]^T \begin{Bmatrix} u \\ v \\ w \end{Bmatrix} \tag{A.4}$$

This is the straightforward transformation rule for the displacement vector, where the transformation matrix $[\mathsf{H}]$ is obtained by considering the ordered array of direction cosines between the (x, y, z) and (r, θ, z) axes, i.e.

$$[\mathsf{H}] = \begin{matrix} & r & \theta & z \\ x & \cos(x0r) & \cos(x0\theta) & \cos(x0z) \\ y & \cos(y0r) & \cos(y0\theta) & \cos(y0z) \\ z & \cos(z0r) & \cos(z0\theta) & \cos(z0z) \end{matrix} \tag{A.5}$$

It is quite important to note the ordering sequence and the positioning of the original coordinate system and the new coordinate system, as indicated in (A.5), *when constructing the transformation matrix*, $[\mathsf{H}]$. If the positioning of the coordinate systems is interchanged for the construction of the transformation matrix (i.e. the (x, y, z) coordinates are now positioned to occupy the *column* directions and the (r, θ, z) coordinates are positioned to occupy the *row* directions), then we obtain a different matrix, say, $[\mathsf{B}]$ and, of course, $[\mathsf{H}]^T = [\mathsf{B}]$ or $[\mathsf{H}] = [\mathsf{B}]^T$. It is also important to note that the coordinate systems under discussion are orthogonal coordinate systems and that

$$[\mathsf{H}]^T[\mathsf{H}] = [\mathsf{B}]\,[\mathsf{B}]^T = [\mathbf{I}]; \quad [\mathsf{H}]\,[\mathsf{H}]^T = [\mathsf{B}]^T[\mathsf{B}] = [\mathbf{I}] \tag{A.6}$$

where $[\mathbf{I}]$ is the *identity* or *unit* matrix. We have explained the transformation rule for the displacement components by considering a specific coordinate system; the procedure can, of course, be applied to any two sets of orthogonal coordinate systems, curvilinear or otherwise.

The next step involves the kinematics of deformation as described by the infinitesimal strains, the matrix of which referred to the Cartesian components is denoted by $[\boldsymbol{\varepsilon}]$. We have, from (1.5) and (1.7),

$$[\boldsymbol{\varepsilon}] = \begin{bmatrix} \varepsilon_{xx} & \varepsilon_{xy} & \varepsilon_{xz} \\ \varepsilon_{yx} & \varepsilon_{yy} & \varepsilon_{yz} \\ \varepsilon_{zx} & \varepsilon_{zy} & \varepsilon_{zz} \end{bmatrix}$$

$$= \begin{bmatrix} \dfrac{\partial u}{\partial x} & \dfrac{1}{2}\left(\dfrac{\partial u}{\partial y} + \dfrac{\partial v}{\partial x}\right) & \dfrac{1}{2}\left(\dfrac{\partial u}{\partial z} + \dfrac{\partial w}{\partial x}\right) \\ \dfrac{1}{2}\left(\dfrac{\partial u}{\partial y} + \dfrac{\partial v}{\partial x}\right) & \dfrac{\partial v}{\partial y} & \dfrac{1}{2}\left(\dfrac{\partial v}{\partial z} + \dfrac{\partial w}{\partial y}\right) \\ \dfrac{1}{2}\left(\dfrac{\partial u}{\partial z} + \dfrac{\partial w}{\partial x}\right) & \dfrac{1}{2}\left(\dfrac{\partial v}{\partial z} + \dfrac{\partial w}{\partial y}\right) & \dfrac{\partial w}{\partial z} \end{bmatrix} = [\boldsymbol{\varepsilon}]^T \tag{A.7}$$

The objective here is to express the components of strains at a point referred to the rectangular Cartesian coordinate system in reference to the cylindrical polar coordinate system, which not only makes use of the components of the displacement vector referred to the cylindrical polar coordinate system, but also in terms of the coordinates (r, θ, z) and the derivatives of these coordinates. There are many ways in which this can be accomplished. One obvious possibility is to obtain a geometrical interpretation of the strain components referred to the cylindrical polar coordinate system in terms of the

(a)

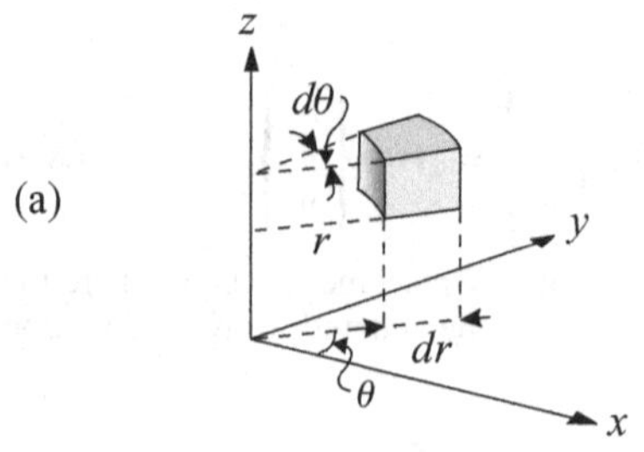

(b)

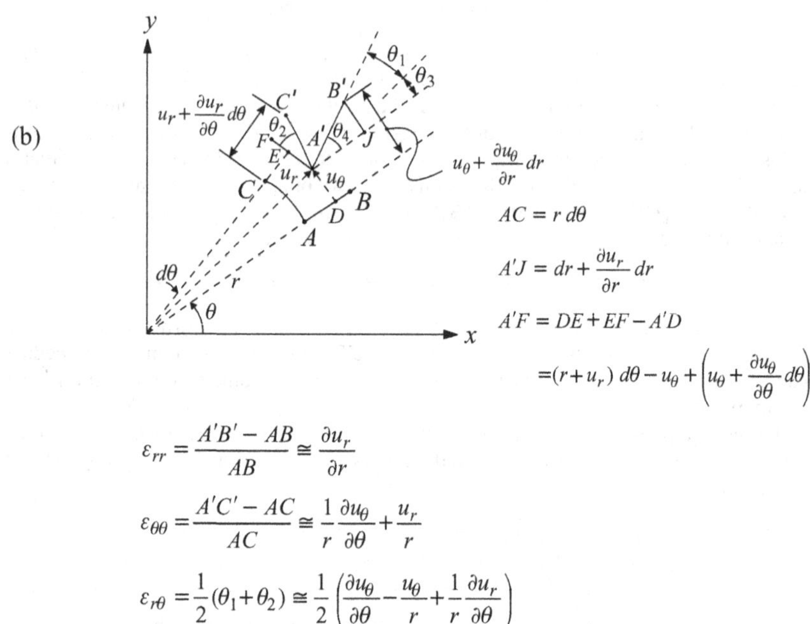

Figure A.9. Geometric representation of small strain in polar coordinates.

displacement components (u_r, u_θ, u_z), by considering the deformations of an element in the shape of a segment of an annular region (Figure A.9). The visualization of these deformations in their entirety is a difficult exercise. The customary approach is to treat the deformations by considering the deformations of a sector that corresponds to an element in the (r, θ)-plane and then to consider separately the deformations referred to the (r, z)- and (θ, z)-planes. Even this is not a trivial exercise in both graphical visualization and calculation. The description of the procedure applicable to the deformations in the (r, θ)-plane is given in the references cited at the end of this Appendix.

An alternative but more expedient approach is to take full advantage of mathematical methods available through consideration of operations in linear algebra and differential calculus. To proceed with such manipulations it is necessary, at the outset, to identify a matrix of strains, which are referred to the cylindrical polar coordinate system. Let us denote this strain matrix by $[\boldsymbol{\varepsilon}^{cp}]$, with the superscript indicating its appropriate reference. Let us denote the components of this matrix as follows:

$$[\boldsymbol{\varepsilon}^{cp}] = \begin{bmatrix} \varepsilon_{rr} & \varepsilon_{r\theta} & \varepsilon_{rz} \\ \varepsilon_{\theta r} & \varepsilon_{\theta\theta} & \varepsilon_{\theta z} \\ \varepsilon_{zr} & \varepsilon_{z\theta} & \varepsilon_{zz} \end{bmatrix} = [\boldsymbol{\varepsilon}^{cp}]^T \tag{A.8}$$

The property of symmetry of the strain matrix in the cylindrical polar coordinate system is a direct consequence of the symmetry of the strain matrix in the rectangular Cartesian coordinate system, since we are describing the state of strain at the same location. Now let us treat the problem of expressing the components of $[\boldsymbol{\varepsilon}^{cp}]$ in terms of the components of $[\boldsymbol{\varepsilon}]$ purely as a matrix transformation where the $[\mathbf{H}]$ matrix is defined by (A.5). Accordingly, we obtain the relationships

$$[\boldsymbol{\varepsilon}^{cp}] = [\mathbf{H}]^T[\boldsymbol{\varepsilon}][\mathbf{H}] \quad \text{or} \quad [\boldsymbol{\varepsilon}^{cp}] = [\mathbf{B}][\boldsymbol{\varepsilon}][\mathbf{B}]^T \tag{A.9}$$

We can evaluate the components of the matrix of strains referred to the cylindrical polar coordinate system in terms of the components in the rectangular Cartesian coordinate system. For the purposes of illustration we note that

$$\begin{aligned} \varepsilon_{rr} &= \varepsilon_{xx}\cos^2\theta + \varepsilon_{yy}\sin^2\theta + 2\varepsilon_{xy}\sin\theta\cos\theta \\ \varepsilon_{\theta\theta} &= \varepsilon_{xx}\sin^2\theta + \varepsilon_{yy}\cos^2\theta - 2\varepsilon_{xy}\sin\theta\cos\theta \\ \varepsilon_{r\theta} &= (\varepsilon_{yy} - \varepsilon_{xx})\sin\theta\cos\theta + \varepsilon_{xy}(\cos^2\theta - \sin^2\theta) \\ &\vdots \end{aligned} \tag{A.10}$$

etc., which could be identified as the Mohr transformation of the strains in the plane. Now we need to express the expressions for strain components ε_{xx}, ε_{yy}, etc., in terms of the displacement components u_r, u_θ, etc., and the derivatives with respect to the variables r, θ and z. The procedure can be adequately illustrated by simply examining the operations applicable to transform the result for ε_{rr}.

Considering the fact that $x = x(r, \theta)$ and $y = y(r, \theta)$, we have

$$\frac{\partial}{\partial x} = \frac{\partial}{\partial r}\frac{\partial r}{\partial x} + \frac{\partial}{\partial \theta}\frac{\partial \theta}{\partial x}; \quad \frac{\partial}{\partial y} = \frac{\partial}{\partial r}\frac{\partial r}{\partial y} + \frac{\partial}{\partial \theta}\frac{\partial \theta}{\partial y} \tag{A.11}$$

From (A.1), we have

$$\begin{aligned} \frac{\partial r}{\partial x} &= \frac{x}{r} = \cos\theta; & \frac{\partial r}{\partial y} &= \frac{y}{r} = \sin\theta \\ \frac{\partial \theta}{\partial x} &= -\frac{y}{r^2} = -\frac{\sin\theta}{r}; & \frac{\partial \theta}{\partial y} &= \frac{x}{r^2} = \frac{\cos\theta}{r} \end{aligned} \tag{A.12}$$

Combining (A.11) and (A.12) gives

$$\frac{\partial}{\partial x} = \cos\theta\frac{\partial}{\partial r} - \frac{\sin\theta}{r}\frac{\partial}{\partial \theta}; \quad \frac{\partial}{\partial y} = \sin\theta\frac{\partial}{\partial r} + \frac{\cos\theta}{r}\frac{\partial}{\partial \theta} \tag{A.13}$$

Also from (A.2) we obtain

$$u = u_r\cos\theta - u_\theta\sin\theta; \quad v = u_r\sin\theta + u_\theta\cos\theta \tag{A.14}$$

From the first equation of (A.10) we have

$$\begin{aligned}\varepsilon_{rr} &= \cos^2\theta\left(\cos\theta\frac{\partial}{\partial r}-\frac{\sin\theta}{r}\frac{\partial}{\partial\theta}\right)(u_r\cos\theta-u_\theta\sin\theta)\\&+\sin^2\theta\left(\sin\theta\frac{\partial}{\partial r}+\frac{\cos\theta}{r}\frac{\partial}{\partial\theta}\right)(u_r\sin\theta+u_\theta\cos\theta)\\&+\sin\theta\cos\theta\left(\sin\theta\frac{\partial}{\partial r}+\frac{\cos\theta}{r}\frac{\partial}{\partial\theta}\right)(u_r\cos\theta-u_\theta\sin\theta)\\&+\sin\theta\cos\theta\left(\cos\theta\frac{\partial}{\partial r}-\frac{\sin\theta}{r}\frac{\partial}{\partial\theta}\right)(u_r\sin\theta+u_\theta\cos\theta)\end{aligned}\tag{A.15}$$

Evaluating this expression it can be shown that

$$\varepsilon_{rr}=\frac{\partial u_r}{\partial r}\tag{A.16}$$

This procedure can be repeated to determine the remaining components of the strain matrix referred to the cylindrical polar coordinate system. The process is tedious, but the effort and the drudgery can be reduced significantly by employing symbolic mathematical manipulation techniques offered by codes such as MATHEMATICA®, MAPLE® or MACSYMA®. We can now summarise the results generated by the above procedures to develop the strain matrix, which is referred to the cylindrical polar coordinate system, i.e.

$$\begin{aligned}[\boldsymbol{\varepsilon}^{cp}] &= \begin{bmatrix}\dfrac{\partial u_r}{\partial r} & \dfrac{1}{2}\left(\dfrac{\partial u_\theta}{\partial r}+\dfrac{1}{r}\dfrac{\partial u_r}{\partial\theta}-\dfrac{u_\theta}{r}\right) & \dfrac{1}{2}\left(\dfrac{\partial u_z}{\partial r}+\dfrac{\partial u_r}{\partial z}\right)\\ \dfrac{1}{2}\left(\dfrac{\partial u_\theta}{\partial r}+\dfrac{1}{r}\dfrac{\partial u_r}{\partial\theta}-\dfrac{u_\theta}{r}\right) & \dfrac{1}{r}\dfrac{\partial u_\theta}{\partial\theta}+\dfrac{u_r}{r} & \dfrac{1}{2}\left(\dfrac{\partial u_\theta}{\partial z}+\dfrac{1}{r}\dfrac{\partial u_z}{\partial\theta}\right)\\ \dfrac{1}{2}\left(\dfrac{\partial u_z}{\partial r}+\dfrac{\partial u_r}{\partial z}\right) & \dfrac{1}{2}\left(\dfrac{\partial u_\theta}{\partial z}+\dfrac{1}{r}\dfrac{\partial u_z}{\partial\theta}\right) & \dfrac{\partial u_z}{\partial z}\end{bmatrix}\\&= [\boldsymbol{\varepsilon}^{cp}]^T\end{aligned}\tag{A.17}$$

A further possibility is to develop the expressions for the strain components by making use of the result given by (1.4), which is already in a form applicable to any coordinate system. This is, of course, easy in principle but requires some familiarity with vector calculus and associated operations applicable to generalized curvilinear coordinates. It is instructive to briefly outline the salient aspects of this procedure and, again, the reader is referred to the texts cited at the end of this Appendix for a more in-depth treatment.

Consider a point P with rectangular Cartesian coordinates (x, y, z) and any curvilinear coordinate system $(\alpha_1, \alpha_2, \alpha_3)$ that admits a representation

$$x = x(\alpha_1,\alpha_2,\alpha_3);\quad y = y(\alpha_1,\alpha_2,\alpha_3);\quad z = z(\alpha_1,\alpha_2,\alpha_3)\tag{A.18}$$

If we hold α_2 and α_3 constant, then as P varies, the position vector $\Re = x\boldsymbol{i} + y\boldsymbol{j} + z\boldsymbol{k}$ describes the curve called the α_1 coordinate through P. Similarly, we can define the coordinate curves α_2 and α_3 through P. The tangent vectors to the coordinate curves are given by the partial derivatives $\partial\Re/\partial\alpha_1$, $\partial\Re/\partial\alpha_2$ and $\partial\Re/\partial\alpha_3$. We can define the unit tangent vectors $\boldsymbol{e}_1$, $\boldsymbol{e}_2$ and $\boldsymbol{e}_3$ applicable to the curves α_1, α_2 and α_3 such that

$$\frac{\partial\Re}{\partial\alpha_1}=h_1\boldsymbol{e}_1;\quad \frac{\partial\Re}{\partial\alpha_2}=h_2\boldsymbol{e}_2;\quad \frac{\partial\Re}{\partial\alpha_3}=h_3\boldsymbol{e}_3\tag{A.19}$$

where h_1, h_2 and h_3 are defined as the scale factors

$$h_n = \left| \frac{\partial \mathfrak{R}}{\partial \alpha_n} \right|; \quad n = 1, 2, 3 \tag{A.20}$$

If $\boldsymbol{e}_1, \boldsymbol{e}_2$ and $\boldsymbol{e}_3$ are orthogonal then the curvilinear coordinate system is said to be orthogonal. The advantage of this formulation is that we can also define an element of arc length ds in terms of these scale factors, such that

$$(ds)^2 = d\mathfrak{R} \cdot d\mathfrak{R} \tag{A.21}$$

where

$$d\mathfrak{R} = \frac{\partial \mathfrak{R}}{\partial \alpha_1} d\alpha_1 + \frac{\partial \mathfrak{R}}{\partial \alpha_2} d\alpha_2 + \frac{\partial \mathfrak{R}}{\partial \alpha_3} d\alpha_3 = h_1 d\alpha_1 \boldsymbol{e}_1 + h_2 d\alpha_2 \boldsymbol{e}_2 + h_3 d\alpha_3 \boldsymbol{e}_3 \tag{A.22}$$

or

$$(ds)^2 = h_1^2 (d\alpha_1)^2 + h_2^2 (d\alpha_2)^2 + h_3^2 (d\alpha_3)^2 \tag{A.23}$$

Avoiding details it can be shown that the gradient operator takes the form

$$\nabla = \frac{1}{h_1} \boldsymbol{e}_1 \frac{\partial}{\partial \alpha_1} + \frac{1}{h_2} \boldsymbol{e}_2 \frac{\partial}{\partial \alpha_2} + \frac{1}{h_3} \boldsymbol{e}_3 \frac{\partial}{\partial \alpha_3} \tag{A.24}$$

In the special case of the cylindrical polar coordinate system

$$h_r = 1; \quad h_\theta = r; \quad h_z = 1$$

$$\boldsymbol{e}_r = \boldsymbol{i}_r; \quad \boldsymbol{e}_\theta = \boldsymbol{i}_\theta; \quad \boldsymbol{e}_z = \boldsymbol{i}_z \tag{A.25}$$

We need to observe the fact that there are also the derivatives of the unit base vectors; for the cylindrical polar coordinate system the only non-zero ones are

$$\frac{\partial \boldsymbol{e}_r}{\partial \theta} = \boldsymbol{e}_\theta; \quad \frac{\partial \boldsymbol{e}_\theta}{\partial \theta} = -\boldsymbol{e}_r \tag{A.26}$$

Let us now consider the operation designated by $\nabla \boldsymbol{u}$ with the understanding that this now signifies the gradient of a vector which gives rise to a *dyadic*. The mathematical physicist Josiah Willard Gibbs is credited with the introduction of the convention, which can be used to display the elements of the operation in terms of components that have two unit vectors associated with each component, i.e.

$$\nabla \boldsymbol{u} = \frac{1}{h_r} \boldsymbol{i}_r \frac{\partial}{\partial r} (u_r \boldsymbol{i}_r + u_\theta \boldsymbol{i}_\theta + u_z \boldsymbol{i}_z) + \frac{1}{h_\theta} \boldsymbol{i}_\theta \frac{\partial}{\partial \theta} (u_r \boldsymbol{i}_r + u_\theta \boldsymbol{i}_\theta + u_z \boldsymbol{i}_z)$$

$$+ \frac{1}{h_z} \boldsymbol{i}_z \frac{\partial}{\partial z} (u_r \boldsymbol{i}_r + u_\theta \boldsymbol{i}_\theta + u_z \boldsymbol{i}_z) \tag{A.27}$$

Performing the operation, we obtain

$$\nabla \boldsymbol{u} = \left(\frac{\partial u_r}{\partial r} \right) \boldsymbol{i}_r \boldsymbol{i}_r + \left(\frac{\partial u_\theta}{\partial r} \right) \boldsymbol{i}_r \boldsymbol{i}_\theta + \left(\frac{\partial u_z}{\partial r} \right) \boldsymbol{i}_r \boldsymbol{i}_z$$

$$+ \left(\frac{1}{r} \frac{\partial u_r}{\partial \theta} - \frac{u_r}{r} \right) \boldsymbol{i}_\theta \boldsymbol{i}_r + \left(\frac{1}{r} \frac{\partial u_\theta}{\partial \theta} + \frac{u_\theta}{r} \right) \boldsymbol{i}_\theta \boldsymbol{i}_\theta + \left(\frac{1}{r} \frac{\partial u_z}{\partial \theta} \right) \boldsymbol{i}_\theta \boldsymbol{i}_z$$

$$+ \left(\frac{\partial u_r}{\partial z} \right) \boldsymbol{i}_z \boldsymbol{i}_r + \left(\frac{\partial u_\theta}{\partial z} \right) \boldsymbol{i}_z \boldsymbol{i}_\theta + \left(\frac{\partial u_z}{\partial z} \right) \boldsymbol{i}_z \boldsymbol{i}_z \tag{A.28}$$

The ordering of the unit vectors is quite important, and the conjugate dyadic $(\nabla \boldsymbol{u})^T$ can also be defined in the same way. We can now construct an array or matrix of physical components of the gradient of the displacement in such a way that $\nabla \boldsymbol{u}$ is signified by

$$\nabla \boldsymbol{u} = \begin{bmatrix} \left(\dfrac{\partial u_r}{\partial r}\right) & \left(\dfrac{\partial u_\theta}{\partial r}\right) & \left(\dfrac{\partial u_z}{\partial r}\right) \\ \left(\dfrac{1}{r}\dfrac{\partial u_r}{\partial \theta} - \dfrac{u_r}{r}\right) & \left(\dfrac{1}{r}\dfrac{\partial u_\theta}{\partial \theta} + \dfrac{u_\theta}{r}\right) & \left(\dfrac{1}{r}\dfrac{\partial u_z}{\partial \theta}\right) \\ \left(\dfrac{\partial u_r}{\partial z}\right) & \left(\dfrac{\partial u_\theta}{\partial z}\right) & \left(\dfrac{\partial u_z}{\partial z}\right) \end{bmatrix} \tag{A.29}$$

with

$$\boldsymbol{u}\nabla = (\nabla \boldsymbol{u})^T \tag{A.30}$$

It can be easily verified that using the result (A.29) and the definition for the strain matrix defined by

$$[\boldsymbol{\varepsilon}^{cp}] = \frac{1}{2}[\nabla \boldsymbol{u} + (\nabla \boldsymbol{u})^T] \tag{A.31}$$

one obtains a result that exactly corresponds to the expression (A.17).

Let us now consider the equations of equilibrium. Take, for example, the first of equations (1.33), where

$$\frac{\partial \sigma_{xx}}{\partial x} + \frac{\partial \sigma_{xy}}{\partial y} + \frac{\partial \sigma_{xz}}{\partial z} - b_x = 0 \tag{A.32}$$

It is a relatively easy matter first to obtain the expressions for the stress components σ_{xx}, σ_{xy} and σ_{xz} in terms of their cylindrical polar coordinate equivalents and to follow this up by converting the partial derivatives of the stress components, which are expressed in terms of their Cartesian components, to their cylindrical polar equivalents. This is a perfectly valid operation, which will certainly give a mathematically correct alternative result for the equation of equilibrium in the x-direction in terms of σ_{rr}, $\sigma_{r\theta}, \sigma_{rz}$, etc., and r, $\theta$and z. In doing these operations we have, in fact, gained nothing. This is of little value when dealing with problems associated with the cylindrical polar coordinate system, where the expressions for traction boundary conditions and problem formulations in general are usually expressed in terms of radial, circumferential and axial directions corresponding to coordinate directions r, θ and z, respectively. For this reason, we have to resort to alternative procedures. The easiest is to consider the variation of stresses over a segment of an annular region (Figure A.10). Considering the equilibrium of forces in the radial direction we have

$$\begin{aligned} &\left[\left(\sigma_{rr} + \frac{\partial \sigma_{rr}}{\partial r}dr\right)(r + dr) - \sigma_{rr}r\right] d\theta\, dz - \left[\sigma_{\theta\theta} + \frac{\partial \sigma_{\theta\theta}}{\partial \theta}d\theta + \sigma_{\theta\theta}\right] \\ &\quad \times dr\, dz \sin\left(\frac{d\theta}{2}\right) + \left[\sigma_{r\theta} + \frac{\partial \sigma_{r\theta}}{\partial \theta}d\theta - \sigma_{r\theta}\right] dr\, dz \cos\left(\frac{d\theta}{2}\right) \\ &\quad + \left[\sigma_{rz} + \frac{\partial \sigma_{rz}}{\partial z}dz - \sigma_{rz}\right] r\, dr\, d\theta - b_r r\, dr\, d\theta\, dz = 0 \end{aligned} \tag{A.33}$$

By making the approximation, $\sin(d\theta/2) \approx d\theta/2$ and $\cos(d\theta/2) \approx 1$, and noting that since the elemental volume considered is arbitrary and non-zero, (A.33) reduces to

$$\frac{\partial \sigma_{rr}}{\partial r} + \frac{1}{r}\frac{\partial \sigma_{r\theta}}{\partial r} + \frac{\sigma_{rr} - \sigma_{\theta\theta}}{r} + \frac{\partial \sigma_{rz}}{\partial z} - b_r = 0 \tag{A.34}$$

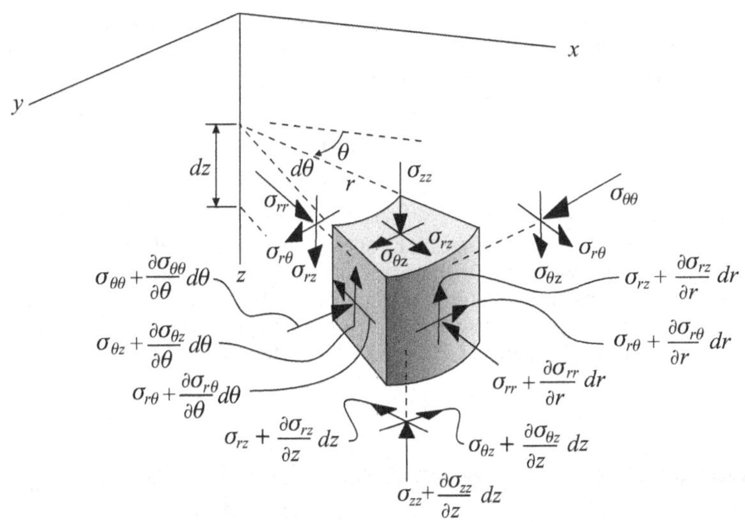

Figure A.10. Stresses acting on an element referred to a cylindrical coordinate system.

We can repeat the procedure to develop the relevant equations of equilibrium appropriate for the θ- and z-directions. This is left as an exercise for the reader.

The alternative to this procedure is to utilise the general vector equation for static equilibrium

$$\nabla\cdot\boldsymbol{\sigma}-\boldsymbol{b}=\mathbf{0} \tag{A.35}$$

and to convert this to its counterpart applicable to the cylindrical polar coordinate system. We note that, in (A.35), $\boldsymbol{\sigma}$ refers to the stress dyadic appropriate for the cylindrical polar coordinate system, i.e.

$$\begin{aligned}\boldsymbol{\sigma} &= \sigma_{rr}\boldsymbol{i}_r\boldsymbol{i}_r+\sigma_{r\theta}\boldsymbol{i}_r\boldsymbol{i}_\theta+\sigma_{rz}\boldsymbol{i}_r\boldsymbol{i}_z+\sigma_{\theta r}\boldsymbol{i}_\theta\boldsymbol{i}_r+\sigma_{\theta\theta}\boldsymbol{i}_\theta\boldsymbol{i}_\theta+\sigma_{\theta z}\boldsymbol{i}_\theta\boldsymbol{i}_z\\ &\quad+\sigma_{zr}\boldsymbol{i}_z\boldsymbol{i}_r+\sigma_{z\theta}\boldsymbol{i}_z\boldsymbol{i}_\theta+\sigma_{zz}\boldsymbol{i}_z\boldsymbol{i}_z\end{aligned} \tag{A.36}$$

with the matrix of physical components of $\boldsymbol{\sigma}$ satisfying the requirement $\boldsymbol{\sigma}=\boldsymbol{\sigma}^T$. We can evaluate the divergence by noting that certain derivatives of the base vectors are non-zero and that the divergence of a dyadic is a vector (see, e.g., (A.26)). This gives

$$\begin{aligned}\nabla\cdot\boldsymbol{\sigma}-\boldsymbol{b} &= \left[\frac{\partial\sigma_{rr}}{\partial r}+\frac{1}{r}\frac{\partial\sigma_{r\theta}}{\partial\theta}+\frac{\sigma_{rr}-\sigma_{\theta\theta}}{r}+\frac{\partial\sigma_{rz}}{\partial z}\right]\boldsymbol{i}_r\\ &\quad+\left[\frac{\partial\sigma_{r\theta}}{\partial r}+\frac{1}{r}\frac{\partial\sigma_{\theta\theta}}{\partial\theta}+\frac{\sigma_{r\theta}}{r}+\frac{\partial\sigma_{\theta z}}{\partial z}\right]\boldsymbol{i}_\theta\\ &\quad+\left[\frac{\partial\sigma_{zz}}{\partial z}+\frac{\partial\sigma_{rz}}{\partial r}+\frac{\sigma_{zr}}{r}+\frac{1}{r}\frac{\partial\sigma_{\theta z}}{\partial\theta}\right]\boldsymbol{i}_z-\boldsymbol{b}=\mathbf{0}\end{aligned} \tag{A.37}$$

As can be seen, the three equations of equilibrium in the three orthogonal coordinate directions r, θ and z are present in the form of the relevant vector components.

Other equations such as the constitutive equations applicable to elastic and plastic behaviour of the material, compatibility equations and boundary conditions can be obtained similarly by considering the general form of the equations applicable to the cylindrical polar coordinate system. In most cases, the equations simply transform in a straightforward manner, because of the orthogonality of the coordinate systems. For example, the equations of elasticity for an isotropic elastic solid, expressed in the cylindrical polar coordinate system take the form

$$\begin{aligned} E\varepsilon_{rr} &= \sigma_{rr} - \nu(\sigma_{\theta\theta} + \sigma_{zz}) \\ E\varepsilon_{\theta\theta} &= \sigma_{\theta\theta} - \nu(\sigma_{rr} + \sigma_{zz}) \\ E\varepsilon_{zz} &= \sigma_{zz} - \nu(\sigma_{\theta\theta} + \sigma_{rr}) \\ 2G\varepsilon_{r\theta} &= \sigma_{r\theta}; \quad 2G\varepsilon_{\theta z} = \sigma_{\theta z}; \quad 2G\varepsilon_{rz} = \sigma_{rz} \end{aligned} \tag{A.38}$$

which is identical in form to the equations of elasticity in rectangular Cartesian coordinates. With regard to yield criteria, the representation in terms of invariants assures their development in terms of the appropriate components of the stresses, which are referred to the cylindrical polar coordinate system. For example, for *axial symmetry*, the stress matrix referred to the cylindrical polar coordinate system is

$$\boldsymbol{\sigma} = \begin{bmatrix} \sigma_{rr} & 0 & \sigma_{rz} \\ 0 & \sigma_{\theta\theta} & 0 \\ \sigma_{rz} & 0 & \sigma_{zz} \end{bmatrix} \tag{A.39}$$

and the appropriate form of the Coulomb failure criterion for a purely granular material where either $\sigma_{rr} > \sigma_{\theta\theta} > \sigma_{zz}$ or $\sigma_{zz} > \sigma_{\theta\theta} > \sigma_{rr}$ takes the form

$$f(\boldsymbol{\sigma}) = (\sigma_{rr} - \sigma_{zz})^2 + (\sigma_{rr} + \sigma_{zz})^2 \sin^2\phi + 4\sigma_{rz}^2 = 0 \tag{A.40}$$

and the associative flow rule (4.12) gives the plastic strain rate components as

$$\begin{aligned} \dot{\boldsymbol{\varepsilon}}^p &= \begin{bmatrix} \dot{\varepsilon}^p_{rr} & 0 & \dot{\varepsilon}^p_{rz} \\ 0 & \dot{\varepsilon}^p_{\theta\theta} & 0 \\ \dot{\varepsilon}^p_{rz} & 0 & \dot{\varepsilon}^p_{zz} \end{bmatrix} \\ &= 2\lambda \begin{bmatrix} \left\{ \begin{matrix} \sigma_{rr}(1+\sin^2\phi) \\ -\sigma_{zz}(1-\sin^2\phi) \end{matrix} \right\} & 0 & 4\sigma_{rz} \\ 0 & 0 & 0 \\ 4\sigma_{rz} & 0 & \left\{ \begin{matrix} \sigma_{zz}(1+\sin^2\phi) \\ -\sigma_{rr}(1-\sin^2\phi) \end{matrix} \right\} \end{bmatrix} \end{aligned} \tag{A.41}$$

Expressions similar to (A.41) can be obtained for other yield criteria with lesser constraints than those invoked in the above due to the assumptions of axial symmetry and the consideration of a relatively simple form of a failure criterion applicable to purely granular material with no dependence on the intermediate principal stress.

Further reading

P.-C. Chou and N.J. Pagano, *Elasticity: Tensor, Dyadic and Engineering Approaches*, Dover Publications, New York, 1967.

T.J. Chung, *Continuum Mechanics*, Prentice-Hall, Englewood Cliffs, NJ, 1988.

R.O. Davis and A.P.S. Selvadurai, *Elasticity and Geomechanics*, Cambridge University Press, Cambridge, 1996.

A.C. Eringen, *Mechanics of Continua*, John Wiley, New York, 1967.

Y.C. Fung, *Foundations of Solid Mechanics*, Prentice-Hall, Englewood Cliffs, NJ, 1965.

S.C. Hunter, *Mechanics of Continuous Media*, Ellis-Horwood-Wiley, New York, 1983.
R.W. Little, *Elasticity*, Prentice-Hall, Englewood Cliffs, NJ, 1973.
L.E. Malvern, *Introduction to the Mechanics of a Continuous Medium*, Prentice-Hall, Englewood Cliffs, NJ, 1969.
J.B. Martin, *Plasticity: Fundamentals and General Results*, MIT Press, Cambridge, MA, 1975.
G.E. Mase and G.T. Mase, *Continuum Mechanics for Engineers*, CRC Press, Boca Raton, FL, 1991.
A.P.S. Selvadurai, *Partial Differential Equations in Mechanics. Vol. 2. The Biharmonic Equation, Poisson's Equation*, Springer-Verlag, Berlin, 2000.
A.J.M. Spencer, *Continuum Mechanics*, Longman, London, 1980.
S.P. Timoshenko and J.N. Goodier, *Theory of Elasticity*, McGraw-Hill, New York, 1970.
E. Volterra and J.H. Gaines, *Advanced Strength of Materials*, Prentice-Hall, Englewood Cliffs, NJ, 1971.

Appendix B
Mohr circles

The graphical construction for the representation of the state of stress at a point within a continuum region is generally attributed to the German engineer Otto Christian Mohr. Although the use of graphical techniques in structural and solid mechanics has been an important area of activity both for engineering calculations and stress analysis, particularly in the eighteenth and nineteenth centuries (see, e.g., Todhunter and Pearson (1886, 1893) and Timoshenko (1953)), the contributions of Karl Culmann and Otto Mohr to the development of this area are regarded as being particularly significant. Despite the passage of time these graphical constructions have continued to serve as efficient educational tools for the visualisation of difficult concepts related to the representation of three-dimensional states of stress, particularly in relation to the description of failure states in materials. The fact that the techniques developed in relation to the stress state at a point that can be represented in terms of a stress matrix of rank two or a second-order tensor implies that the procedures are equally applicable to the description of other properties and states in continua, which can be described in a similar manner. Examples include the description of moments of inertia of solids, flexural characteristics of plates and the hydraulic conductivity characteristics of porous media, etc. The purpose of this Appendix is to present a brief outline of the significant features of Mohr circles and to develop the basic equations applicable to the three-dimensional graphical representation of the stress state at a point. The naming of the graphical procedures for the representation of the state of stress at a point, in honour of Otto Mohr is very much in recognition of his formal development of the procedures through archival publications. There are earlier references to techniques resembling a graphical method in the work of Karl Culmann, although they are in a form that is perhaps less well developed than the presentations of Mohr.

As a prelude to the development of the relevant equations, we first consider Cauchy's relationship, which deals with the stress state at a point within the medium and the traction vectors that act on an arbitrary plane either through or located at an infinitesimal distance from the point. Before doing this it is worthwhile making some remarks with regard to sign conventions that are used to identify particular stress states. From an engineer's perspective, sign conventions are crucial to identifying the 'sense' of stress components accurately. This is clearly not the case if we were to treat the stress state as a '*matrix*' or a '*tensor*', which is amenable to purely mathematical operations. In this context we are not concerned as to whether the normal stresses are compressive or tensile or whether the shear stresses are positive or negative. These are simply elements of a matrix or a tensor; we can transform the matrix, calculate its eigenvalues and perform all the legitimate operations of linear algebra without ever worrying about the physical significance of the manipulations. Also, for example when a stress transformation rule such as

$$[\sigma^{cp}] = [\mathbf{H}]^T[\sigma][\mathbf{H}] \tag{B.1}$$

which involves the transformation of the stress matrix from the Cartesian to the cylindrical polar equivalent, the sign convention adopted in the definition of the stress matrix referred to the rectangular Cartesian coordinate system, $[\sigma]$, simply translates to the definition of the sign convention for the stress matrix $\lfloor\sigma^{cp}\rfloor$ referred to the cylindrical polar coordinate system. There are of course 'bonuses' that arise from these mathematical operations, such as the fact that the eigenvalues of a symmetric matrix must always be real, which straight away translates to the deduction that the principal stresses must always be real, but this is a secondary issue. From an engineering perspective, sign conventions are crucial to the proper physical understanding of the 'mechanics' of the manipulations.

Sign conventions for the description of the stresses are many and varied and they are, at the same time, a vexation to expert and novice alike. There are many possible sign conventions that are found in the literature. The fair advice is to suggest that if a particular sign convention works for you, by all means use it. The purpose of this commentary is to outline the limitations of some commonly adopted sign conventions and to suggest a sign convention that will be user-friendly in most circumstances. First, let us consider the sign convention normally associated with the axial stresses. In solid mechanics in general, tensile stresses are usually considered to be positive whereas in geomechanics, and in this text in particular, compressive stresses are considered to be positive. This is not a major area of concern since we can associate some differences in the physical actions that will result from the applications of either a tensile stress or a compressive stress. Line elements can either extend or shorten depending on the nature of the axial stress. What about the shear stresses? The usual procedure is to consider first the shear stress acting on a surface of interest and to select a point just outside the region. If the shear stresses cause *clockwise moments* then, in solid mechanics, the shear stresses are considered to be *positive.* In geomechanics counterclockwise replaces clockwise in their definition (Figure B.1a). (In doing this we have also, by deduction, introduced the definition of a negative shear stress.) Performing a simple operation, however, unravels this definition. Let us draw Figure B.1(a) on an acetate transparency and look at the figure from the opposite side. The view will correspond to that shown in Figure B.1(b). We now have the same shear stress but it appears to have a negative magnitude. Regrettably, this definition of the shear stress becomes dependent on the point of view of the observer. The previous definition is perfectly acceptable so long as you do not move outside of the plane of the paper. This is obviously somewhat restrictive when three-dimensional

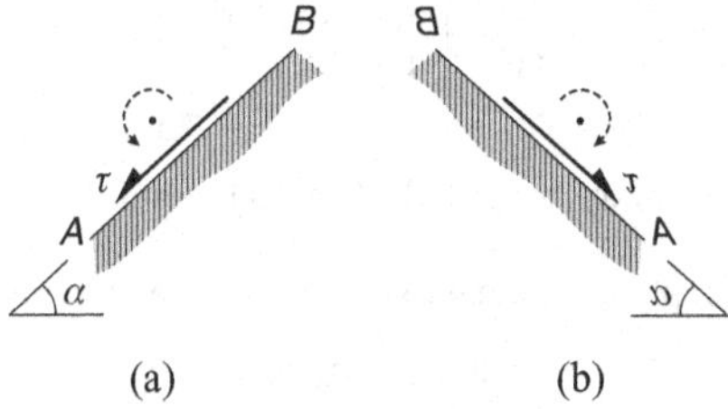

Figure B.1. Sign convention for shear stresses based on a clockwise and a counterclockwise sense of the moment induced by the shear stress about an exterior point.

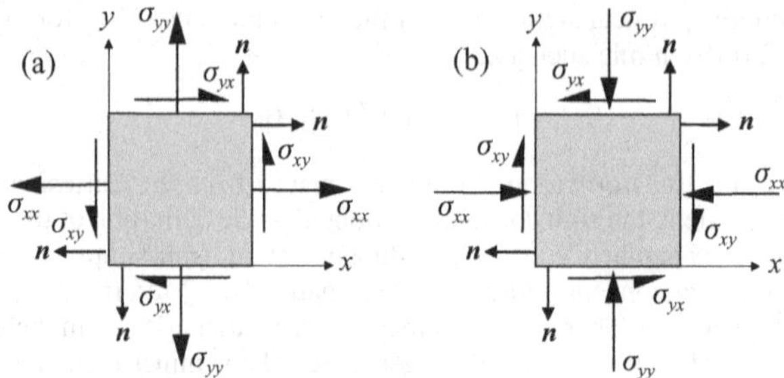

Figure B.2. Sign convention for normal stresses and shear stresses in association with coordinate directions and alignment of unit normals.

states of stress are encountered, and the dependence on a change in the perspective only compounds the problem. Naturally, there are means of overcoming this deficiency and the simplest is to attach a frame of reference to the definition of the sign convention. For example, consider the definition of tensile stresses as being positive in the context of solid mechanics. The appropriate definition of a positive tensile stress is one in which the traction vector acts in a positive (or negative) coordinate direction and on planes the outward normal of which is also oriented in a positive (or negative) coordinate direction. A similar definition can be adopted when defining a positive shear stress consistent with this definition of a positive tensile stress. Referring to Figure B.2(a), all the stresses shown there are *positive stresses* in the context of *solid mechanics*. We can draw Figure B.2(a) on an acetate transparency and look through from the reverse side and still all the stresses will be positive according to our definition. We have eliminated the observer dependence by attaching a system of coordinates to the element, the positive directions of which will remain observer invariant. In the same way, we can now proceed to define a consistent set of definitions to account for the geomechanics convention of considering compressive stresses as being positive. So, the definition of a positive compressive or shear stress is one where the traction vector acts in the negative (or positive) coordinate direction and on planes the outward normal to which acts in the positive (or negative) direction. Referring to Figure B.2(b), all the stresses shown there are considered to be *positive stresses* in the context of *geomechanics*. As has been demonstrated, a certain consistency is necessary in assigning a sign convention for the stress components defining the state of stress at a point. Several such possibilities exist and the prudent option is to select one with the minimum number of limitations.

With the above comments in mind let us restrict our attention to a system of rectangular Cartesian coordinates and a stress state defined by

$$[\sigma] = \begin{bmatrix} \sigma_{xx} & \sigma_{xy} & \sigma_{xz} \\ \sigma_{yx} & \sigma_{yy} & \sigma_{yz} \\ \sigma_{zx} & \sigma_{zy} & \sigma_{zz} \end{bmatrix} \tag{B.2}$$

where in the absence of body couples, $[\sigma] = [\sigma]^T$, and the superscript refers to the transpose.

The stress vectors on planes normal to the axes x, y and z (Figure B.3) are given by

$$\boldsymbol{T}_x = \sigma_{xx}\boldsymbol{i}_x + \sigma_{xy}\boldsymbol{i}_y + \sigma_{xz}\boldsymbol{i}_z \tag{B.3}$$

$$\boldsymbol{T}_y = \sigma_{yx}\boldsymbol{i}_x + \sigma_{yy}\boldsymbol{i}_y + \sigma_{yz}\boldsymbol{i}_z \tag{B.4}$$

$$\boldsymbol{T}_z = \sigma_{zx}\boldsymbol{i}_x + \sigma_{zy}\boldsymbol{i}_y + \sigma_{zz}\boldsymbol{i}_z \tag{B.5}$$

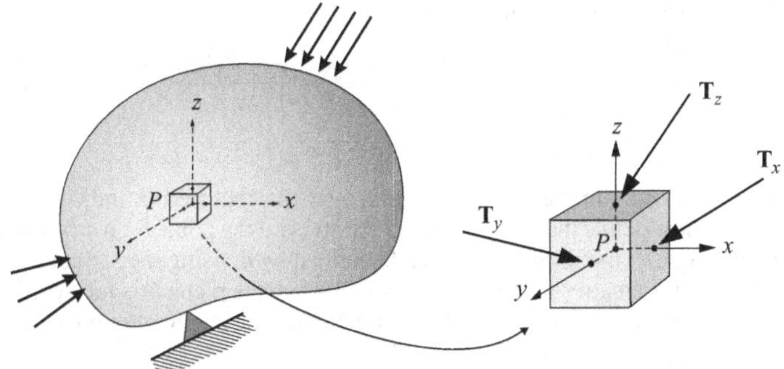

Figure B.3. Traction vectors on a cuboidal element encompassing location P.

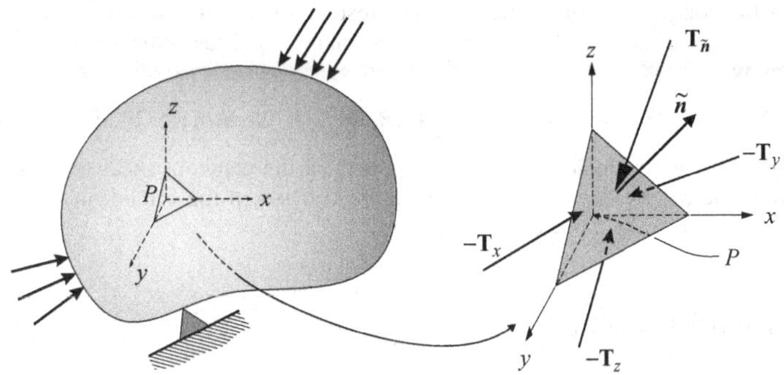

Figure B.4. Traction vectors on a tetrahedral element located at an infinitesimal distance from P.

where $\boldsymbol{i}_x, \boldsymbol{i}_y$ and $\boldsymbol{i}_z$ are the unit vectors in the x-, y- and z-directions, respectively. Let us now consider an oblique plane S located at an infinitesimal distance from point P (Figure B.4) such that the unit normal to the plane $\tilde{\boldsymbol{n}}$ is defined by

$$\tilde{\boldsymbol{n}} = \tilde{n}_x \boldsymbol{i}_x + \tilde{n}_y \boldsymbol{i}_y + \tilde{n}_z \boldsymbol{i}_z \tag{B.6}$$

where the components of $\tilde{\boldsymbol{n}}$ in the x-, y- and z-directions are implied. We can consider equilibrium of forces acting on the tetrahedral element with infinitesimal dimensions and show that (see, e.g., Davis and Selvadurai 1996)

$$\boldsymbol{T}_{\tilde{n}} = \tilde{n}_x \boldsymbol{T}_x + \tilde{n}_y \boldsymbol{T}_y + \tilde{n}_z \boldsymbol{T}_z \tag{B.7}$$

Upon substituting (B.3)–(B.5) in (B.7) we have

$$\begin{aligned} \boldsymbol{T}_{\tilde{n}} = {} & (\tilde{n}_x \sigma_{xx} + \tilde{n}_y \sigma_{yx} + \tilde{n}_z \sigma_{zx}) \boldsymbol{i}_x + (\tilde{n}_x \sigma_{xy} + \tilde{n}_y \sigma_{yy} + \tilde{n}_z \sigma_{zy}) \boldsymbol{i}_y \\ & + (\tilde{n}_x \sigma_{xz} + \tilde{n}_y \sigma_{yz} + \tilde{n}_z \sigma_{zz}) \boldsymbol{i}_z \end{aligned} \tag{B.8}$$

Note that, although the stress matrix is symmetric, we will retain the designations for the components of the stress matrix defined by (B.2) primarily to illustrate the development of a sequence. We can now define the stress vector $\boldsymbol{T}_{\tilde{n}}$ on the oblique plane in terms of projections along the x, y and z axes such that

$$\boldsymbol{T}_{\tilde{n}} = \sigma_{\tilde{n}x} \boldsymbol{i}_x + \sigma_{\tilde{n}y} \boldsymbol{i}_y + \sigma_{\tilde{n}z} \boldsymbol{i}_z \tag{B.9}$$

where the following scalar definitions apply:

$$\sigma_{\tilde{n}x} = \tilde{n}_x\sigma_{xx} + \tilde{n}_y\sigma_{yx} + \tilde{n}_z\sigma_{zx} = \tilde{n}_\alpha\sigma_{\alpha x} \tag{B.10}$$

$$\sigma_{\tilde{n}y} = \tilde{n}_x\sigma_{yx} + \tilde{n}_y\sigma_{yy} + \tilde{n}_z\sigma_{yz} = \tilde{n}_\alpha\sigma_{\alpha y} \tag{B.11}$$

$$\sigma_{\tilde{n}z} = \tilde{n}_x\sigma_{xz} + \tilde{n}_y\sigma_{yz} + \tilde{n}_z\sigma_{zz} = \tilde{n}_\alpha\sigma_{\alpha z} \tag{B.12}$$

In these equations a summation takes place over the repeated α indices. Equations (B.10)–(B.12) now define the components of the stress at the point P on an oblique plane S passing through P, the normal of which is defined by $\tilde{\boldsymbol{n}}$, using the six components of the symmetric stress matrix $[\boldsymbol{\sigma}]$. It is evident that when the plane S is located directly at the point P, the relationships (B.10)–(B.12) are, in fact, the traction boundary conditions given by

$$\tilde{\boldsymbol{T}} = \tilde{\boldsymbol{n}}[\boldsymbol{\sigma}] \text{ or, using the summation convention,} \quad \tilde{T}_i = \tilde{n}_j\sigma_{ij} \tag{B.13}$$

There are two other results that can be deduced from expressions (B.10)–(B.12); we can express the components of the traction as a resultant of a stress normal to the oblique plane $\sigma_{\tilde{n}\tilde{n}}$ and a shear stress tangent the oblique plane, $\sigma_{\tilde{n}\tilde{t}}$. The normal stress to the plane is given by

$$\sigma_{\tilde{n}\tilde{n}} = \tilde{\boldsymbol{n}} \cdot \boldsymbol{T}_{\tilde{\boldsymbol{n}}} = \tilde{n}_x^2\sigma_{xx} + \tilde{n}_y^2\sigma_{yy} + \tilde{n}_z^2\sigma_{zz} + 2\tilde{n}_x\tilde{n}_y\sigma_{xy} + 2\tilde{n}_x\tilde{n}_z\sigma_{xz} + 2\tilde{n}_y\tilde{n}_z\sigma_{yz} \tag{B.14}$$

Similarly, we can evaluate the shear component of the resultant shear traction on the oblique plane $\sigma_{\tilde{n}\tilde{t}}$ by considering its relationship between the Euclidean norm of the vector $\boldsymbol{T}_{\tilde{\boldsymbol{n}}}$ and $\sigma_{\tilde{n}\tilde{n}}$, which takes the form

$$\|\boldsymbol{T}_{\tilde{\boldsymbol{n}}}\|^2 = \sigma_{\tilde{n}\tilde{n}}^2 + \sigma_{\tilde{n}\tilde{t}}^2 \tag{B.15}$$

Evaluating (B.15) we have

$$\sigma_{\tilde{n}\tilde{t}}^2 = \sigma_{\tilde{n}x}^2 + \sigma_{\tilde{n}y}^2 + \sigma_{\tilde{n}z}^2 - \sigma_{\tilde{n}\tilde{n}}^2 \tag{B.16}$$

We can now proceed to discuss the graphical representations associated with Mohr circles of stress. While several aspects of these graphical representations can be discussed we shall restrict attention to the following: the first deals with the use of Mohr circles as a graphical interpretation of the transformation rule applicable to stresses, strains and other dependent variables encountered in geomechanics and the second deals with the use of Mohr circles as a means of identifying admissible states of stress acting on arbitrary planes located through a point at which the stress matrix is defined in terms of the principal components. For the discussion of the first aspect of Mohr circles, it is convenient to further restrict one's attention to a two-dimensional state of stress characterized by a state of plane stress defined by

$$[\boldsymbol{\sigma}] = \begin{bmatrix} \sigma_{xx} & \sigma_{xy} & 0 \\ \sigma_{yx} & \sigma_{yy} & 0 \\ 0 & 0 & 0 \end{bmatrix} \tag{B.17}$$

We consider this particular state of stress, which may now be referred to a new set of rectangular Cartesian coordinates (X, Y, Z), obtained by rotating the (x, y, z) coordinate system about the z-axis by an angle θ in the anticlockwise direction (Figure B.5). Following developments given in Appendix A, the transformation matrix $[\mathbf{H}]$ is given by

$$[\mathbf{H}] = \begin{bmatrix} \cos\theta & -\sin\theta & 0 \\ \sin\theta & \cos\theta & 0 \\ 0 & 0 & 1 \end{bmatrix} \tag{B.18}$$

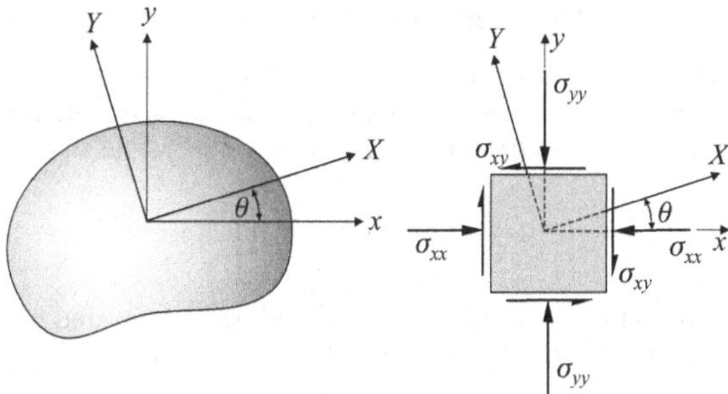

Figure B.5. State of stress and rotation of the reference coordinate system.

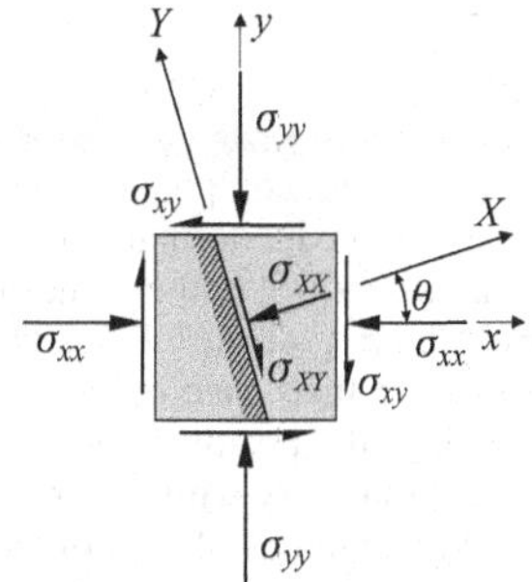

Figure B.6. Stresses on an oblique plane.

The stress matrix $[\boldsymbol{\Sigma}]$ referred to the rotated coordinate system is given by

$$[\boldsymbol{\Sigma}] = \begin{bmatrix} \sigma_{XX} & \sigma_{XY} & 0 \\ \sigma_{YX} & \sigma_{YY} & 0 \\ 0 & 0 & 0 \end{bmatrix} = [\mathbf{H}]^T[\boldsymbol{\sigma}][\mathbf{H}] \tag{B.19}$$

The non-zero components of (B.19) are given by

$$\sigma_{XX} = \sigma_{xx}\cos^2\theta + \sigma_{yy}\sin^2\theta + 2\sigma_{xy}\sin\theta\cos\theta \tag{B.20}$$

$$\sigma_{YY} = \sigma_{xx}\sin^2\theta + \sigma_{yy}\cos^2\theta - 2\sigma_{xy}\sin\theta\cos\theta \tag{B.21}$$

$$\sigma_{XY} = (\sigma_{yy} - \sigma_{xx})\sin\theta\cos\theta + (\cos^2\theta - \sin^2\theta)\sigma_{xy} \tag{B.22}$$

We can, for example, consider these expressions when $\theta = 0$, which gives

$$\sigma_{XX} = \sigma_{xx}; \quad \sigma_{YY} = \sigma_{yy}; \quad \sigma_{XY} = \sigma_{xy} \tag{B.23}$$

Similarly when $\theta = \pi/2$, (B.20)–(B.22) give

$$\sigma_{XX} = \sigma_{yy}; \quad \sigma_{YY} = \sigma_{xx}; \quad \sigma_{XY} = -\sigma_{xy} \tag{B.24}$$

the negative sign of the shear stress indicating that the positive shear stress acts in a direction opposite to that indicated by σ_{XY} (see, e.g., Figure B.6).

Mohr circles in two dimensions

Since the description of the state of stress referred to a coordinate system is dependent on the orientation of that coordinate system, we can choose an orientation in the (x, y, z) configuration such that the matrix $[\boldsymbol{\sigma}]$ corresponds to a principal state of stress with

$$[\boldsymbol{\sigma}] = \begin{bmatrix} \sigma_1 & 0 & 0 \\ 0 & \sigma_2 & 0 \\ 0 & 0 & 0 \end{bmatrix} \tag{B.25}$$

where the convention that $\sigma_1 > \sigma_2$ is used and both stresses σ_1 and σ_2 are assumed to be compressive. Using this result in (B.20)–(B.22) we obtain

$$\sigma_{XX} = \frac{1}{2}(\sigma_1 + \sigma_2) + \frac{1}{2}(\sigma_1 - \sigma_2)\cos 2\theta \tag{B.26}$$

$$\sigma_{YY} = \frac{1}{2}(\sigma_1 + \sigma_2) - \frac{1}{2}(\sigma_1 - \sigma_2)\cos 2\theta \tag{B.27}$$

$$\sigma_{XY} = -\frac{1}{2}(\sigma_1 - \sigma_2)\sin 2\theta \tag{B.28}$$

If we interpret these transformed stress components in relation to the new set of axes (X, Y, Z), we note that the stress σ_{XX} is the normal stress acting along the X-direction and σ_{XY} is the positive shear stress acting on the same plane. Similar interpretations can be given to the stresses σ_{YY} and σ_{XY} (Figure B.7). The negative sign for the shear stress σ_{XY} is consistent with the fact that since $\sigma_1 > \sigma_2$, the actual direction of σ_{XY} will be opposite to that indicated by the positive sign convention. The magnitudes of the stresses σ_{XX}, σ_{YY} and σ_{XY} will thus vary with the choice of the angle of orientation θ. Therefore this variation in the magnitudes of the components of the stress matrix in the transformed configuration can be illustrated graphically by constructing a diagram in which either the set σ_{XX} and σ_{XY} or σ_{YY} and σ_{XY} are taken as coordinates. To obtain such a relationship we square (B.26) and (B.28); the addition of these gives

$$\left[\sigma_{XX} - \frac{1}{2}(\sigma_1 + \sigma_2)\right]^2 + (\sigma_{XY})^2 = \frac{1}{2}(\sigma_1 - \sigma_2)^2 \tag{B.29}$$

This represents the equation of a circle in the σ_{XX} vs. σ_{XY} plane with its centre at $(\frac{1}{2}(\sigma_1 + \sigma_2), 0)$ and radius $\frac{1}{2}(\sigma_1 - \sigma_2)$. This circle is referred to as the Mohr circle (Figure B.8).

At this point we need to talk about sign conventions once again. There is no difficulty with the normal stresses σ_{XX} and σ_{YY}. Compressive normal stresses are taken as positive in geomechanics. But another question arises concerning the shear stress σ_{XY}. In the

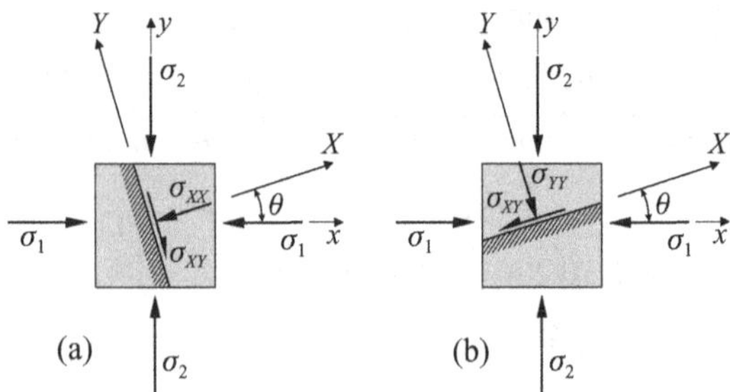

Figure B.7. Transformation of stresses for the principal plane stress state; $\sigma_1 \geq \sigma_2$.

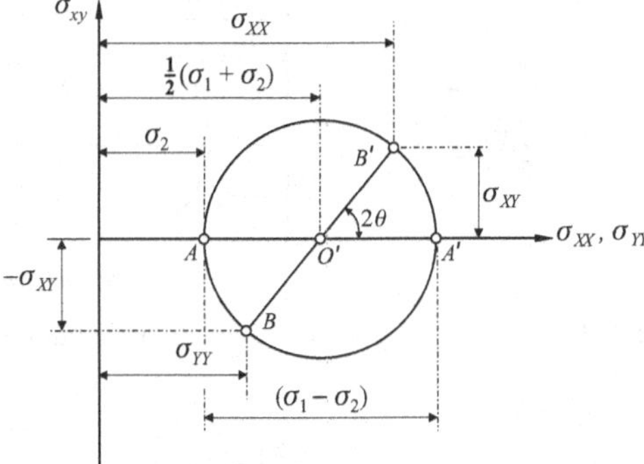

Figure B.8. The Mohr circle.

context of our user-friendly sign convention, σ_{XY} is positive when it acts in the negative Y-direction on the surface whose outward normal points in the positive X-direction, as shown in Figure B.6. The exact same stress acts on the surface with normal pointing in the Y-direction as shown in the same figure. Nevertheless, when we look at the Mohr circle in Figure B.8 it appears that the two shear stresses have opposite signs. This seems to pose a serious contradiction. In fact, it is not serious; it is simply the result of the fact that σ_{XY} enters (B.29) as a squared quantity. The equation cannot distinguish between positive and negative shear stress and the resulting Mohr diagram cannot either.

It will still be useful, however, to be able to *interpret* the sign of the shear stress *strictly within the context* of Mohr circles. We will see where the utility arises in a moment, but we must first return to our original sign convention. *Only* for the purpose of Mohr circles, we will interpret the sign of the shear stress as follows. If the shear stress induces a counterclockwise moment about the point P in Figure B.1(a), then we will plot the stress on the Mohr diagram as positive. If a clockwise moment is indicated then we plot the stress as negative. Looking at Figure B.7(a) we see that σ_{XY} there would be plotted as positive on a Mohr diagram. In Figure B.7(b), σ_{XY} would be plotted as negative. This convention gives exactly the result shown in Figure B.8. Note that this applies *only* for interpretation of the Mohr diagram.

In one sense this is all a storm in a teacup. The sign of the shear stress does not have a physical significance similar to compressive and tensile normal stress. If a material fails due to excessive shear stress it makes little difference whether that stress was positive of negative; failure is just as inconvenient in either case. But there is one more feature we can associate with the Mohr diagram that makes it useful for us to introduce our special sign convention. That is the existence of a unique point on the circumference of any Mohr circle called the *pole*. The concept of the pole, sometimes called the origin of planes, is especially useful in developing a graphical understanding of any stress state, and we must use a special sign convention to make it work.

To understand the concept of the pole we begin by noting from (B.26) and (B.28) that the stress point $(\sigma_{XX}, \sigma_{XY})$ makes a central angle of 2θ with the horizontal axis in Figure B.8. If we alter θ the stress point moves around the circumference of the circle to some new point. Let $(\sigma_{XX}, \sigma_{XY})$ and $(\bar{\sigma}_{XX}, \bar{\sigma}_{XY})$ be two stress states acting on two surfaces. If the orientations of the surfaces differ by an angle α, then the *central* angle on the Mohr circle between two stress points will be 2α as shown in Figure B.9. This doubling of angle applies for any surfaces we wish to consider. Next, suppose we draw on the Mohr diagram two lines: one line through the point $(\sigma_{XX}, \sigma_{XY})$ oriented parallel to the surface

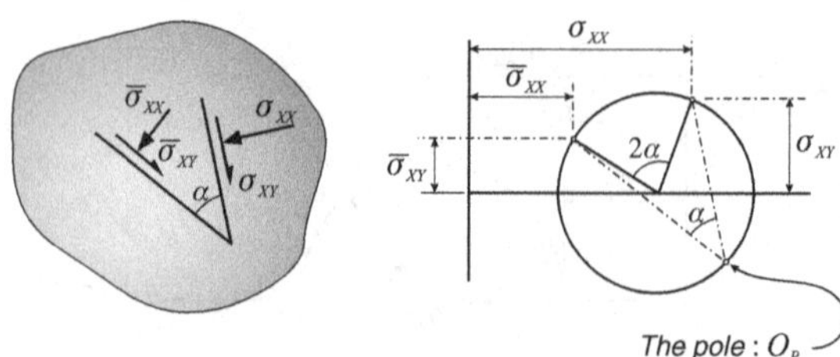

Figure B.9. Location of the pole on the Mohr circle circumference.

on which those stresses act and the other through $(\bar{\sigma}_{XX}, \bar{\sigma}_{XY})$ oriented parallel to the surface on which the second stress state acts. It is a fact that the two lines will always intersect at a point on the circumference of the circle. This follows from a theorem of geometry stating that the *central* angle between any two points on a circle will always be twice the corresponding *inscribed* angle. Note that the angle between our two lines must be α since they were drawn parallel to the two surfaces. Then on the Mohr diagram the two lines must intersect at some point on the circumference of the circle. We call this point the pole, denoted by O_P. These ideas are all summarised in Figure B.9.

Note that the pole is a unique point. The two surfaces used in the discussion above were totally arbitrary and therefore every line drawn through the stress point corresponding to any surface will intersect the circle at the pole. Conversely, any line drawn through the pole must intersect the circle at the stress state which acts on the surface parallel to that line. This is an extremely powerful tool for visualisation of the stresses associated with any surface. Once the location of the pole is determined, the stress on any surface is found by simply drawing a line parallel to that surface. We can find the pole so long as we know the stresses acting on a surface of known orientation. In all of this the special sign convention concerning shear stress applies. Stress points on the upper half of any Mohr circle imply a positive shear stress producing a counterclockwise moment about the point nearby the surface.

Mohr circles in three dimensions

We can extend Mohr circle construction to three dimensions. As with the two-dimensional case the procedure is most conveniently demonstrated using the principal stress state where σ_1 is the maximum principal stress, σ_2 is the intermediate principal stress and σ_3 is the minimum principal stress with the result, $\sigma_1 > \sigma_2 > \sigma_3$. The result (B.14) concerning the normal stress acting on an oblique plane and referred to a generalized state of stress is equally valid for the principal stress state. If we consider the principal stress state shown in Figure B.10 and consider the obliquely oriented plane with a unit normal having components

$$\boldsymbol{n} = n_1 \boldsymbol{i}_1 + n_2 \boldsymbol{i}_2 + n_3 \boldsymbol{i}_3 \tag{B.30}$$

where the unit base vectors in the principal directions are implied, the normal stress acting on the oblique plane is given by

$$\sigma_{nn} = n_1^2 \sigma_1 + n_2^2 \sigma_2 + n_3^2 \sigma_3 \tag{B.31}$$

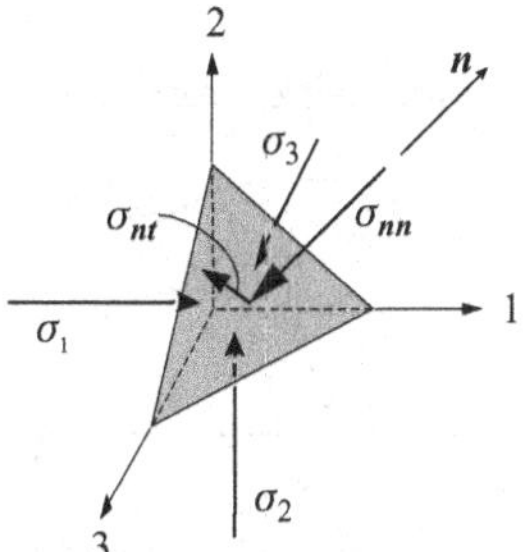

Figure B.10. Traction vectors on an oblique plane referred to the principal stress space.

We can also use the reduced version of the result (B.16) to define the shear stress that acts on the oblique plane. This gives

$$\sigma_{nt}^2 = (n_1\sigma_1)^2 + (n_2\sigma_2)^2 + (n_3\sigma_3)^2 - \left(n_1^2\sigma_1 + n_2^2\sigma_2 + n_3^2\sigma_{31}\right)^2 \tag{B.32}$$

Using both (B.31) and (B.32) we can obtain two equations

$$\sigma_{nn}^2 + \sigma_{nt}^2 = (n_1\sigma_1)^2 + (n_2\sigma_2)^2 + (n_3\sigma_3)^2 \tag{B.33}$$

$$(\sigma_{nn})^2 = \left(n_1^2\sigma_1 + n_2^2\sigma_2 + n_3^2\sigma_3\right)^2 \tag{B.34}$$

which can be combined with the consistency condition for the direction cosines,

$$n_1^2 + n_2^2 + n_3^2 = 1 \tag{B.35}$$

to give a set of equations for the squares of the three direction cosines. The set of equations (B.33)–(B.35) has a non-trivial solution, which can be evaluated using symbolic mathematical manipulation programs such as MATHEMATICA®, MACSYMA® or MAPLE®. Imposing the constraints

$$n_1^2 \geq 0; \quad n_2^2 \geq 0; \quad n_3^2 \geq 0 \tag{B.36}$$

The relevant solutions are

$$n_1^2 = \frac{\sigma_{nt}^2 + (\sigma_{nn} - \sigma_2)(\sigma_{nn} - \sigma_3)}{(\sigma_1 - \sigma_2)(\sigma_1 - \sigma_3)} \geq 0 \tag{B.37}$$

$$n_2^2 = \frac{\sigma_{nt}^2 + (\sigma_{nn} - \sigma_1)(\sigma_{nn} - \sigma_3)}{(\sigma_2 - \sigma_3)(\sigma_2 - \sigma_1)} \geq 0 \tag{B.38}$$

$$n_3^2 = \frac{\sigma_{nt}^2 + (\sigma_{nn} - \sigma_1)(\sigma_{nn} - \sigma_2)}{(\sigma_3 - \sigma_1)(\sigma_3 - \sigma_2)} \geq 0 \tag{B.39}$$

Since the principal stresses are in a ranked order, the equations (B.37)–(B.39) are equivalent to

$$\sigma_{nt}^2 + (\sigma_{nn} - \sigma_2)(\sigma_{nn} - \sigma_3) \geq 0 \tag{B.40}$$

$$\sigma_{nt}^2 + (\sigma_{nn} - \sigma_1)(\sigma_{nn} - \sigma_3) \leq 0 \tag{B.41}$$

$$\sigma_{nt}^2 + (\sigma_{nn} - \sigma_1)(\sigma_{nn} - \sigma_2) \geq 0 \tag{B.42}$$

and the different directions of the inequalities should be noted. We can rewrite

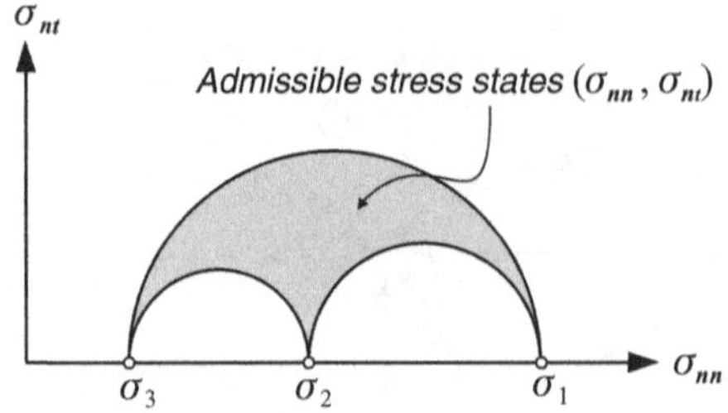

Figure B.11. Mohr circle in three dimensions.

(B.40)–(B.42) in the following forms:

$$\sigma_{nt}^2 + \left[\sigma_{nn} - \frac{(\sigma_2 + \sigma_3)}{2}\right]^2 \geq \left[\frac{1}{2}(\sigma_2 - \sigma_3)\right]^2 \tag{B.43}$$

$$\sigma_{nt}^2 + \left[\sigma_{nn} - \frac{(\sigma_1 + \sigma_3)}{2}\right]^2 \leq \left[\frac{1}{2}(\sigma_3 - \sigma_1)\right]^2 \tag{B.44}$$

$$\sigma_{nt}^2 + \left[\sigma_{nn} - \frac{(\sigma_1 + \sigma_2)}{2}\right]^2 \geq \left[\frac{1}{2}(\sigma_1 - \sigma_2)\right]^2 \tag{B.45}$$

The similarity between equation (B.29) and equations (B.43)–(B.45) is abundantly evident, the only difference being the inequalities that appear in the latter equations. How can we interpret the result? We can do so graphically and assign σ_{nn} as the axis corresponding to σ_{XX} and σ_{nt} as the axis corresponding to σ_{XY} used in the Mohr circle plot shown in Figure B.8. For the sake of convenience let us restrict our attention only to the region where both σ_{nn} and σ_{nt} are considered to be positive. We can plot the circular boundaries defined by the three Mohr circles (B.43)–(B.45) and identify the region to which the inequalities apply. This is shown in Figure B.11. Every combination of σ_{nn} and σ_{nt} shown in the shaded area is an admissible state of stress in a generalized sense. This is a real bonus when it comes to identifying states of stress that are responsible for failure of materials. For example, if failure of the material is governed by the maximum shear stress, the three-dimensional stress state can be converted to its principal components and the maximum shear stress is given by

$$\tau_{max} = \frac{1}{2}(\sigma_1 - \sigma_3) \tag{B.46}$$

The relationship between the principal stresses can also be expressed in terms of Lode's parameter

$$\mu = 2\frac{(\sigma_2 - \sigma_3)}{(\sigma_1 - \sigma_3)} - 1 \tag{B.47}$$

which determines the position of the intermediate principal stress σ_2 in relation to the other principal stresses as μ varies from –1 to +1,

for pure compression: $\sigma_1 > 0;\ \sigma_2 = \sigma_3 = 0$ and $\mu = -1$

for pure tension: $\sigma_1 = \sigma_2 = 0;\ \sigma_3 < 0$ and $\mu = +1$

for pure shear: $\sigma_1 > 0; \sigma_2 = 0;\ \sigma_3 = -\sigma_1$ and $\mu = 0$

The graphical illustration shown in Figure B.11 assumes that all the principal stresses are

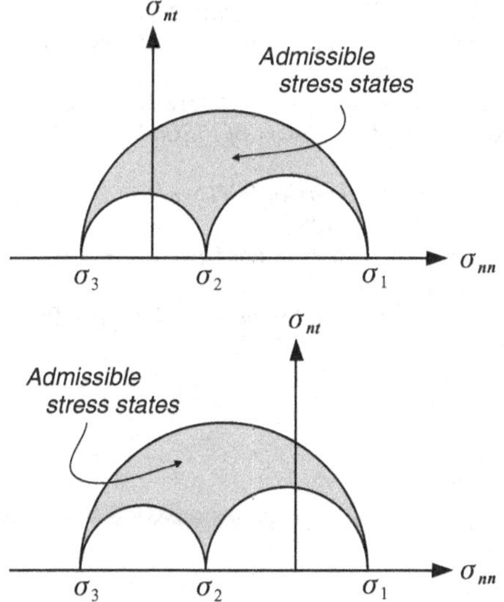

Figure B.12. Alternative admissible stress states.

compressive; this is not a requirement for the identification of the domain of admissible stress states. Figure B.12 shows alternative representations.

This Appendix summarizes some basic attributes of the Mohr circle. Other features associated with its development are covered within the context of the chapters in this volume, which utilise such features extensively. Further valuable discussions can also be found in the suggested reading.

Further reading

A.P. Boresi and P.P. Lynn, *Elasticity in Engineering Mechanics*, Prentice-Hall, Englewood Cliffs, NJ, 1974.

K. Culmann, *Die Graphische Statik*, Meyer & Zeller, Zurich, 1875.

R.O. Davis and A.P.S. Selvadurai, *Elasticity and Geomechanics*, Cambridge University Press, Cambridge, 1996.

D.C. Drucker, *An Introduction to the Mechanics of Deformable Solids*, McGraw-Hill, New York, 1966.

M.E. Harr, *Foundations of Theoretical Soil Mechanics*, McGraw-Hill, New York, 1966.

O. Mahrenholtz and L. Gaul, Mechanik und technik , *Zeit. TU Hannover.*, H.2, 8–34 (1980).

L.E. Malvern, *Introduction to the Mechanics of a Continuous Medium*, Prentice-Hall, Englewood Cliffs, NJ, 1969.

G.E. Mase and G.T. Mase, *Continuum Mechanics for Engineers*, CRC Press, Boca Raton, FL, 1991.

O. Mohr, Uber die Darstellung der Spannungzustandes und der Deformations-Zustandes eines Koerperelements, *Zivilingenieur*, **28**, 113–156 (1882).

O. Mohr *Abhandlungen aus dem Gebiete der Technischen Mechanik*, Wilhelm Ernst & Sohn, Berlin, 1914.

A. Nadai, *Theory of Flow and Fracture of Solids*, McGraw-Hill, New York, 1963.
R.H.G. Parry, *Mohr Circles, Stress Paths and Geotechnics*, E. & F.N. Spon, London, 1995.
S.P. Timoshenko, *History of the Strength of Materials*, McGraw-Hill, New York, 1953.
S.P. Timoshenko and J.N. Goodier, *Theory of Elasticity*, McGraw-Hill, New York, 1970.
I. Todhunter and K. Pearson, *A History of the Theory of Elasticity,* Vol. 1, Cambridge University Press, Cambridge, 1886.
I. Todhunter and K. Pearson *A History of the Theory of Elasticity,* Vol. 2, Cambridge University Press, Cambridge, 1893.
E. Volterra and J.H. Gaines *Advanced Strength of Materials*, Prentice-Hall, Englewood Cliffs, NJ, 1971.

Appendix C
Principles of virtual work

The principles of virtual work, which bring together the concepts of equilibrium and compatibility, or kinematics, are an important development in the mechanics of solids and in applied mechanics in general. The fact that the principles do not rely on the constitutive behaviour that pertains to the material is a major advantage in their applicability to elastic as well as inelastic materials and to problems that deal with dynamic and stability effects. There are, of course, various versions of the principle of virtual work, the forms of which will depend on the manner in which mechanical and kinematic variables are defined and presented. In the following we shall present a general statement of the principle of virtual work, which is of particular relevance to applications to elastic as well as plastic continua.

Let us consider a continuum region of finite extent V, which is bounded by the surface S. The region is restrained against rigid motions by suitable boundary constraints (Figure C.1). We now apply prescribed values of tractions and displacements T_i^* and u_i^* respectively, which act over separate regions of the boundary S. The displacement and traction boundary conditions applicable to the boundary value problem can be written as

$$u_i = u_i^*; \quad x_i \in S_u \tag{C.1}$$

$$T_i = \sigma_{ij} n_j = T_i^*; \quad x_i \in S_T \tag{C.2}$$

where n_i are the components of the outward unit normal to S. In general we assume that $S = S_T \cup S_u$. It must be noted that this presupposes that there are no regions where both traction and displacement boundary conditions are prescribed simultaneously. There are situations where this is possible, an example being that of the contact between a rigid footing with a smooth frictionless base and the surface of a continuum region such as a halfspace. Here, the component of the normal displacement and the tangential component of the tractions at the contact region are specified. In this case the regions S_u and S_T will have to overlap to account for the mixed nature of the boundary conditions applicable to the same region. For the purposes of the present discussion it is sufficient to assume that the boundary conditions correspond to (C.1) and (C.2), with the understanding that the discussion that follows can equally well be extended to cover this class of mixed–mixed boundary conditions. In addition to these prescribed displacements and tractions, we assume that the region V is also subjected to a body force field defined by the vector b_i and ignore the effects of dynamics, thereby reducing the problem to a boundary value problem rather than an initial boundary value problem. Let us assume that the applications of these boundary displacements, boundary tractions and body forces gives rise to a *kinematically*

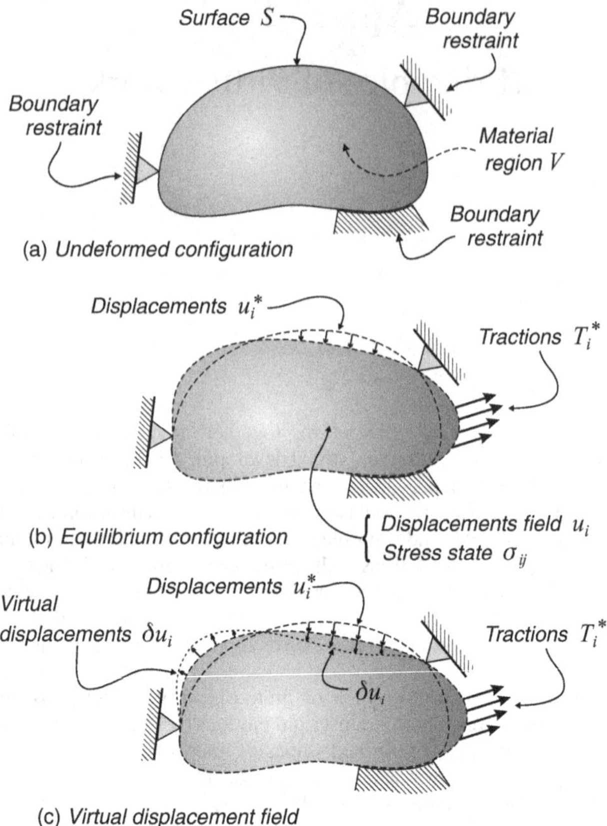

Figure C.1. Undeformed and equilibrium configurations and the virtual displacement field.

admissible set of displacements u_i and stresses σ_{ij} that satisfy boundary conditions (C.1) and (C.2) and the equations of equilibrium

$$\sigma_{ij,j} - b_i = 0 \tag{C.3}$$

where the subscript comma implies partial differentiation with respect to the appropriate spatial variable. By definition, a *kinematically admissible* displacement field is one that satisfies any external constraints, as defined by (C.1), any internal constraints (such as either material incompressibility or inextensibility), and is continuous and piecewise continuously differentiable in the region V, which includes S. Also, we define a stress field σ_{ij} that satisfies the traction boundary conditions (C.2) and the equations of equilibrium (C.3) as being *statically admissible*.

Let us now consider a *virtual displacement field* as defined by the difference between neighbouring kinematically admissible displacement fields. The term 'neighbouring' immediately introduces the notion of the infinitesimal into the definition. If we consider u_i to be one displacement state and $u_i + \delta u_i$ as the neighbouring displacement field then, by definition, δu_i is also a kinematically admissible field with small deformation gradients (i.e. $|\partial(\delta u_i)/\partial x_j| \ll 1$), and the qualifier δ is intended to signify an incremental difference. We can also define the virtual strain field $\delta\varepsilon_{ij}$ associated with the virtual displacements δu_i as

$$\delta\varepsilon_{ij} = \frac{1}{2}(\delta u_{i,j} + \delta u_{j,i}) \tag{C.4}$$

If we consider the variation of δu_i on S then by virtue of (C.1),

$$\delta u_i = 0; \quad x_i \in S_u \tag{C.5}$$

which satisfies the requirement of kinematic admissibility for the second displacement state. We can now compute the work done by this virtual displacement field by considering the body forces and the tractions T_i^*. We can define this 'external work', ΔW_{ext} as[†]

$$\Delta W_{ext} = -\iiint_V b_i \delta u_i \, dV + \iint_{S_T} T_i^* \delta u_i \, dS \tag{C.6}$$

We can also consider the internal work ΔW_{int}, associated with the virtual strains, resulting from the virtual displacements, operating on the stresses σ_{ij}. We have

$$\Delta W_{int} = \iiint_V \sigma_{ij} \delta\varepsilon_{ij} \, dV \tag{C.7}$$

We note that, in the absence of body couples, the stress tensor is symmetric, i.e. $\sigma_{ij} = \sigma_{ji}$. Using this fact, we can write

$$\sigma_{ij}\delta\varepsilon_{ij} = \sigma_{ij}(\delta u_i)_{,j} = [\sigma_{ij}\delta u_i]_{,j} - \sigma_{ij,j}\delta u_i \tag{C.8}$$

Using (C.8) in (C.7) and making use of Green's theorem we have

$$\Delta W_{int} = \iint_S \sigma_{ij} n_j \delta u_i \, dS - \iiint_V \sigma_{ij,j} \delta u_i \, dV \tag{C.9}$$

In view of (C.5) the surface integral in (C.9) can be restricted to S_T rather than S. Combining (C.6) and (C.9) we have

$$\Delta W_{int} - \Delta W_{ext} = -\iiint_V [\sigma_{ij,j} - b_i]\delta u_i \, dV + \iint_{S_T} [\sigma_{ij} n_j - T_i^*]\delta u_i \, dS \tag{C.10}$$

The integrals in (C.10) will vanish for every choice of the set of virtual displacements δu_i, if and only if the terms in the bracketed quantities reduce exactly to zero. This is ensured by the equations of equilibrium (C.3) and the traction boundary conditions (C.2). Hence, a body is in equilibrium under the application of a system of applied forces if and only if the '*principle of virtual work*', defined by

$$\Delta W_{ext} = \Delta W_{int} \tag{C.11}$$

is satisfied identically.

At this point, a comment regarding the expression for the internal work ΔW_{int}, as defined by (C.7), is in order. In the expression for ΔW_{int}, we have not specified the nature of the internal work nor the agencies that would be responsible for generating this internal work. It is only sufficient that such a measure exists. The only apparent requirement is that the body should experience 'virtual straining', under the imposed virtual deformation. We have not even specified whether such internal work is

[†] Although it might appear that the negative sign preceding the first integral in (C.6) is in error, that is not the case. The sign results from our convention that positive displacements act in *negative* coordinate directions while positive body forces act in *positive* coordinate directions.

either conservative or dissipative. This leaves room for choosing the measure of internal work to conform to the dominant internal process associated with the generation of internal work. For example, with elastic materials, this internal work could be associated with the *elastic energy* that is stored in the material during the virtual deformation, with the assumption that for an elastic material the stored energy is indeed fully recoverable. Alternatively, we can assume that the medium under consideration is an ideally plastic solid, in which case the internal work ΔW_{int}, as defined by (C.7) will now correspond to the plastic energy dissipation resulting from the virtual plastic straining, resulting from the application of the virtual displacements δu_i. This ability to interpret the internal work in a manner appropriate to the continuum under consideration makes virtual work principles a powerful tool in mechanics.

Now we proceed to define the second principle of virtual work. The *principle of complementary virtual work* is based on the concept of a *virtual stress field* as opposed to a virtual displacement field. Again we assume two statically admissible stress fields, such that their difference gives the symmetric virtual stress field $\delta\sigma_{ij}$. From (C.2) and (C.3) we note that this virtual stress field should satisfy, respectively,

$$\delta\sigma_{ij}\, n_j = 0; \quad x_i \in S_T \tag{C.12}$$

$$\delta\sigma_{ij,j} = 0; \quad x_i \in V \tag{C.13}$$

In connection with the derivation of the above equations, let us note that the body force field b_i and the applied external tractions T_i^* are exactly the same for two neighbouring states. When we interpret a virtual stress field as the difference between the two neighbouring states, these terms will naturally disappear from the equations governing internal equilibrium in the region V and on the boundary S. We can now define the external complementary virtual work $\overline{\Delta W}^{C}_{ext}$ as

$$\overline{\Delta W}^{C}_{ext} = \iint_{S_u} u_i^* \delta\sigma_{ij} n_j \, dS \tag{C.14}$$

Similarly, we can define an internal complementary virtual work $\overline{\Delta W}^{C}_{int}$ as

$$\overline{\Delta W}^{C}_{int} = \iiint_{V} \varepsilon_{ij} \delta\sigma_{ij} \, dV \tag{C.15}$$

Again we can take the difference between these two measures and apply Green's theorem as well as (C.12) and (C.13) to arrive at the following the result:

$$\overline{\Delta W}^{C}_{int} - \overline{\Delta W}^{C}_{ext} = \iint_{S_u} (u_i - u_i^*)\delta\sigma_{ij}\, n_j \, dS + \iiint_{V} \left[\varepsilon_{ij} - \frac{1}{2}(u_{i,j} + u_{j,i}) \right] \delta\sigma_{ij}\, dV \tag{C.16}$$

Again, the integrals occurring in (C.16) will vanish identically for every choice of the virtual stress field $\delta\sigma_{ij}$ provided the terms within the brackets vanish identically. For this to be satisfied, the strain field ε_{ij} should be compatible with the kinematically admissible displacement field u_i. The principle of complementary virtual work thus gives

$$\overline{\Delta W}^{C}_{int} = \overline{\Delta W}^{C}_{ext} \tag{C.17}$$

Here again, the comments made earlier in relation to the definition of the *internal virtual work* ΔW_{int}, are also applicable to the definition of the *complementary internal virtual work* $\overline{\Delta W}^{C}_{int}$.

When dealing with the application of the principles of virtual work to problems arising from the theory of plasticity, we are dealing with quantities such as *virtual velocities* δv_i and virtual stress rates $\delta\dot{\sigma}_{ij}$, as opposed to virtual displacements and virtual stresses, to indicate primarily the incremental nature of the problem formulation. In this case, the definitions of the internal and external virtual work and their complementary counterparts have to be identified as rates of virtual work. The result (C.11) for the *principle of virtual work* can be restated as a *principle of the rate of virtual work* in the form

$$\Delta \dot{W}_{ext} = \Delta \dot{W}_{int} \tag{C.18}$$

and the result (C.17) for the *principle of complementary virtual work* can be restated as a *principle of the rate of complementary virtual work* in the form

$$\overline{\Delta \dot{W}}^{C}_{int} = \overline{\Delta \dot{W}}^{C}_{ext} \tag{C.19}$$

Both principles are used quite extensively in the development of governing equations, specific solutions and the development of computational schemes for the numerical solution of problems in mechanics. In the context of the theory of plasticity of geomaterials, the principle of the rate of virtual work is used quite extensively for the development of proofs of the upper and lower bound theorems in plasticity and in the development of associated solutions.

Further reading

A.P. Boresi and P.P. Lynn, *Elasticity in Engineering Mechanics*, Prentice-Hall, Englewood Cliffs, NJ, 1974.

R.O. Davis and A.P.S. Selvadurai, *Elasticity and Geomechanics*, Cambridge University Press, Cambridge, 1996.

D.C. Drucker, *An Introduction to the Mechanics of Deformable Solids*, McGraw-Hill, New York, 1966.

C.L. Dym and I.H. Shames, *Solid Mechanics: a Variational Approach*, McGraw-Hill, New York, 1973.

H.L. Langharr, *Energy Methods in Applied Mechanics*, John Wiley, New York, 1962.

L.E. Malvern, *Introduction to the Mechanics of a Continuous Medium*, Prentice-Hall, Englewood Cliffs, NJ, 1969.

J.B. Martin, *Plasticity: Fundamentals and General Results*, MIT Press, Cambridge, MA, 1975.

G.E. Mase and G.T. Mase, *Continuum Mechanics for Engineers*, CRC Press, Boca Raton, FL, 1991.

J.T. Oden, *Mechanics of Elastic Structures*, McGraw-Hill, New York, 1968.

T.J. Tauchert, *Energy Methods in Structural Mechanics*, Addison Wesley, New York, 1974.

K. Washizu, *Variational Methods in Elasticity and Plasticity*, Pergamon Press, Oxford, 1968.

Appendix D
Extremum principles

An extremum principle is basically a mathematical concept that relies on some physical law. In mechanics, extremum principles such as the principle of minimum total potential energy and minimum total complementary energy form an important base of knowledge that has provided the means for obtaining approximate solutions to a variety of problems in engineering. This is particularly the case with the theory of elasticity. The celebrated *principles of least work* attributed to Alberto Castigliano, are also in the realm of extremum principles that have been used extensively in the solution of problems in classical structural mechanics dealing with elastic materials. In general, extremum principles and for that matter variational principles start with the basic premise that the solution to a problem can be represented as a class of functions that would satisfy some but not all of the equations governing the exact solution. It is then shown that a certain functional expression, usually composed of scalar quantities such as the total potential energy, strain energy, energy dissipation rate, etc., that have physical interpretations associated with them and are defined through the use of this class of functions, will yield an extremum (i.e. either a maximum or a minimum) for that function. Moreover, the extremum will satisfy the remaining equations required for the complete solution. For example, the principle of minimum total potential energy states that of all the kinematically admissible displacement fields in an elastic body, which also satisfy the governing constitutive equations, only those that satisfy the equations of equilibrium will give rise to a total potential energy that has a stationary value or an extremum. Furthermore, this stationary value will be a minimum for systems that are in stable equilibrium. The underlying power of extremum principles in elasticity is clearly indicated in their earlier applications to structural mechanics and the recent developments associated with numerical methods such as the Rayleigh–Ritz method, the precursor to and the mathematical basis of the finite-element method. An extremum principle is, however, a stronger principle than a variational one since it establishes the existence of an extremum by considering all admissible functions of a certain class and not restricting it to those that are infinitesimal in the proximity of the extremum. Also, in general, for a variational principle, the existence of even a local extremum is not a requirement; it is only sufficient that the functional satisfying the variational principle has a stationary value. Considering the success these principles have enjoyed in their applications to a wider class of problems in mechanics, it is therefore entirely natural to enquire whether extremum principles can indeed be developed to facilitate the development of solutions for materials that exhibit plasticity effects.

The study of extremum principles and indeed the general area of variational methods is quite a mathematically demanding subject. The purpose of this presentation is not to indulge in rigorous mathematical proofs applicable to all types of elasto-plastic materials, but to give a brief exposé of the basic facets of extremum principles since they constitute the basic foundation upon which the theorems of limit analysis have been developed. We can appreciate the power of the upper and lower bound solutions when we begin to realize that the solution to a plasticity problem is provided with a set of 'bounds' without ever solving the complete set of partial differential equations governing the problem. This is a distinct advantage since these equations are generally non-linear partial differential equations. Excellent accounts of the developments concerning extremum principles are given in the original articles by pioneers of this area of research, notably G. Colonetti, L.M. Kachanov, M.A. Sadowsky, G.H. Handelman, A.A. Markov, H.J. Greenberg, A. Nadai, R. Hill, W. Prager, D.C. Drucker and P.G. Hodge. The references to the articles by these researchers and more complete accounts of developments of extremum principles applicable to elastic–plastic media and those materials experiencing large-strain phenomena can be found in the bibliography cited at the end of this Appendix.

As a prelude to the discussion of extremum principles for elastic–plastic solids it is instructive to illustrate, as an example, the proof of the *principle of minimum potential energy*, bearing in mind that the principle is applicable *only* to elastic solids. In a typical boundary value problem in elasticity, displacements are usually prescribed on a part of the boundary and tractions are prescribed on the remainder. It is also possible to generalise this by considering a part of the boundary where in each of the three independent directions we specify either a component of the displacement or a component of traction. These are the so-called mixed–mixed boundary conditions. An example would be a body in smooth contact with a rigid plane where a single displacement is prescribed and two components of the traction are specified as zero. For the present purposes let us restrict our attention to the specification of the conventional displacement boundary conditions on S_u in the form

$$u_i = \hat{u}_i \quad \text{on} \quad x_i \in S_u \tag{D.1}$$

and traction boundary conditions on the remainder of the boundary such that

$$\sigma_{ij} n_j = T_i = \hat{T}_i \quad \text{on} \quad x_i \in S_T \tag{D.2}$$

where $\hat{u}_i$ and $\hat{T}_i$ are specified functions and n_i are the direction cosines of the outward unit normal to S_T. For the purposes of the discussions that follow, it is sufficient to assume that the region $S = S_u \cup S_T$, and during any deformation $S_u \cap S_T = 0$. Considering the elasticity problem, we assume that the solution to any well-posed boundary value problem can be expressed in terms of the stresses σ_{ij} and strains ε_{ij}, that are required to satisfy certain conditions. For example, any stress state σ_{ij}^0 that satisfies both the equations of internal equilibrium, which in the absence of body forces and dynamic effects reduce to

$$\sigma_{ij,j}^0 = 0 \quad \text{on} \quad x_i \in V \tag{D.3}$$

and the traction boundary conditions

$$\sigma_{ij}^0 n_j = T_i^0 \quad \text{on} \quad x_i \in S \tag{D.4}$$

and where n_i are the components of the outward unit normal to S, is considered to be a *statically admissible stress state*. Also Cauchy's condition (D.4) ensures that at all boundary points where a vector T_i^0 is specified, the internal stress field σ_{ij}^0 satisfies equilibrium between the applied tractions and the internal stresses.

The strain field ε_{ij}, on the other hand, must be determined from a displacement vector u_i, such that given ε_{ij}, we should be able to determine u_i, at least to within a set of rigid body displacements. If we now consider a displacement field u_i^*, which satisfies all the boundary conditions applicable to the displacements (i.e. of the type (D.1)) and ε_{ij}^* the corresponding strains, then these strains are considered to be *kinematically admissible*.

In elasticity, the *statically admissible stresses* σ_{ij}^0 and the corresponding strains ε_{ij}^0 are related through Hooke's law, as follows:

$$\varepsilon_{ij}^0 = C_{ijkl}\sigma_{kl}^0 \tag{D.5}$$

where C_{ijkl} is the generalised elasticity matrix. The inversion of (D.5) is assured by the positive definiteness of the generalized elasticity matrix. Similarly, the *kinematically admissible strains* ε_{ij}^* and the stresses σ_{ij}^* derived from these strains are also related to each other through Hooke's law as follows:

$$\varepsilon_{ij}^* = C_{ijkl}\sigma_{kl}^* \tag{D.6}$$

In general, the strains ε_{ij}^0 cannot be integrated to obtain the displacements u_i^0 and the stresses σ_{ij}^* generally do not satisfy equilibrium.

Since we have a kinematically admissible set of displacements u_i^* and a statically admissible set of stresses σ_{ij}^0 applicable to the same region V with boundary S, we can apply the principle of virtual work to the region; combining (C.6) and (C.7) and setting $b_i = 0$, we have

$$\iiint_V \sigma_{ij}^0\varepsilon_{ij}^*\, dV = \iint_S T_i^0 u_i^*\, dS \tag{D.7}$$

The internal energy per unit volume associated with any kinematically admissible state is

$$U^* = \int \sigma_{ij}^*\, d\varepsilon_{ij}^* \tag{D.8}$$

Since we are considering linear elastic behaviour (and isothermal or adiabatic deformations) we have from (D.6) and (D.8)

$$U^* = \frac{1}{2}C_{ijkl}\sigma_{ik}^*\sigma_{jl}^* \tag{D.9}$$

Hence the total potential energy for the kinematically admissible state of deformation is

$$\Pi^* = \frac{1}{2}\iiint_V C_{ijkl}\sigma_{ik}^*\sigma_{jl}^*\, dV - \iint_{S_T} T_i u_i^*\, dS \tag{D.10}$$

The equivalent expression for the total potential energy associated with the exact solution takes the form

$$\Pi = \frac{1}{2}\iiint_V C_{ijkl}\sigma_{ik}\sigma_{jl}\, dV - \iint_{S_T} T_i u_i\, dS \tag{D.11}$$

• • •

Theorem D1. The theorem of minimum total potential energy states that, of all the kinematically admissible states of deformation in an elastic body, only the true one will minimise the total potential energy.

Proof. Considering (D.10) and (D.11), the theorem is equivalent to the statement

$$\Delta\Pi = \Pi^* - \Pi \geq 0 \tag{D.12}$$

with the equality sign applicable when $u_i^* \equiv u_i$. Using (D.10) and (D.11) we have

$$\Delta\Pi = \frac{1}{2}\iiint_V C_{ijkl}(\sigma_{ik}^*\sigma_{jl}^* - \sigma_{ik}\sigma_{jl})\,dV - \iint_{S_T} T_i(u_i^* - u_i)\,dS \tag{D.13}$$

Since u_i and u_i^* have to satisfy the same prescribed displacement boundary conditions on S_u of the type (D.1), we must have

$$\iint_{S_u} T_i(u_i^* - u_i)\,dS = 0 \tag{D.14}$$

Hence

$$\iint_{S_T} T_i(u_i^* - u_i)\,dS = \iint_S T_i(u_i^* - u_i)\,dS = \iint_S \sigma_{ij}n_j\,(u_i^* - u_i)\,dS \tag{D.15}$$

and applying Green's theorem to the above, we can show that since $\sigma_{ij} = \sigma_{ji}$

$$\iint_{S_T} T_i(u_i^* - u_i)\,dS = \iiint_V \sigma_{ij}(\varepsilon_{ij}^* - \varepsilon_{ij})\,dV = \iiint_V C_{ijkl}\sigma_{ik}(\sigma_{jl}^* - \sigma_{jl})\,dV \tag{D.16}$$

Combining (D.13) and (D.16) we have

$$\Delta\Pi = \frac{1}{2}\iiint_V C_{ijkl}(\sigma_{ik}^*\sigma_{jl}^* - 2\sigma_{ik}\sigma_{jl}^* + \sigma_{ik}\sigma_{jl})\,dV \tag{D.17}$$

Note that since C_{ijkl} is symmetric and, since the summation is carried out over the complete set of indices to provide a scalar result, we can interchange the suffixes without altering the final result. We can write (D.17) in the form

$$\Delta\Pi = \frac{1}{2}\iiint_V C_{ijkl}(\sigma_{ik}^* - \sigma_{ik})(\sigma_{jl}^* - \sigma_{jl})\,dV \tag{D.18}$$

Since C_{ijkl} is positive definite, the integrand of (D.17) is positive definite at each $x_i \in V$. Hence $\Delta\Pi \geq 0$ with the equality being applicable if and only if $\sigma_{ij}^* \equiv \sigma_{ij}$. This latter condition implies that $\varepsilon_{ij}^* \equiv \varepsilon_{ij}$ and $u_i^* = u_i$ to within a rigid body displacement. This proves the assertion that, of the kinematically admissible sets of displacement fields, the exact one, which also satisfies the equations of equilibrium, renders the total potential energy a minimum.

• • •

We can use the principle of minimum complementary energy to develop a similar proof for any *statically admissible stress field*; i.e. of all the statically admissible stress fields only the stress state that will also give compatible strain fields will render the complementary energy a minimum. Both of these extremum principles and their mixed versions feature prominently in aspects related to the development of procedures for obtaining approximate computational solutions to problems in elasticity. These aspects are discussed in detail in works cited in the bibliography at the end of this Appendix.

Let us now focus attention on the discussion of the extremum principles that are applicable to elastic–plastic materials. First, in keeping with the developments consistent with the theory of plasticity, we will consider velocities, strain rates and stress rates as opposed to displacements, strains and stresses, with the understanding that the specification of the rate is to account for the incremental nature of the developments. Analogous to (D.1), we can define a region S_v on which velocities are prescribed: i.e.

$$v_i = \hat{v}_i \quad \text{on} \quad x_i \in S_v \tag{D.19}$$

Similarly, for a surface on which the traction rate is defined we have

$$\dot{\sigma}_{ij} n_j = \dot{T}_i = \hat{\dot{T}}_i \quad \text{on} \quad x_i \in S_T \tag{D.20}$$

The class of boundary value problems to be solved usually assumes that at a certain time t, the displacements and stresses are known throughout V and the traction rates and velocities are prescribed on S in relation to (D.19) and/or (D.20). The objective here is to determine the stress rates and velocities within V. In keeping with the decomposition rule applicable to small-strain rates, we now assume that the total strain rate $\dot{\varepsilon}_{ij}$ consists of the summation of the elastic and plastic strain rates $\dot{\varepsilon}_{ij}^{(el)}$ and $\dot{\varepsilon}_{ij}^{(pl)}$, respectively. We further assume that the elastic strain rates are derived from Hooke's law and the plastic strain rates are obtained through the specification of a yield criterion and a flow rule. We shall restrict attention to only the class of materials that satisfy the *associated flow rule*. We also assume that, given a yield criterion, $\dot{\varepsilon}_{ij}^{(pl)}$ can be determined uniquely. This is, of course, not the case with yield functions with edge surfaces such as those encountered in the Tresca yield surface or even for that matter the vertex point in the Drucker–Prager conical yield surface. This restriction can be removed from the presentation that follows by adopting a discussion to include edges or points where, conventionally, the orientation and magnitude of the plastic strain rate is undetermined. These aspects can be further studied in references cited in the bibliography provided at the end of this Appendix.

We can write

$$\dot{\varepsilon}_{ij} = C_{ijkl}\dot{\sigma}_{kl} + \lambda \frac{\partial f}{\partial \sigma_{ij}} \tag{D.21}$$

where, at a plastic point

$$\lambda \geq 0 \quad \text{if} \quad f = k; \quad \dot{f} = 0 \tag{D.22}$$

and at an elastic point

$$\lambda = 0 \quad \text{if either} \quad f < k \quad or \text{ if} \quad f = k \quad \text{and} \quad \dot{f} < 0 \tag{D.23}$$

We now define a statically admissible field of stress rates $\dot{\sigma}_{ij}^0$ such that they satisfy the equilibrium equations in V and boundary traction rates $\hat{\dot{T}}_i$ on the surface S_T as defined through (D.20), and do not violate the plasticity conditions (D.22). The requirement concerning non-violation of the plasticity conditions (D.22) is automatically satisfied if $f < k$, but imposes the additional constraint that if

$$f = k \quad \text{then} \quad \dot{f}^0 \leq 0 \tag{D.24}$$

Here, the superscript 0 refers to the quantity evaluated at the stress state corresponding to the statically admissible state. The strain rates corresponding to (D.21) applicable to the value of the statically admissible stress state are now given by

$$\dot{\varepsilon}_{ij}^0 = C_{ijkl}\dot{\sigma}_{kl}^0 + \lambda^0 \frac{\partial f}{\partial \sigma_{ij}} \tag{D.25}$$

Considering (D.23) and (D.24) , the above expression is subject to the following constraints:

$$\lambda^0 \geq 0 \quad \text{if} \quad f = k \quad \text{and} \quad \dot{f}^0 = 0, \tag{D.26}$$

$$\lambda^0 = 0 \quad \text{if either} \quad f < k \quad \text{or if} \quad f = k \quad \dot{f}^0 < 0 \tag{D.27}$$

A point to note here is that we have chosen a statically admissible stress state that will specifically *exclude* yield in the material, which should be present if plastic strain rates are to manifest. At the outset it would appear that the third condition of (D.26) implies that there may be plastic energy dissipation. However the specification of the additional constraints (D.24) along with (D.26) and (D.27) safeguards the non-violation of the yield condition which is necessary for any stress state σ_{ij}^0 to be considered statically admissible (see e.g. Hodge (1958) and Koiter (1960)).

The analogous kinematically admissible velocity field v_i^* is one that satisfies the velocity boundary conditions of the type (D.19) on S_v. The strain rates $\dot{\varepsilon}_{ij}^*$ are derived directly from the velocity vector v_i^*. The related stress rates are any solution satisfying

$$\dot{\varepsilon}_{ij}^* = C_{ijkl}\dot{\sigma}_{kl}^* + \lambda^* \frac{\partial f}{\partial \sigma_{ij}} \tag{D.28}$$

with the constraints

$$\text{if} \quad f = k \quad \text{and} \quad \dot{f}^* = 0 \quad \text{then} \quad \lambda^* \geq 0 \tag{D.29}$$

$$\text{if} \quad f < k \quad \text{or} \quad \dot{f}^* < 0 \quad \text{then} \quad \lambda^* = 0 \tag{D.30}$$

In (D.24)–(D.30), it should be noted that quantities such as f and $\partial f/\partial\sigma_{ij}$ depend only on the stress rather than the stress rate and are evaluated for the actual given stress state.

Since $\dot{\varepsilon}_{ij}^*$ represents any kinematically admissible strain rate derived from a velocity field that satisfies the velocity boundary conditions, the corresponding analogy to the energy per unit volume of the material is the energy production rate per unit volume of the material, which is given by

$$\dot{W}^* = \int \dot{\sigma}_{ij}^*\, d\dot{\varepsilon}_{ij}^* = \int \left(C_{ijkl}\dot{\sigma}_{ik}^*\, d\dot{\sigma}_{jl}^* + \dot{\sigma}_{ij}^* \frac{\partial f}{\partial \sigma_{ij}}\, d\lambda^* \right) \tag{D.31}$$

With regard to the last term on the right-hand side of (D.31), the differential of $\partial f/\partial\sigma_{ij}$ depends solely on the stresses and not the stress rates. Also considering (D.28)–(D.30) it follows that since either $\dot{f}^*$ or λ^* is identically zero, we have

$$\left(\frac{\partial f}{\partial \sigma_{ij}} \dot{\sigma}_{ij}^* \right) d\lambda^* = \dot{f}^*\, d\lambda^* = 0 \tag{D.32}$$

Therefore

$$\dot{W}^* = \frac{1}{2} C_{ijkl}\dot{\sigma}_{ik}^*\, \dot{\sigma}_{jl}^* \tag{D.33}$$

and the total energy rate is given by

$$\dot{\Lambda}^* = \frac{1}{2} \iiint_V C_{ijkl}\dot{\sigma}_{ik}^*\dot{\sigma}_{jl}^*\, dV - \iint_{S_T} \dot{T}_i v_i^*\, dS \tag{D.34}$$

We can now use this functional to develop the first of two extremum principles applicable to elastic–plastic materials.

• • •

Theorem D2. The first minimum principle states that, of all the kinematically admissible velocity fields in an elastic plastic material, the true velocity field will minimise $\dot{\Lambda}^*$.

Proof. The procedure for developing the proof is similar to that outlined in connection with the principle of minimum total potential energy for an elastic material. We consider the total energy rate associated with the exact result, which is given by

$$\dot{\Lambda} = \frac{1}{2} \iiint_V C_{ijkl}\dot{\sigma}_{ik}\dot{\sigma}_{jl}\, dV - \iint_{S_T} \dot{T}_i v_i\, dS \tag{D.35}$$

and construct the difference between the total energy rate (D.34) associated with the assumed kinematically admissible velocity field v_i^* and the result (D.35). This gives (after converting the resulting surface integral in the expression to a volume integral)

$$\Delta\dot{\Lambda} = \dot{\Lambda}^* - \dot{\Lambda} = \frac{1}{2}\iiint_V C_{ijkl}(\dot{\sigma}_{ik}^* - \dot{\sigma}_{ik})(\dot{\sigma}_{jl}^* - \dot{\sigma}_{jl})\,dV + \iiint_V \frac{\partial f}{\sigma_{ij}}\dot{\sigma}_{ij}(\lambda - \lambda^*)\,dV \tag{D.36}$$

We need to prove that $\Delta\dot{\Lambda}$ is positive definite. The integrand of the first integral in (D.36) is always positive in view of the fact that C_{ijkl} is positive definite and the remaining term is in a quadratic form. Considering (D.34), the integrand in the second term can be written as $\dot{f}(\lambda - \lambda^*)$ and, in view of (D.22), this term will vanish at every plastic point. If, on the other hand, the material is elastic it follows from (D.23) that $\lambda = 0$ with the result that the integrand is equal to $-\dot{f}\,\lambda^*$. Now if $f < k$, then no finite stress rates can make the neighbourhood of a stress state immediately plastic, so that from (D.30) we have $\lambda^* = 0$; if, on the other hand, $f = k$, then we require from (D.23), $\dot{f} < 0$ and from (D.29) we have $\lambda^* \geq 0$. Hence $-\dot{f}\lambda^*$ is always positive and the integrand of the second integral is also positive. Consequently, $\Delta\dot{\Lambda}$ is positive definite. Implicit in this positive definiteness assumption is the requirement that the material is elastic–perfectly plastic and is *non-softening*, in order to satisfy the constraint.

• • •

The analysis can be extended to the consideration of the total complementary energy rate defined by

$$\dot{\Lambda}_c^0 = \frac{1}{2}\iiint_V C_{ijkl}\sigma_{ik}^0\sigma_{jl}^0\,dV - \iint_{S_v} \dot{T}_i^0 v_i\,dS \tag{D.37}$$

where the superscripts ^{0} are associated with the statically admissible stress states, to develop a second extremum principle.

• • •

Theorem D3. The second minimum principle states that among all the statically admissible states of stress rates, the true one will minimise $\dot{\Lambda}_c^0$.

Proof. Again by considering the difference between the integral expressions for the complementary energy rate applicable to a statically admissible state of stress rate and the complementary energy rate applicable to the exact solution we obtain

$$\begin{aligned}\Delta\dot{\Lambda}_c^0 = \dot{\Lambda}_c^0 - \dot{\Lambda}_c = \frac{1}{2}\iiint_V C_{ijkl}\left(\dot{\sigma}_{ik}^0 - \dot{\sigma}_{ik}\right)\left(\dot{\sigma}_{jl}^0 - \dot{\sigma}_{jl}\right)dV \\ + \iiint_V \lambda\frac{\partial f}{\sigma_{ij}}\left(\dot{\sigma}_{ij} - \dot{\sigma}_{ij}^0\right)dV\end{aligned} \tag{D.38}$$

and we need to prove that $\Delta\dot{\Lambda}_c^0$ is positive definite. Since the integrand of the first integral in (D.38) is positive definite, attention can be directed to proving that the integrand of the second integral is always positive definite. We can rewrite the second integrand as $\lambda(\dot{f} - \dot{f}^0)$. In the case of elastic behaviour, in view of (D.23), this quantity will be zero. For plastic behaviour, from (D.22), $\lambda \geq 0$ and $\dot{f} = 0$ and from (D.27) and (D.28),

$\dot{f}^0 \leq 0$. As a consequence $\lambda(\dot{f} - \dot{f}^0) \geq 0$, and the integrand is positive definite, which proves the second extremum principle.

• • •

The two theorems presented here can be combined to give upper and lower bounds on either $\dot{\Lambda}$ or $\dot{\Lambda}_c$ as follows:

$$-\dot{\Lambda}_c^0 \leq -\dot{\Lambda}_c = \dot{\Lambda} \leq \dot{\Lambda}^* \tag{D.39}$$

This represents the basis for the development of a number of important relationships associated with not only the upper and lower bound theorems but also to address the question of uniqueness of solution. Let us also not overlook the fact that the extremum principles for elastic–plastic materials, experiencing small strains presented here, have as their basis the requirement concerning the applicability of the *associated flow rule* for the determination of the plastic strain increments. This indirectly provides the proof for the necessity of the associated flow rule and the normality condition as minimum requirements for the valid application of limit analysis techniques in the development of approximate solutions to problems in soil plasticity.

Further reading

Certain original articles dealing with extremum principles are:

G. Colonetti, De l'equilibre des systems elastiques dans lesquels se produisant des deformations plastiques, *J. Math. Pures Appl.*, **17**, 233–255 (1938).

D.C. Drucker, Plasticity, in *Structural Mechanics*, (eds. J.N. Goodier and N.J. Hoff) Pergamon Press, Oxford, pp. 407–455, 1960.

H.J. Greenberg, Complementary minimum principles for an elastic–plastic stress, *Quart. Appl. Math.*, **7**, 85–95 (1949).

G.H. Handelman, A variational principle for a state of combined plastic stress, *Quart. Appl. Math.*, **1**, 351–353 (1944).

R. Hill, A variational principle of maximum plastic work in classical plasticity, *Quart. J. Mech. Appl. Math.*, **1**, 18–28 (1948).

R. Hill, A comparative study of some variational principles in the theory of plasticity, *J. Mech. Phys. Solids*, **5**, 66–74 (1950).

P.G. Hodge Jr., The mathematical theory of plasticity, in *Elasticity and Plasticity. Surveys in Applied Mathematics*, Vol. 1 (eds. J.N. Goodier and P.G. Hodge Jr.) John Wiley, New York, pp. 49–144, 1958.

P.G. Hodge Jr. and W. Prager, A variational principle for plastic materials with strain hardening, *J. Math. Phys.*, **27**, 1–10 (1948).

L.M. Kachanov, Variational principles for elastic plastic solids (in Russian), *Prikl. Math. Mech.*, **6**, 187–194 (1942).

W.T. Koiter, Stress–strain relations, uniqueness and variational theorems for elastic–plastic materials with a singular yield surface, *Quart. Appl. Math.*, **11**, 350–361 (1953).

W.T. Koiter, General theorems for elastic–plastic solids, In *Progress in Solid Mechanics* (eds. I.N.Sneddon and R.Hill) North-Holland, Amsterdam, pp. 165–221, 1960.

A.A. Markov, On variational principles in the theory of plasticity (in Russian), *Prikl. Math. Mech.*, **11**, 339–350 (1947).

A. Nadai, The principle of minimum work applied to states of finite, homogeneous, plane plastic strain, *Proc. 1st U.S. National Congress of Applied Mechanics*, pp. 479–485, 1952.

M.A. Sadowsky, A principle of maximum plastic resistance, *J. Appl. Mech.*, **10**, A65–A68 (1943).

Informative accounts of the general topic of extremum principles and variational methods as applied to problems in mechanics and more specifically to problems in the theory of elastic–plastic solids can be found in the following:

J.M. Ball, The calculus of variations and materials science, *Quart. Appl. Math.*, **56**, 719–740 (1998).

J. Chakrabarty, *Theory of Plasticity*, 2nd edition, McGraw-Hill, New York, 1998.

P.-C. Chou and N.J. Pagano, *Elasticity: Tensor, Dyadic and Engineering Approaches*, Dover Publications, New York, 1967.

R. Courant and D. Hilbert, *Methods of Theoretical Physics*, Vol. 1, Wiley Interscience, New York, 1953.

C.L. Dym and I.H. Shames, *Solid Mechanics. A Variational Approach*, McGraw-Hill, New York, 1973.

R. Hill, *The Mathematical Theory of Plasticity*, Oxford University Press, Oxford, 1950.

L.M. Kachanov, *Fundamentals of the Theory of Plasticity* (Translated from the Russian by M. Konyaeva) Mir Publishers, Moscow, 1974.

J. C. Lubliner, *Plasticity Theory*, Collier-Macmillan, New York, 1990.

J.B. Martin, *Plasticity: Fundamentals and General Results*, MIT Press, Cambridge, MA, 1975.

W. Olszak, Z. Mroz and P. Perzyna, *Recent Trends in the Development of the Theory of Plasticity*, Macmillan Co., New York, 1963.

W. Prager and P.G. Hodge Jr., *Theory of Perfectly Plastic Solids*, John Wiley and Sons, New York, 1951.

J. Salencon, *Applications of the Theory of Plasticity in Soil Mechanics*, John Wiley, New York, 1974.

I.S. Sokolnikoff, *Mathematical Theory of Elasticity*, 2nd edition, McGraw-Hill, New York, 1956.

R.A.C. Slater, *Engineering Plasticity*, John Wiley, New York, 1977.

R. Weinstock, *Calculus of Variations*, McGraw-Hill, New York, 1952.

G.A. Wempner, *Mechanics of Solids with Applications to Thin Bodies*, McGraw-Hill, New York, 1973.

Appendix E
Drucker's stability postulate

An approach to the development of the constitutive equations of plasticity involves the consideration of plastic energy dissipation in an irreversible process. This is somewhat analogous to the determination of the constitutive equations for an elastic material by considering the energy stored during deformation. The notion of material stability is an important aspect of the development of any self-consistent theory of plasticity, which not only includes the constitutive equations governing plastic behaviour but also appropriate uniqueness theorems and procedures for the solution of boundary value problems.

The concept of material stability implies the existence of a one-to-one correspondence in the constitutive equations in the range of small strains. The notion of material stability *in the small* can be illustrated by appeal to the behaviour of a material in uniaxial straining. Figure E.1 shows non-linear stress–strain behaviour where σ is the uniaxial Cauchy stress, ε is the corresponding small strain and $\dot{\varepsilon}$ is the strain rate. Let us consider the situation where the specimen is subjected to an arbitrary stress σ and $\Delta\sigma$ is the increment in stress, which produces a corresponding increment in strain $\Delta\varepsilon$. The material is said to *stable* if

$$\Delta\sigma\,\Delta\varepsilon > 0 \tag{E.1}$$

The result (E.1) implies that in a stable material, strain increments result in positive work from the stresses. We can generalize (E.1) to the following form involving all components of the strain tensor and the stress tensor to give the following requirement for a stable material:

$$\Delta\sigma_{ij}\,\Delta\varepsilon_{ij} > 0 \tag{E.2}$$

The notion of material stability and plastic energy dissipation during yielding is central to Drucker's stability postulate, which applies to geomaterials that exhibit strain hardening phenomena and as a special case can also apply to perfectly plastic materials.

The plastic energy dissipation during a closed cycle of loading can be demonstrated by appeal to the closed path shown in Figure E.2. If the loading that induces a strain increment $\Delta\varepsilon_{ij}$ commences from a reference stress state σ_{ij}^0, the inequality (E.2) can be written as

$$\left(\sigma_{ij} - \sigma_{ij}^0\right)\Delta\varepsilon_{ij}^p > 0 \tag{E.3}$$

and the incremental irreversible plastic strains $\Delta\varepsilon_{ij}^p$ are those that occur subsequent to the application of the reference stress σ_{ij}^0.

Drucker's postulate hinges on this concept of a stable material. Let us consider a stable state in a material, which can experience plastic energy dissipation. We subject the body

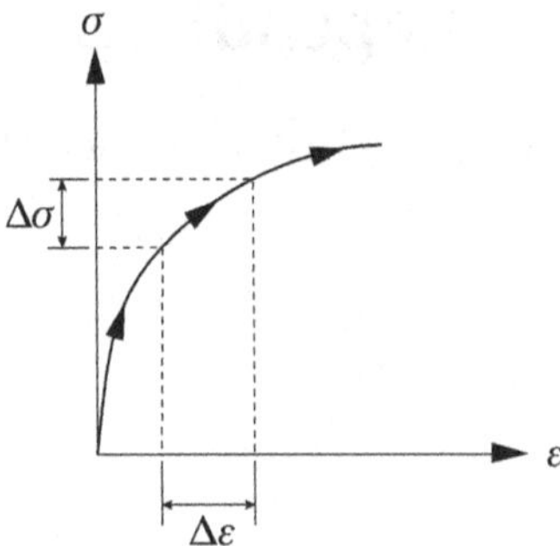

Figure E.1. The uniaxial stress–strain behaviour for a stable material.

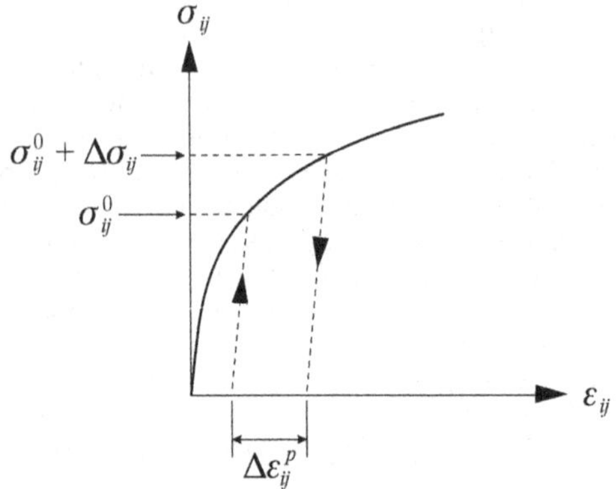

Figure E.2. Schematic representation of a loading cycle in a stable material.

to tractions, body forces and displacements, which result in the stress state σ_{ij}^0. We now slowly alter the tractions, body forces, etc., such that the new stress state is σ_{ij}. Finally, we return slowly to the original reference stress state σ_{ij}^0. If plastic strains develop during this stress cycle then the work done by the stresses is non-negative. Drucker's postulate therefore states that, for any stable material, the rate of work done by the stresses during plastic deformation at a point in the medium, over a closed cycle involving loading and unloading, is non-negative. If we denote this plastic work rate by $\dot{W}_P$, we have

$$W_P = \int_0^t (\sigma_{ij} - \sigma_{ij}^0)\dot{\varepsilon}_{ij}\, dt \geq 0 \tag{E.4}$$

We can use a schematic geometric representation shown in Figure E.3 to illustrate stress cycling in relation to the yield surface $f(\sigma_{ij}) = 0$. Considering the closed stress cycle (i.e. the stress cycle commences from σ_{ij}^0 and returns to σ_{ij}^0), the rate of work done by the stresses on the elastic strain rates $\dot{\varepsilon}_{ij}^e$ is fully recoverable. Therefore we can focus on the representation of the stability postulate in terms of the total work of the stresses done on the plastic strain rates, W_p, over the history of the stress cycle where $t \in (t_1\,,\ t_2)$, and for which the stresses satisfy the yield condition. Therefore in terms of the plastic

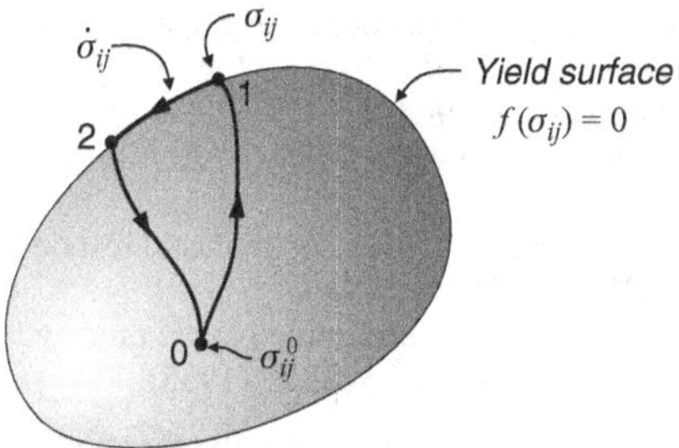

Figure E.3. Closed stress cycle 0–1–2–0 in a generalised stress space.

strain rate, (E.4) can be written as

$$W_p = \int_{t_1}^{t_2} \left(\sigma_{ij} - \sigma_{ij}^0\right)\dot{\varepsilon}_{ij}^p \, dt \geq 0 \tag{E.5}$$

Since plastic strains materialise only at $t = t_1$, we can expand W_p as a Taylor series about the neighbourhood of $\sigma_{ij}(t_1)$; this gives

$$W_p(t) = (t - t_1)\left(\sigma_{ij} - \sigma_{ij}^0\right)\dot{\varepsilon}_{ij}^p + \frac{(t - t_1)^2}{2!}\left[\dot{\sigma}_{ij}\dot{\varepsilon}_{ij}^p + \left(\sigma_{ij} - \sigma_{ij}^0\right)\ddot{\varepsilon}_{ij}^p\right] + \cdots. \tag{E.6}$$

Then, considering the leading terms on the right-hand side of (E.6), if the plastic work rate is to satisfy Drucker's postulate we must require (since $(t - t_1) > 0$)

$$\begin{aligned} \left(\sigma_{ij} - \sigma_{ij}^0\right)\dot{\varepsilon}_{ij}^p \geq 0 \quad &\text{if} \quad \sigma_{ij} \neq \sigma_{ij}^0 \\ \dot{\sigma}_{ij}\dot{\varepsilon}_{ij}^p \geq 0 \quad &\text{if} \quad \sigma_{ij} = \sigma_{ij}^0 \end{aligned} \tag{E.7}$$

The first requirement of (E.7) is associated with the energy dissipation and the second requirement refers to the stability of the material since, if $\dot{\sigma}_{ij}\dot{\varepsilon}_{ij}^p \geq 0$ and $\dot{\varepsilon}_{ij}^p \neq 0$, material stability is assured. The stability postulate is a key feature in the development of associated flow rules in the theory of plasticity and in the development of uniqueness theorems for perfectly plastic behaviour. Several investigators including H. Ziegler, A.E. Green and P.M. Naghdi have discussed the stability postulate in the context of thermodynamics of continua. Ziegler's work shows that Drucker's stability postulate is a special case of maximum entropy production. Green and Naghdi's work has shown that the postulate implies constraints on the flow rule that do not necessarily follow from laws of thermodynamics. This makes the concept of stability as constitutive assumption valid for certain classes of materials.

Further reading

The key works of Drucker, which introduce the concept of stability postulate as applied to plastic materials, can be found in:

D.C. Drucker, The definition of a stable inelastic material, *J. Appl. Mech., Trans ASME*, **26**, 101–106 (1958).

D.C. Drucker, Plasticity, In *Structural Mechanics* (eds. J.N. Goodier and N.J. Hoff) Pergamon Press, Oxford, pp. 407–455, 1960.

D.C. Drucker, On the postulate of stability of material in the mechanics of continua, *J. Mecanique*, **3**, 235–249 (1964).

D.C. Drucker, Concepts of path independence and material stability in soils. In *Rheology and Mechanics of Soils, Proc. IUTAM Symposium* (eds. J. Kravtchenko and P.M. Sirieys) pp. 23–43, 1964.

D.C. Drucker, *An Introduction to the Mechanics of Deformable Solids*, McGraw-Hill, New York, 1966.

D.C. Drucker, Conventional and unconventional plastic response and representation, *Appl. Mech. Rev.*, **41**, 151–167 (1988).

D.C. Drucker and M. Li, Stable response in the plastic range with local instability, *J. Appl. Math. Phys (ZAMP)*, **46**, S375–S385 (1995).

Further discussions of the stability concept from the thermodynamical viewpoint can be found in:

A.E. Green and P.M. Naghdi, A general theory for an elastic plastic continuum, *Arch. Rational Mech. Anal.*, **18**, 251–281 (1965).

H. Ziegler, Some extremum principles in irreversible thermodynamics, with applications to continuum mechanics. In *Progress in Solid Mechanics*, Vol. 4 (eds. I.N. Sneddon and R. Hill) North-Holland, Amsterdam, pp. 91–193, 1963.

Additional references include

W.-F. Chen, *Limit Analysis and Soil Plasticity*, Elsevier, Amsterdam, 1975.

J.N. Goodier and P.G. Hodge Jr., *Elasticity and Plasticity. Surveys in Applied Mathematics*, Vol. 1, John Wiley, New York, 1958.

L.E. Malvern, *Introduction to the Mechanics of a Continuous Medium*, Prentice-Hall, Englewood Cliffs, NJ, 1969.

A. Sawczuk, *Mechanics and Plasticity of Structures*, Ellis Horwood, New York, 1989.

Appendix F

The associated flow rule

In Appendix E we have examined Drucker's postulate for the stability of the material undergoing plastic deformations. To develop the plastic constitutive equations or the associated flow rule it is necessary to assume that a yield function exists, i.e.

$$f(\sigma_{ij}) = k \tag{F.1}$$

As discussed in Chapter 3, when referred to the multi-dimensional stress space, the *convex yield function* with a *unique normal at each point* identifies the boundary between elastic states in the material for which $f(\sigma_{ij}) < k$ and plastic states for which $f(\sigma_{ij}) = k$. For the present purposes we shall restrict attention to non-strain hardening materials. Consider the inequalities given by (E.7) in relation to a vector space consisting of the stress tensor and the strain rate vector. The expression related to plastic energy dissipation rate can be visualised as the scalar product of two vectors $(\sigma_{ij} - \sigma_{ij}^0)$ and $\dot{\varepsilon}_{ij}^p$. In order for the first inequality of (E.7) to be satisfied, the included angle between the vectors $(\sigma_{ij} - \sigma_{ij}^0)$ and the plastic strain rate vector $\dot{\varepsilon}_{ij}^p$ should be acute. This condition will hold for any σ_{ij}^0 located either within the yield surface or on the yield surface itself.

Consider the point B in Figure F.1, which is located on the yield surface $f(\sigma_{ij}) = k$, and assume that the associated flow rule with the governing normality condition gives the plastic strain rate vector, which will therefore be normal to the yield surface. Now consider the tangent plane to the yield surface at this point. We are assured, by the convexity of the yield surface, that any stress point σ_{ij} will lie to one side of this tangent plane. The line of action of the vector $(\sigma_{ij} - \sigma_{ij}^0)$ must therefore make an *acute angle* with $\dot{\varepsilon}_{ij}^p$. Therefore for the rate of plastic energy dissipation to be positive definite we must have

$$\left(\sigma_{ij} - \sigma_{ij}^0\right) \frac{\partial f}{\partial \sigma_{ij}} \geq 0 \tag{F.2}$$

We can also restate these operations in the following way. Consider a material for which the yield surface in the generalized stress space is convex and possesses a unique normal at each point. If plastic deformations are to occur such that material stability is preserved, then the plastic strain rate should satisfy the normality condition or the plastic strain rate must always be normal to the yield surface, i.e.

$$\dot{\varepsilon}_{ij}^p = \lambda \frac{\partial f}{\partial \sigma_{ij}} \tag{F.3}$$

where λ is an undetermined scalar multiplier which is *non-negative*. This associated form of the flow rule was first postulated by R. von Mises without any formal proof.

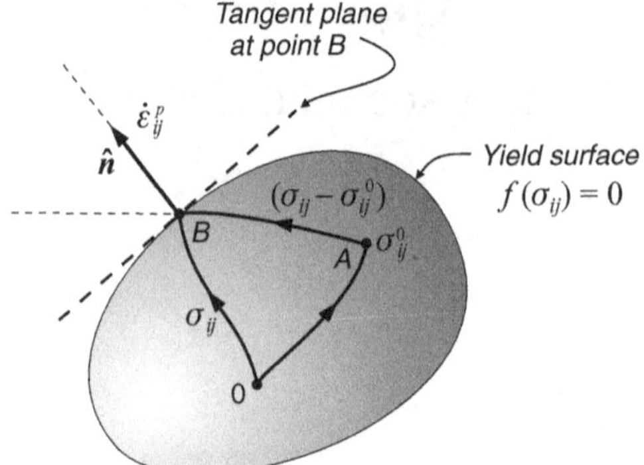

Figure F.1. Geometrical representation of the stability postulate.

Other arguments for this form of the associated flow rule are also given by R. Hill and T.Y. Thomas.

The second inequality of (E.7) dealing with the requirement for material stability can be written as

$$\dot{\sigma}_{ij}\dot{\varepsilon}_{ij}^{p} = \lambda \frac{\partial f}{\partial \sigma_{ij}} \dot{\sigma}_{ij} \geq 0 \tag{F.4}$$

The state of either loading, neutral loading or unloading of the plastic medium can be defined in relation to the direction of the stress increments in the following manner:

$$\frac{\partial f}{\partial \sigma_{ij}} \dot{\sigma}_{ij} \begin{cases} > 0; & \text{loading} \\ = 0; & \text{neutral loading} \\ < 0; & \text{unloading} \end{cases} \tag{F.5}$$

In the case of neutral loading, the loading path follows the yield surface itself and the unloading process results in no plastic deformation. The condition (F.4) implies that for plastic deformations to occur the scalar multiplier must be non-negative; i.e.

$$\lambda \geq 0 \tag{F.6}$$

If the stress state satisfies the yield condition but with $\lambda = 0$, then there is no plastic deformation. Also for a perfectly plastic material, there is no essential difference between the processes of loading and neutral loading since in the stress space $(\partial f / \partial \sigma_{ij})\dot{\sigma}_{ij} = 0$, whenever $\dot{\sigma}_{ij}$ lies on the yield surface.

In the preceding we have focused on the application of Drucker's stability postulate to the development of the associated flow rule for failure surfaces that have a unique normal at each point on the surface. Let us now focus attention on situations where the failure surface can have either edges or corners along which the orientation of the unit normal is not determined uniquely. Examples of such failure surfaces can include the Tresca and Coulomb failure criteria. We can extend the definition of the associated flow rule to cover such non-singular boundaries (Figure F.2). When considering non-singular failure surfaces, the associative flow rule should be modified to include several (say n) intersecting surfaces at a point. In such a case, the associated flow rule can be written as

$$\dot{\varepsilon}_{ij}^{p} = \sum_{\alpha=1,}^{n} \lambda_{\alpha} \frac{\partial f_{\alpha}}{\partial \sigma_{ij}} \quad \text{with} \quad \lambda_{\alpha} > 0 \tag{F.7}$$

where the derivatives $\partial f_{\alpha} / \partial \sigma_{ij}$ are linearly independent in view of the fact that the failure

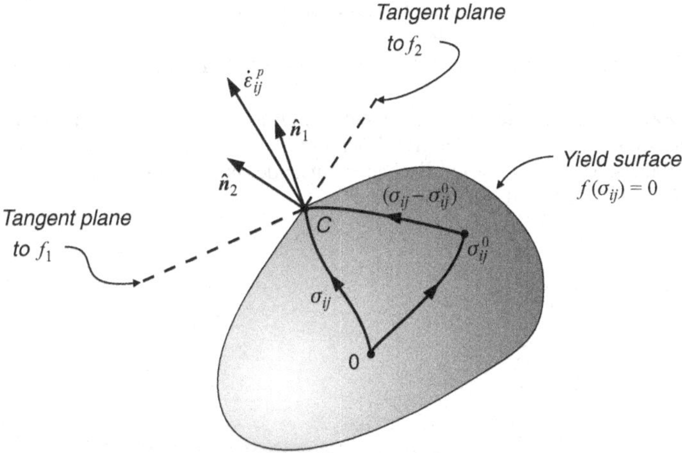

Figure F.2. Convex failure surface with non-singular points.

surfaces themselves are independent. The complete plastic strain rate is found according to (F.7) and the resultant of these plastic strain rates will be contained within the region obtained by surfaces spanning the unit normals to the segments of the yield surfaces, which intersect at the singular points. For example, referring to Figure F.2, where $n = 2$, the plastic strain rate at the corner C is given by

$$\dot{\varepsilon}_{ij}^p = \lambda_1 \frac{\partial f_1}{\partial \sigma_{ij}} + \lambda_2 \frac{\partial f_2}{\partial \sigma_{ij}} = \lambda \left[\phi \frac{\partial f_1}{\partial \sigma_{ij}} + (1 - \phi) \frac{\partial f_2}{\partial \sigma_{ij}} \right] \tag{F.8}$$

where $0 \leq \phi \leq 1$. It should be noted that although the direction of the plastic strain rate is not unique, the energy dissipation rate is uniquely determined, since for a given plastic strain rate the corresponding stress is unique. For example, considering a perfectly plastic material, which obeys the associated flow rule, the energy dissipation rate is given by

$$\dot{D} = \sigma_{ij} \dot{\varepsilon}_{ij}^p \tag{F.9}$$

If we realize that, at failure, the stresses are uniquely determined by the failure criterion, then the dissipation function can be expressed solely in terms of the plastic strain rate, i.e. $\dot{D} = \dot{D}(\dot{\varepsilon}_{ij}^p)$, and such a representation can be used to present an inverse form of the associated flow rule (F.3), where now the stresses can be expressed in terms of the strain rates.

Further reading

J.F.W. Bishop and R. Hill, A theory of plastic distortion of a polycrystalline aggregate under combined stress, *Phil. Mag.*, **42**, 414–427 (1951).

W.-F. Chen, *Limit Analysis and Soil Plasticity*, Elsevier, Amsterdam, 1975

T.J. Chung, *Applied Continuum Mechanics*, Cambridge University Press, Cambridge, 1996.

Y.C. Fung, *Foundations of Solid Mechanics*, Prentice-Hall, Englewood Cliffs, NJ, 1965.

L.E. Malvern, *Introduction to the Mechanics of a Continuous Medium*, Prentice-Hall, Englewood Cliffs, NJ, 1969.

J.B. Martin, *Plasticity: Fundamentals and General Results*, MIT Press, Cambridge, MA, 1975.

T.Y. Thomas, Interdependence of the yield condition and the stress–strain relations for plastic flow, *Proc. U.S. Nat. Acad. Sci.*, **40**: 593–597 (1954).

R. von Mises, Mechanik der plastischen Formaenderung von Kristallen, *Zeit. Angew. Math. Mech.*, **8**: 161–185 (1928).

Appendix G

A uniqueness theorem for elastic–plastic deformation

The concept of the uniqueness of a solution is an essential requirement to the well-posed nature of a boundary value problem. A uniqueness theorem assures us that there is only one solution possible for the governing set of equations subject to appropriate boundary conditions. In *EG* we have discussed a uniqueness theorem in the context of the linear theory of elasticity. With linear theories in mechanics and physics, the development of a proof of uniqueness of solutions to boundary value problems and initial boundary value problems is well established. Comprehensive discussions of these topics are given in many texts on mathematical physics and on the theory of partial differential equations and also discussed in recent volumes by Selvadurai (2000a,b). The question that arises in the context of plasticity focuses on the development of a uniqueness theorem for what is basically a non-linear problem. This is not a straightforward issue, even with regard to certain situations involving non-linear behaviour of linear elastic materials. Examples that illustrate the concept of non-uniqueness of elasticity solutions can be readily found in problems dealing with elastic buckling of structural elements such as beam-columns and shallow shells under lateral loads. In these categories of problem the structure can exhibit multiple equilibrium states corresponding to the same level of loading. The purpose of the discussion given below is then to address the basic question of what constraints should be imposed, specifically regarding plastic stress–strain relations, in order that the solution to a particular boundary value problem is unique.

Theorem G1. In keeping with the formulation of many approaches to proof of uniqueness, let us consider a finite region V of an elastic–plastic material that is bounded by a surface S. The region is subjected to tractions, displacements and body forces as follows:

$$T_i = T_i^0; \quad x_i \in S_T \tag{G.1}$$

$$u_i = u_i^0; \quad x_i \in S_u \tag{G.2}$$

$$b_i = b_i^0; \quad x_i \in V \tag{G.3}$$

where S_T and S_u are complementary subsets of S; i.e. $S_T \cup S_u = S$ and $S_T \cap S_u = 0$, and T_i^0, u_i^0 and b_i^0 are prescribed functions (Figure G.1a). The state of stress and strain in the medium due to the application of the prescribed tractions, displacements and body

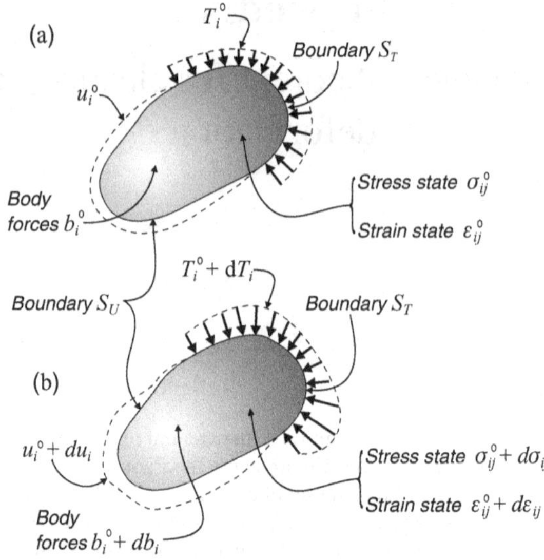

Figure G.1. Initial and final states of stress and strain in an elasto-plastic body.

forces in the respective regions is given by

$$\sigma_{ij} = \sigma_{ij}^0; \quad x_i \in V \tag{G.4}$$

$$\varepsilon_{ij} = \varepsilon_{ij}^0; \quad x_i \in V \tag{G.5}$$

We now alter the boundary tractions, boundary displacements and body forces by their corresponding incremental values as follows:

$$T_i = T_i^0 + dT_i; \quad x_i \in S_T \tag{G.6}$$

$$u_i = u_i^0 + du_i; \quad x_i \in S_u \tag{G.7}$$

$$b_i = b_i^0 + db_i; \quad x_i \in V \tag{G.8}$$

We assume that the corresponding states of stress and strain in the body (Figure G.1b) can be expressed in the forms

$$\sigma_{ij} = \sigma_{ij}^0 + d\sigma_{ij}; \quad x_i \in V \tag{G.9}$$

$$\varepsilon_{ij} = \varepsilon_{ij}^0 + d\varepsilon_{ij}; \quad x_i \in V \tag{G.10}$$

where the incremental values of the stresses and strains are implied. The examination of the question of uniqueness of the solution reduces to the assessment of whether the stress increments $d\sigma_{ij}$ and the strain increments $d\varepsilon_{ij}$ are uniquely determined by the increments of change in the surface tractions dT_i, the increment in the surface displacements du_i and the increment in the body forces db_i. In this connection we hope to prove that the associated flow rule is both necessary and sufficient for the condition for uniqueness of the stress increments $d\sigma_{ij}$ and the strain increments $d\varepsilon_{ij}$.

Proof. As in many proofs of this kind we adopt an approach, based on *'proof by contradiction'*. Let us assume that the state of stress and strain resulting from the application of the tractions, displacements and body forces defined by (G.6)–(G.8) results in *two* states of stress and strain in the following forms:

$$\sigma_{ij} = \sigma_{ij}^0 + d\sigma_{ij}^{(1)}; \quad x_i \in V \tag{G.11}$$

$$\varepsilon_{ij} = \varepsilon_{ij}^0 + d\varepsilon_{ij}^{(1)}; \quad x_i \in V \tag{G.12}$$

and

$$\sigma_{ij} = \sigma_{ij}^0 + d\sigma_{ij}^{(2)}; \quad x_i \in V \tag{G.13}$$

$$\varepsilon_{ij} = \varepsilon_{ij}^0 + d\varepsilon_{ij}^{(2)}; \quad x_i \in V \tag{G.14}$$

Now we make use of the principle of virtual work described in Appendix C. Assuming that the displacement field is continuous throughout V, such that incremental tractions $d\tilde{T}_i$ are defined through equilibrium considerations and incremental displacements du_i satisfy kinematic or compatibility constraints we have

$$\iint_{S_T} d\tilde{T}_i\, du_i\, dS + \iint_{S_u} d\tilde{T}_i\, du_i\, dS - \iiint_V db_i\, du_i\, dV = \iiint_V d\tilde{\sigma}_{ij}\, d\varepsilon_{ij}\, dV \tag{G.15}$$

There is, of course, no requirement for the two solutions mentioned above to be in any way related. The difference between the two states given by (G.11)–(G14) gives

$$\Delta(d\sigma_{ij}) = d\sigma_{ij}^{(2)} - d\sigma_{ij}^{(1)}; \quad x_i \in V \tag{G.16}$$

$$\Delta(d\varepsilon_{ij}) = d\varepsilon_{ij}^{(2)} - d\varepsilon_{ij}^{(1)}; \quad x_i \in V \tag{G.17}$$

and with

$$d\tilde{T}_i = dT_i^{(2)} - dT_i^{(1)} = 0; \quad x_i \in S_T \tag{G.18}$$

$$du_i = du_i^{(2)} - du_i^{(1)} = 0; \quad x_i \in S_u \tag{G.19}$$

Substituting (G.16)–(G.19) in (G.15) we obtain the result

$$\iiint_V \Delta(d\sigma_{ij})\, \Delta(d\varepsilon_{ij})\, dV = 0 \tag{G.20}$$

Since V is finite we obtain, from the *Dubois–Reymond lemma*, that the integrand of (G.20) must vanish everywhere in V, i.e.

$$dI = \Delta(d\sigma_{ij})\Delta(d\varepsilon_{ij}) \equiv 0 \tag{G.21}$$

If we could now show that, for a particular elastic–plastic constitutive relation, the quantity dI is positive definite, then we obtain the contradiction we seek. We could then satisfy (G.21) if and only if $\Delta(d\varepsilon_{ij})$ were identically zero. So how can we proceed to show that $\Delta(d\sigma_{ij})\Delta(d\varepsilon_{ij})$ is positive definite? To begin, assume that the difference in the incremental strains can be represented as a linear combination of the elastic and plastic components as follows:

$$\Delta(d\varepsilon_{ij}) = \Delta\left(d\varepsilon_{ij}^e\right) + \Delta\left(d\varepsilon_{ij}^p\right) \tag{G.22}$$

Using (G.22) we can rewrite (G.21) in the form

$$dI = \Delta(d\sigma_{ij})\left[\Delta\left(d\varepsilon_{ij}^e\right) + \Delta\left(d\varepsilon_{ij}^p\right)\right] \equiv 0 \tag{G.23}$$

Considering the elastic behaviour of the material we have for every state of incremental stresses

$$\Delta(d\sigma_{ij})\Delta\left(d\varepsilon_{ij}^e\right) \geq 0 \tag{G.24}$$

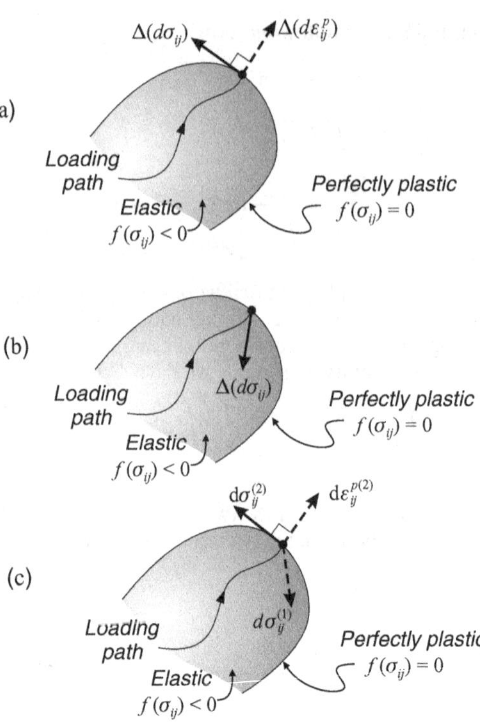

Figure G.2. Schematic representation of loading and unloading paths in relation to the failure surface of an elastic–perfectly plastic solid.

Therefore the problem is reduced to the examination of the conditions under which the scalar product $\Delta(d\sigma_{ij})\,\Delta(d\varepsilon^p_{ij})$ is positive definite. In order to examine this we need to consider separately three possible states, relating to loading and unloading (Figure G.2) associated with $d\sigma^{(1)}_{ij}$, $d\sigma^{(2)}_{ij}$ and $\Delta(d\sigma_{ij})$.

Case 1. Consider the case when both increments $d\sigma^{(1)}_{ij}$ and $d\sigma^{(2)}_{ij}$ correspond to loading paths. In this instance, both $d\sigma^{(1)}_{ij}$ and $d\sigma^{(2)}_{ij}$ lie on the tangent plane at the stress point on the failure surface and by virtue of (G.16), $\Delta(d\sigma_{ij})$ also lies on the tangent plane. It is evident that for the scalar product $\Delta(d\sigma_{ij})\,\Delta(d\varepsilon^p_{ij})$ to be positive definite for all vector increments $\Delta(d\sigma_{ij})$ that are tangent to the failure surface, the plastic strain vectors $d\varepsilon^{p(1)}_{ij}$ and $d\varepsilon^{p(2)}_{ij}$, and consequently $\Delta(d\varepsilon^p_{ij})$, must be normal to the failure surface. So the associated flow rule ensures uniqueness in this case.

Case 2. Consider the situation where both loading increments correspond to unloading. In this case by definition (see also Appendix F) $\Delta(d\varepsilon^p_{ij}) = 0$. Hence by virtue of (G.24), dI is positive definite and uniqueness is a trivial consequence.

Case 3. The remaining possibility deals with the situation when one stress increment corresponds to loading and the other corresponds to unloading. It is immaterial which one we would choose as loading and which as unloading. For purposes of analysis, let us assume that $d\sigma^{(1)}_{ij}$ is an unloading path and $d\sigma^{(2)}_{ij}$ corresponds to a loading path.

Owing to this assumption we have $d\varepsilon_{ij}^{p(1)} \equiv 0$ and the part of dI whose positive definiteness needs to be investigated is

$$dI = \left(d\sigma_{ij}^{(2)} - d\sigma_{ij}^{(1)}\right) d\varepsilon_{ij}^{p(2)} = d\sigma_{ij}^{(2)}\, d\varepsilon_{ij}^{p(2)} - d\sigma_{ij}^{(1)}\, d\varepsilon_{ij}^{p(2)} \tag{G.25}$$

Since $d\sigma_{ij}^{(2)}$ corresponds to a loading path, this stress increment is located on the tangent plane through the stress point. By virtue of the associated flow rule the plastic strain increment vector $d\varepsilon_{ij}^{p(2)}$ is orthogonal to $d\sigma_{ij}^{(2)}$. Hence the scalar product $d\sigma_{ij}^{(2)}\, d\varepsilon_{ij}^{p(2)}$ is identically equal to zero.

Let us now focus on the second term on the right-hand side of (G.25). By definition $d\sigma_{ij}^{(1)}$ is an unloading path and the vector is directed to the interior of the failure surface (Figure G.2c), commencing from the stress point. The plastic strain increment vector $d\varepsilon_{ij}^{p(2)}$, for the loading path, is on the other hand directed away from the failure surface normal to the tangent plane at the stress point. This means that so long as the increment $d\sigma_{ij}^{(1)}$ is an unloading path and the increment $d\sigma_{ij}^{(2)}$ is a loading path, the included angle between $d\sigma_{ij}^{(1)}$ and $d\varepsilon_{ij}^{p(2)}$ will be an obtuse angle. Hence the scalar product, $-d\sigma_{ij}^{(1)}\, d\varepsilon_{ij}^{p(2)}$, will always be positive definite, for any convex failure surface. Therefore the positive definiteness of the integral is assured and we can conclude that uniqueness of the boundary value problem is assured, provided the failure surface is convex and the plastic strains are determined through the associated flow rule.

Further reading

Discussions of the development of uniqueness theorems for elastic plastic solids can be found in the following:

W.-F. Chen, *Limit Analysis and Soil-Plasticity*, Elsevier, Amsterdam, 1975.

W.-F. Chen, *Plasticity in Reinforced Concrete*, McGraw-Hill, New York, 1982.

R. Hill, *The Mathematical Theory of Plasticity*, Oxford University Press, Oxford, 1950.

L.M. Kachanov, *Foundations of the Theory of Plasticity*, North-Holland, Amsterdam, 1971.

W.T. Koiter, Stress–strain relations, uniqueness and variational theorems for elastic–plastic materials with a singular yield surface, *Quart. Appl. Math.*, **11**, 350–354 (1953).

J.B. Martin, *Plasticity: Fundamentals and General Results*, MIT Press, Cambridge, MA, 1975.

A. Sawczuk, *Mechanics and Plasticity of Structures*, Ellis-Horwood, Chichester, 1989.

General discussions pertaining to the development of uniqueness theorems for problems in mathematical physics and mechanics can be found in the following references:

R. Courant and D. Hilbert, *Methods of Mathematical Physics*, Vols. 1 and 2, Wiley-Interscience, New York, 1962.

P.M. Morse and H. Feshbach, *Methods of Theoretical Physics*, Parts 1 and 2, McGraw-Hill, New York, 1953.

A.P.S. Selvadurai, *Partial Differential Equations in Mechanics. Vol. 1 Fundamentals, Laplace's Equation, Diffusion Equation, Wave Equation.* Springer-Verlag, Berlin, 2000a.

A.P.S. Selvadurai, *Partial Differential Equations in Mechanics. Vol. 2. The Biharmonic Equation, Poisson's Equation,* Springer-Verlag, Berlin, 2000b.
I.N. Sneddon, *Elements of Partial Differential Equations*, McGraw-Hill, New York, 1957.
I. Stakgold, *Boundary Value Problems in Mathematical Physics*, Vols. I and II, Macmillan, New York, 1957.
E. Zauderer, *Partial Differential Equations of Applied Mathematics*, 2nd edition, John Wiley, New York.

Appendix H

Theorems of limit analysis

When a continuum region consists of either a rigid strain hardening or an elastic strain hardening material, the strains and displacements of the region for a given history of loading can be determined. If, on the other hand, the continuum region is made of either a rigid perfectly plastic or an elastic perfectly plastic material the situation is quite different. A qualitative picture of the behaviour of such a material was discussed at the very start of this volume. For example, at sufficiently small loads the region can either remain rigid or experience small elastic deformations. As the loading increases parts of the continuum region can become plastic, but the region as a whole can withstand collapse due to the restraining effect of elastic regions. As the loads increase, larger regions of the continuum can experience plastic yield and eventually the continuum region can undergo 'indefinite' plastic deformations leading to what we term 'collapse' of the region. Two interesting examples that illustrate the definition of a 'collapse' state in the context of limit analysis are given by Drucker *et al.* (1952). In this process we implicitly assume that the deformations experienced by the continuum region are small enough so that changes in the geometry of the region may be neglected and that all the deformations take place in a quasi-static fashion so that any dynamic effects can be ignored. The study of the behaviour of continua that exhibit 'collapse' marks an important milestone in the development of the theory of plasticity. The availability of procedures for the calculation of collapse loads for elastic plastic and rigid plastic solids provided a valuable design tool, which could be utilised to calculate the loads required either to initiate collapse or to prevent collapse of the structure or a component made of an elastic–perfectly plastic material. The availability of limit analysis techniques also shifted the attention from the mathematical solution of the governing non-linear partial differential equations. These partial differential equations are difficult to solve through analytical procedures except in the simplest of situations involving elementary states of deformation. Often recourse is made to methods involving numerical computations. The development of the theorems of limit analysis for rigid plastic materials is attributed to A.A. Gvozdev and to R. Hill. The generalization of these concepts to include elastic–plastic solids was given in benchmark papers by D.C. Drucker, W. Prager and H.J. Greenberg (1952, 1957). The formulation of the limit theorems within the context of extremum principles (Appendix D) added a new dimension to the mathematical theory underlying the calculation of 'collapse' and to the development of the theorems necessary to ensure the accuracy of the calculations. In Chapter 5 we applied the limit theorems to develop upper and lower bound solutions to several problems of interest to geomechanics. The purpose of this Appendix is to provide certain proofs that will establish the validity of the limit theorems. The presentation is not meant to be mathematically all encompassing;

the interested reader is encouraged to consult the bibliography presented at the end of this Appendix.

The development of the proofs of the limit theorems is facilitated by first considering the following Lemma:

In the case of elastic–perfectly plastic materials, upon attainment of the limit load, the stress state experiences no change and the increments of strain are only due to the development of plastic strains in the medium. As a result, the application of elastic–perfectly plastic relations is formally similar to the application of rigid perfectly plastic relations. With continued plastic deformations, the elastic deformations can be neglected from the analysis.

To prove the lemma, let us consider a finite region V of an elastic–perfectly plastic material with surface S. Also, S_T and S_v are, respectively, subsets of S on which tractions and velocities can be prescribed. The tractions T_i act on the surface S_T and body forces act in V and satisfy the conditions for static equilibrium

$$\sigma_{ij} n_j = T_i; \quad x_i \in S_T \tag{H.1}$$

$$\sigma_{ij,j} - b_i = 0; \quad x_i \in V \tag{H.2}$$

the comma denotes the partial derivative with respect to the spatial variable and n_i are the direction cosines to the outward unit normal to S. In addition to this statically admissible set of stresses, v_i^* represents the kinematically admissible velocity field in the region V and $\dot{\varepsilon}_{ij}^*$ the corresponding strain rates that satisfy the kinematic relations

$$2\dot{\varepsilon}_{ij}^* = v_{i,j}^* + v_{j,i}^* \tag{H.3}$$

Neither the equilibrium set T_i, b_i and σ_{ij} nor the kinematically admissible set v_i^* and $\dot{\varepsilon}_{ij}^*$ need be the actual state nor in any way related to each other. We shall proceed to develop the rate form of the virtual work equation where we have

$$\dot{W}_{ext} = -\iiint_V b_i v_i^* \, dV + \iint_{S_T} T_i v_i^* \, dS + \iint_{S_v} T_i v_i^* \, dS \tag{H.4}$$

and

$$\dot{W}_{int} = \iiint_V \sigma_{ij} \dot{\varepsilon}_{ij}^* \, dV \tag{H.5}$$

The result (H.5) can also be extended to cover the rate of work of internal forces associated with discontinuities that are encountered in plastically deforming materials; this is not, however, central to the discussion that pertains to the proofs of the limit theorems. The rate form of the equation of virtual work gives

$$-\iiint_V b_i v_i^* \, dV + \iint_{S_T} T_i v_i^* \, dS + \iint_{S_v} T_i v_i^* \, dS = \iiint_V \sigma_{ij} \dot{\varepsilon}_{ij}^* \, dV \tag{H.6}$$

This result is applicable to *any* equilibrium set of tractions, body forces and stresses. Therefore we can choose the set as *increments* of tractions, body forces and stresses. Substituting the increments or rates of the equilibrium set we obtain from (H.6)

$$-\iiint_V \dot{b}_i v_i^* \, dV + \iint_{S_T} \dot{T}_i v_i^* \, dS + \iint_{S_v} \dot{T}_i v_i^* \, dS = \iiint_V \dot{\sigma}_{ij} \dot{\varepsilon}_{ij}^* \, dV \tag{H.7}$$

We can now utilise (H.7) to prove the Lemma. Let $\dot{T}_i^c$, $\dot{b}_i^c$ and $\dot{\sigma}_{ij}^c$ be the rates or increments of the body forces, surface tractions and stresses and v_i^c and $\dot{\varepsilon}_{ij}^c$ the corresponding velocity and strain rates associated with a collapse state, such that, analogously to (H.7), we obtain

$$-\iiint_V \dot{b}_i^c v_i^c \, dV + \iint_{S_T} \dot{T}_i^c v_i^c \, dS + \iint_{S_v} \dot{T}_i^c v_i^c \, dS = \iiint_V \dot{\sigma}_{ij}^c \dot{\varepsilon}_{ij}^c \, dV \quad \text{(H.8)}$$

and the superscript $(\ )^c$ is intended to signify the fact that all the quantities are associated with the collapse state. Let us note that in the development of the proof of this Lemma, we will be considering only elastic–perfectly plastic materials. For such a material, at limiting or collapse conditions, the peak values of all the forces, tractions (either applied over S_T or induced at S_v) and stresses have been reached giving rise to the requirements

$$\dot{b}_i^c \equiv 0; \quad x_i \in V \quad \text{(H.9)}$$

$$\dot{T}_i^c \equiv 0; \quad x_i \in S \quad \text{(H.10)}$$

In view of (H.9) and (H.10), the entire left-hand side of equation (H.8) is zero. If we further assume that the strain rates $\dot{\varepsilon}_{ij}^c$ admit an additive decomposition of the form

$$\dot{\varepsilon}_{ij}^c = \dot{\varepsilon}_{ij}^{c(el)} + \dot{\varepsilon}_{ij}^{c(pl)} \quad \text{(H.11)}$$

where the superscripts (el) and (pl) refer, respectively, to the elastic and plastic strain rates, (H.8) gives

$$\iiint_V \left(\dot{\varepsilon}_{ij}^{c(el)} + \dot{\varepsilon}_{ij}^{c(pl)}\right) \dot{\sigma}_{ij}^c \, dV = 0 \quad \text{(H.12)}$$

From considerations of the stability postulate of Drucker (Appendix E) that utilises the concepts of the convexity of the yield surface, the associated flow rule (Appendix F) and the normality condition, it can be shown that at collapse the increment of the plastic energy dissipation rate is zero: i.e. the plastic strain increment vector is orthogonal to the stress increment vector. This gives

$$\iiint_V \dot{\varepsilon}_{ij}^{c(pl)} \dot{\sigma}_{ij}^c \, dV = 0 \quad \text{(H.13)}$$

which reduces (H.12) to

$$\iiint_V \dot{\varepsilon}_{ij}^{c(el)} \dot{\sigma}_{ij}^c \, dV = 0 \quad \text{(H.14)}$$

Since, for any elastic material, the integrand of (H.14) is positive definite for $\dot{\sigma}_{ij}^c \neq 0$, the vanishing of the integral in (H.14) implies that

$$\dot{\sigma}_{ij}^c = 0; \quad x_i \in V \quad \text{(H.15)}$$

This leads to the conclusion that, at collapse, there is no incremental change in the stress and accordingly there is no incremental change in the elastic strain during deformations occurring at the collapse load. In other words, the elastic deformations play no role in defining the collapse load. With this important result we can proceed to provide proofs of the limit theorems as originally postulated in the classic papers by D.C. Drucker, W. Prager and H.J. Greenberg.

Theorem H1. Consider a region V of a perfectly plastic material with surface S. If an equilibrium distribution of stress denoted by σ_{ij}^E applicable to V can be found such that

σ_{ij}^E satisfies the traction boundary conditions

$$\sigma_{ij}^E n_j = T_i \quad \text{for} \quad x_i \in S_T \tag{H.16}$$

and is everywhere below yield, i.e.

$$f\left(\sigma_{ij}^E\right) < k \quad \text{for} \quad x_i \in V \tag{H.17}$$

then collapse will not occur under the action of the loads.

Proof. The basic approach to proving this theorem involves a proof by contradiction, which makes the assumption that the theorem is false and that such an assumption will lead to a contradiction. Let us consider the finite region V with surface S that is subjected to tractions T_i on S_T a subset of S and body forces b_i in V. At some values of these loads we assume that collapse of the body occurs, resulting in an actual state of stress σ_{ij}^E, strain rates $\dot{\varepsilon}_{ij}^E$ and corresponding velocities v_i^E. This state of collapse will obviously correspond to tractions

$$\sigma_{ij}^E n_j = T_i \quad \text{on} \quad x_i \in S_T \tag{H.18}$$

and body forces b_i that satisfy

$$\sigma_{ij,j}^E - b_i = 0 \quad \text{in} \quad x_i \in V \tag{H.19}$$

but with the requirement that, at collapse, the region under consideration is suitably constrained to eliminate any undetermined rigid body movement.

There are two equilibrium states; one corresponding to $\{T_i, b_i, \sigma_{ij}^E\}$ and the second corresponding to the assumed state of collapse $\{T_i, b_i, \sigma_{ij}^c\}$. We can apply the virtual rate of work equation to each of these states. Using (H.6) we can write

$$-\iiint_V b_i v_i^c \, dV + \iint_{S_T} T_i v_i^c \, dS + \iint_{S_v} T_i v_i^c \, dS = \iiint_V \sigma_{ij}^c \dot{\varepsilon}_{ij}^c \, dV \tag{H.20}$$

and

$$-\iiint_V b_i v_i^c \, dV + \iint_{S_T} T_i v_i^c \, dS + \iint_{S_v} T_i v_i^c \, dS = \iiint_V \sigma_{ij}^E \dot{\varepsilon}_{ij}^c \, dV \tag{H.21}$$

Subtracting (H.21) from (H.20) and noting that at collapse the influence of elastic deformations can be neglected, we obtain the result

$$\iiint_V \left(\sigma_{ij}^c - \sigma_{ij}^E\right) \dot{\varepsilon}_{ij}^{c(pl)} \, dV = 0 \tag{H.22}$$

From assumptions of convexity of the yield surface the normality of the plastic strain increment vector to the yield surface implied by the associated flow rule, we require

$$\left(\sigma_{ij}^c - \sigma_{ij}^E\right) \dot{\varepsilon}_{ij}^{c(pl)} > 0 \quad \text{if} \quad f\left(\sigma_{ij}^E\right) < k \tag{H.23}$$

and the resulting sum of positive terms cannot be zero. As a result (H.22) must be false and the lower bound theorem is proved. That is, if

$$f\left(\sigma_{ij}^E\right) < k \tag{H.24}$$

collapse is not possible and if

$$f\left(\sigma_{ij}^E\right) = k \tag{H.25}$$

then the body may be in an *imminent state of collapse*. The result (H.25) can then be used to compute the external loads that are necessary just to initiate collapse in the body, which provides the lower bound of the capacity of the body. In terms of the extrema, this is the absolute minimum carrying capacity of the body, and the *real collapse load is expected to be greater than the lower bound.*

• • •

We can now proceed to provide, in a similar manner, a theorem concerning the upper limit of the carrying capacity of an elastic–perfectly plastic body.

Theorem H2. Collapse must occur for any compatible mechanism of plastic deformation for which the rate of working of the external forces either equals or exceeds the rate of internal energy dissipation. The collapse load which is obtained by considering the balance between the rate of working of the external forces and the rate of internal energy dissipation will either be greater than or equal to the true collapse load.

Proof. The procedure for the proof is again to assume that the theorem is false and to show that the assumption leads to a contradiction. Consider the region V of an elastic–rigid plastic material with surface S, which is under the action of surface tractions T_i acting on S_T and body forces b_i in the volume. Suppose the loads computed by equating the rate of working of the tractions and body forces to the rate of internal dissipation are *less* than the actual collapse load. Then the body will *not* experience collapse at this load and we can obtain an equilibrium state of stress σ_{ij}^E such that

$$\sigma_{ij}^E n_j = T_i; \quad x_i \in S_T \tag{H.26}$$

$$\sigma_{ij,j}^E - b_i = 0; \quad x_i \in V \tag{H.27}$$

and

$$f(\sigma_{ij}^E) < k; \quad x_i \in V_E \tag{H.28}$$

$$f(\sigma_{ij}^E) = k; \quad x_i \in V_F \tag{H.29}$$

where $V_F \cup V_E = V$. At this point we should clarify the conditions implied by (H.28) and (H.29) in the light of comments made both at the beginning of this Appendix and in the introductory paragraphs of Chapter 5. It is implicitly assumed that failure may occur in restricted regions within V, without attainment of collapse conditions in the body.

Let us now consider a compatible mechanism of plastic deformation corresponding to a collapse mechanism, which is defined by a plastic component of the velocity $v_i^{c(pl)}$ in the region S_v, a resulting plastic strain rate $\dot{\varepsilon}_{ij}^{c(pl)}$ and a corresponding stress state $\sigma_{ij}^{c(pl)}$ in the region V. Considering the stress state at collapse and the velocity field at collapse, the rate form of the virtual work equation (H.6) can be written as

$$-\iiint_V b_i v_i^{c(pl)}\, dV + \iint_{S_T} T_i v_i^{c(pl)}\, dS + \iint_{S_v} T_i v_i^{c(pl)}\, dS = \iiint_V \sigma_{ij}^{c(pl)} \dot{\varepsilon}_{ij}^{c(pl)}\, dV \tag{H.30}$$

Similarly, considering the equilibrium stress state and the velocity field at collapse we can write

$$-\iiint_V b_i v_i^{c(pl)}\, dV + \iint_{S_T} T_i v_i^{c(pl)}\, dS + \iint_{S_v} T_i v_i^{c(pl)}\, dS = \iiint_V \sigma_{ij}^E \dot{\varepsilon}_{ij}^{c(pl)}\, dV \tag{H.31}$$

Subtracting (H.31) from (H.30) we obtain

$$\iiint_V \left(\sigma_{ij}^{c(pl)} - \sigma_{ij}^E\right) \dot{\varepsilon}_{ij}^{c(pl)}\, dV = 0 \tag{H.32}$$

Considering (H.28) and (H.29), we can rewrite (H.32) as

$$\iiint_{V_F} \left(\sigma_{ij}^{c(pl)} - \sigma_{ij}^E\right) \dot{\varepsilon}_{ij}^{c(pl)}\, dV + \iiint_{V_E} \left(\sigma_{ij}^{c(pl)} - \sigma_{ij}^E\right) \dot{\varepsilon}_{ij}^{c(pl)}\, dV = 0 \tag{H.33}$$

We are aware from Appendices E and F that, for a convex yield surface and an associated flow rule obeying the normality condition, the quantity $(\sigma_{ij}^{c(pl)} - \sigma_{ij}^E)\dot{\varepsilon}_{ij}^{c(pl)}$ can be no smaller than zero. Moreover, if σ_{ij}^E is not a yield state (as is the case in the region V_E), then $(\sigma_{ij}^{c(pl)} - \sigma_{ij}^E)\dot{\varepsilon}_{ij}^{c(pl)}$ must be strictly positive. Therefore the first integral in (H.33) is non-negative and the second integral is positive definite. This contradicts our initial assumption and we conclude that the theorem must be true. The upper bound theorem assures us of the fact that the true collapse load must either be less than or at most equal to the collapse load obtained by equating the rate working of the external loading with the rate of internal energy dissipation for a compatible mechanism of plastic deformation.

• • •

In the preceding sections we have presented the two basic limit theorems that provide distinct upper and lower limits to the actual collapse loads, which could be obtained by solving the complete set of partial differential equations governing a problem in the theory of perfectly plastic solids. Furthermore, if the upper and lower bounds coincide then this would also correspond to the exact solution of the problem. There are also several corollaries that arise from the classical lower bound theorem, since the original state of stress is also applicable in the modified situation. These extensions are described in detail by Drucker *et al.* (1952) and in the text by Chen (1975).

It is also worth re-iterating the important role that requirements such as convexity of the yield surface and the associated flow rule play in the developments of the theorems of limit equilibrium. Indeed the theorems cannot be proved in a general sense without the aid of the convexity and associativity arguments. Despite these advantages, it is worth recognising the fact that the flow laws for many geomaterials point specifically to their non-associated character. The uniqueness of solutions and the validity of the upper and lower bound theorems do not, in a general sense, extend to such materials. Several investigators have examined the conditions under which the bounding techniques can be applied to geomaterials, which obey non-associated flow rules. Examples of these are given, among others, by A.D. Cox, Z. Mroz, G. de Josselin de Jong, A.C. Palmer, G. Maier, E.H. Davis, T. Hueckel, J.L. Dais, I.F. Collins and J. Salencon. Other approximate procedures for the calculation of limit loads for geomaterials exhibiting non-associated flow laws have also been proposed recently by A. Drescher and E. Detournay and R.L. Michalowski.

Further reading

The original articles dealing with limit theorems are due to:

D.C. Drucker, W. Prager and H.J. Greenberg, Extended limit design theorems for continuous media, *Quart. Appl. Math.*, **9**, 381–389 (1952).

D.C. Drucker, H.J. Greenberg and W. Prager, The safety factor for an elastic–plastic body in plane strain, *J. Appl . Mech.*, **73**, 371–378 (1957).

A.A. Gvozdev, Determination of the collapse load for statically indeterminate structures subjected to plastic deformations (in Russian), *Proc. Conf. Plastic Deformations*, Moscow, 1938.

A.A. Gvozdev, The determination of the value of the collapse load for statically determinate systems undergoing plastic deformations (Translation by R.M. Haythornthwaite), *Int. J. Mech. Sci.*, **1**, 322–335 (1960).

R. Hill, On the state of stress in a plastic rigid body at the yield point, *Phil. Mag.*, **7**: 868–875 (1951).

A.A. Markov, On variational principles in the theory of plasticity (in Russian), *Prikl. Math. Mech.*, **11**, 339–350 (1947).

W. Prager, Limit analysis: the development of a concept, in *Problems in Plasticity* (ed. A. Sawczuk), Noordhoff International Publ., Leyden, 3–24, 1974.

Excellent accounts of the developments of the limit theorems applicable to ideally plastic solids are given by

J. Chakrabarty, *Theory of Plasticity*, 2nd edition, McGraw-Hill, New York, 1998.

W.-F. Chen, *Limit Analysis and Soil Plasticity*, Elsevier, Amsterdam, 1975.

D.C. Drucker, Plasticity, in *Structural Mechanics. Proceedings of the First Symposium on Naval Structural Mechanics* (eds. J.N. Goodier and N.J. Hoff), Pergamon Press, New York, pp. 407–448, 1960.

L.M. Kachanov, *Fundamentals of the Theory of Plasticity* (Translated from the Russian by M. Konyaeva) Mir Publishers, Moscow, 1974.

J. C. Lubliner, *Plasticity Theory*, Collier-Macmillan, New York, 1990.

W. Olszak, Z. Mroz and P. Perzyna, *Recent Trends in the Development of the Theory of Plasticity*, Macmillan, New York, 1963.

W. Prager and P.G. Hodge Jr., *Theory of Perfectly Plastic Solids*, John Wiley and Sons, New York, 1951.

I.M. Rabinovich, (Ed.) Structural Mechanics in the USSR 1915–1957 (Transl. Ed. G. Herrmann) Pergamon Press, Oxford, 1960.

J. Salencon, *Applications of the Theory of Plasticity in Soil Mechanics*, John Wiley, New York, 1974.

R.A.C. Slater, *Engineering Plasticity*, John Wiley, New York, 1977.

Aspects of limit theorems applicable to materials that obey non-associated flow rules are given in the following articles and books:

I.F. Collins, The upper bound theorem for rigid–plastic solids generalized to include Coulomb friction, *J. Mech. Phys. Solids*, **17**, 323–338 (1969).

A.D. Cox, *The Use of Non Associated Flow Rules in Soil Plasticity*, RARDE (UK) Report B2/63 (1963).

J.L. Dais, Non-uniqueness of collapse load for a frictional material, *Int. J. Solids Struct.*, **6**, 1315–1319 (1970).

E.H. Davis, Theories of plasticity and failure of soil masses, in *Soil Mechanics, Selected Topics* (ed. I.K. Lee) American Elsevier, New York, 341–354, 1968.

de Josselin de Jong, G. Lower bound collapse theorem and lack of normality of strain rate to the yield surface, in *Rheology and Soil Mechanics. Proc. IUTAM Symposium*, Grenoble (eds. J. Kravtchenko and P.M. Sirieys) Springer-Verlag, Berlin, 69–75 (1964).

A. Drescher and E. Detournay, Limit load in translational failure mechanisms for associative and non-associative materials, *Geotechnique*, **43**, 443–456 (1993).

G. Maier and T. Hueckel, Nonassociated and coupled flow rules of elastoplasticity for rock-like materials, *Int. J. Rock Mech. Min. Sci.*, **16**, 77–92 (1979).

R.L. Michalowski, An estimate of the influence of soil weight on bearing capacity using limit analysis, *Soils and Foundations*, **37**, 57–64 (1997).

Z. Mroz, Non-associated flow laws in plasticity, *J. Mecanique*, **2**, 21–42 (1963).

A.C. Palmer, A limit theorem for materials with non-associated flow laws, *J. Mecanique*, **5**, 217–222 (1966).

J. Salencon, *Applications of the Theory of Plasticity in Soil Mechanics*, John Wiley, New York, 1974.

Appendix I
Limit analysis and limiting equilibrium

Coulomb's retaining wall analysis was based on equilibrium of forces acting upon a wedge of soil isolated behind the retaining wall. His method is generally referred to as a *limiting equilibrium* analysis. It gives exactly the same result as the energy balance method used in the upper bound theorem for any translational collapse mechanism.* Moreover the equivalence of the two methods holds regardless of what material model we choose, particularly the choice of flow rule. To see why this is so consider the rigid triangular soil element with area A shown in Figure I.1.

The soil element in Figure I.1 could have more than three sides, but a triangle will be the most simple geometry for our purposes. The element itself is numbered 1 while the surrounding elements are numbered 2, 3 and 4. The velocities of each element with respect to some common stationary point 0 are denoted $\boldsymbol{v}_{0k}$, $k = 1, \ldots, 4$. Relative velocities are shown on the hodograph and are denoted $\boldsymbol{v}_{1k}$, $k = 2, 3, 4$. Let the three sides of element 1 be numbered L_{12}, L_{13}, L_{14} and let the traction vectors which act on those sides be $\boldsymbol{T}_{12}$, $\boldsymbol{T}_{13}$, $\boldsymbol{T}_{14}$. Then we can write out the external rate of work $\mathbb{R}$ associated with our element as follows:

$$\mathbb{R} = \boldsymbol{v}_{02} \cdot \int_{L_{12}} \boldsymbol{T}_{12}\, dL + \boldsymbol{v}_{03} \cdot \int_{L_{13}} \boldsymbol{T}_{13}\, dL + \boldsymbol{v}_{04} \cdot \int_{L_{14}} \boldsymbol{T}_{14}\, dL - \boldsymbol{v}_{01} \cdot \iint_{A} \boldsymbol{b}\, dA \qquad \text{(I.1)}$$

Here $\boldsymbol{b}$ denotes the body force vector. Equation (I.1) would apply to a unit thickness of soil measured perpendicular to the plane of the Figure I.1. We can also write out the corresponding expression for the rate of dissipation $\mathbb{D}$.

$$\mathbb{D} = \boldsymbol{v}_{12} \cdot \int_{L_{12}} \boldsymbol{T}_{12}\, dL + \boldsymbol{v}_{13} \cdot \int_{L_{13}} \boldsymbol{T}_{13}\, dL + \boldsymbol{v}_{14} \cdot \int_{L_{14}} \boldsymbol{T}_{14}\, dL \qquad \text{(I.2)}$$

When we use the upper bound theorem we set $\mathbb{D}$ equal to $\mathbb{R}$.

Note from the velocity hodograph in Figure I.1 how the relative velocity vectors are associated with the element velocities.

$$\boldsymbol{v}_{02} = \boldsymbol{v}_{01} + \boldsymbol{v}_{12}, \quad \boldsymbol{v}_{03} = \boldsymbol{v}_{01} + \boldsymbol{v}_{13}, \quad \boldsymbol{v}_{04} = \boldsymbol{v}_{01} + \boldsymbol{v}_{14} \qquad \text{(I.3)}$$

* By a translational collapse mechanism we mean a plane system of rigid soil blocks separated by thin shear bands exactly as used throughout Chapter 5.

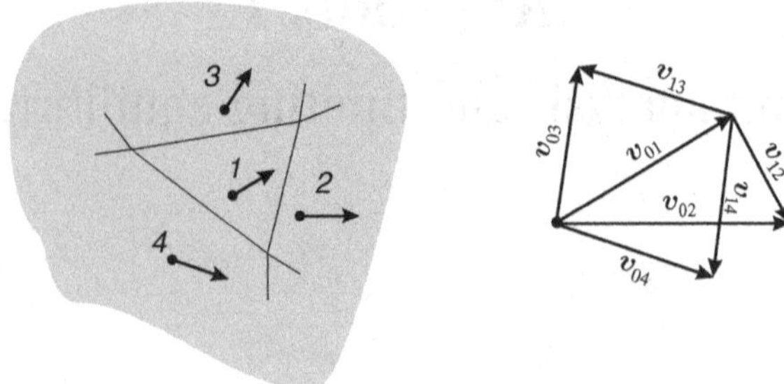

Figure I.1. A triangular material element with velocities of surrounding elements.

Using these relationships in (I.1) and setting $\mathbb{R} = \mathbb{D}$ shows that (I.1) reduces to

$$0 = \boldsymbol{v}_{01} \cdot \int_{L_{12}} \boldsymbol{T}_{12}\, dL + \boldsymbol{v}_{01} \cdot \int_{L_{13}} \boldsymbol{T}_{13}\, dL + \boldsymbol{v}_{01} \cdot \int_{L_{14}} \boldsymbol{T}_{14}\, dL - \boldsymbol{v}_{01} \cdot \iint_{A} \boldsymbol{b}\, dA \qquad \text{(I.4)}$$

Note that the vector $\boldsymbol{v}_{01}$ can be factored from this expression. Then since $\boldsymbol{v}_{01}$ is arbitrary we conclude that

$$\int_{L_{12}} \boldsymbol{T}_{12}\, dL + \int_{L_{13}} \boldsymbol{T}_{13}\, dL + \int_{L_{14}} \boldsymbol{T}_{14}\, dL = \iint_{A} \boldsymbol{b}\, dA \qquad \text{(I.5)}$$

Equation (I.5) is a statement of equilibrium of forces for the triangular element. Thus we see that, for a translational collapse mechanism, the energy balance equation from the upper bound theorem is equivalent to the equations of equilibrium. Note that no reference has made to the material that composes the triangular element.

Further reading

I.F. Collins, A note on the interpretation of Coulomb's analysis of the thrust on a rough retaining wall in terms of the limit theorems of plasticity theory, *Geotechnique*, **24**, 442–447 (1973).

A. Drescher and E. Detournay, Limit load in translational failure mechanisms for associative and non-associative materials, *Geotechnique*, **43**, 443–456 (1993).

J. Heyman, *Coulomb's Memoir on Statics – an Essay in the History of Civil Engineering*, Cambridge University Press, Cambridge, 1972.

Index